ROYAL HISTORICAL SOCIETY

ANNUAL BIBLIOGRAPHY OF BRITISH AND IRISH HISTORY

ROYAL HISTORICAL SOCIETY

ANNUAL BIBLIOGRAPHY OF BRITISH AND IRISH HISTORY

PUBLICATIONS OF 1994

GENERAL EDITORS:
BARBARA ENGLISH & J.J.N. PALMER

Glenn Burgess Virginia Davis C.C. Eldridge
Barbara English Stanley Ireland James Kirk
V.J. Morris Mary O'Dowd D.J. Orton
J.J.N. Palmer Michael S. Partridge Huw Pryce
Stephen Taylor John Turner A. Williams

For the Royal Historical Society and
in association with the Institute of
Historical Research

OXFORD UNIVERSITY PRESS
1995

Oxford University Press, Walton Street, Oxford OX2 6DP
Oxford New York
Athens Auckland Bangkok Bombay
Calcutta Cape Town Dar es Salaam Delhi
Florence Hong Kong Istanbul Karachi
Kuala Lumpur Madras Madrid Melbourne
Mexico City Nairobi Paris Singapore
Taipei Tokyo Toronto
and associated companies in
Berlin Ibadan

Oxford is a trade mark of Oxford University Press

Published in the United States
by Oxford University Press Inc., New York

British Library Cataloguing in Publication Data
Data available

Library of Congress Cataloging in Publication Data
Data applied for
ISBN 0–19–820488–4
1 3 7 9 10 8 6 4 2

Set by Hope Services (Abingdon) Ltd.
Printed in Great Britain
on acid-free paper by
Biddles Ltd.
Guildford & King's Lynn

CONTENTS

Contents

PREFACE

The Annual Bibliography of the Royal Historical Society is designed in the first place to serve the urgent needs of scholars, which has meant subordinating total coverage and refinements of arrangement to speed of production. Nevertheless, it aims to be as comprehensive as possible: this year's volume includes 1507 books and 2038 articles by 4034 authors. There is some variety in the subdivisions of sections, which are those approved by section editors. Searchers are advised to use these subdivisions in conjunction with the four indices. The indices form an important part of the book and cover, as far as is practicable, all major subjects of the books and articles listed, whether or not they figure in the titles. The counties used in identifying places are, in general, those of the pre-1974 counties. A list of the county abbreviations employed is given at the head of the Index of Places.

Within each subsection (except Bc) the items are arranged in alphabetical order of authors. All collective works, including collections of reprints, are placed in the section Bc. Reprinted articles and brief reports (in, for example, archaeological reports or exhibition catalogues) are not normally itemised separately, and where this is the case the fact is noted in the Bc section. All other articles contained in these collective works are individually listed in the appropriate subsection and there referred to by their Bc numbers in lieu of the usual publication details. Apart from the short list of standard abbreviations which follows this preface, these Bc numbers are the only abbreviated form of publication details.

Items covering more than two sections are listed in Ba or Bb; any items that extend over two sections may appear in the first section and be cross-referenced at the head of the second. Items in the section 'Empire and Commonwealth post 1783' are included only if they contribute substantially to British or Irish history: for example, 'British teaplanters in Ceylon' would be included whereas 'The social structure of Ceylon' would not.

This is the last year in which the Annual Bibliography has been edited at the University of Hull. The general editors have been helped over the years by many members of academic staff, both at Hull and elsewhere, in their varied requests asking for translations of everything from Welsh to Swahili; for identifications of persons and places; for assistance with pagination and places of publication. The section editors carry a heavy burden of searching and checking a wide range of publications, in an era where the numbers of new books and periodicals in university libraries are falling (though the number published continues to grow). The problems are to some extent ameliorated by the increased usefulness of electronic bibliographical networks, and the effectiveness of communication by fax machine. Publishers are very generous in checking and correcting material sent to them.

The Director and library staff of the Institute of Historical Research have, as in every year, undertaken the searches without which the compilation of the Bibliography would be impracticable; at the Institute, we are particularly indebted to Steven Smith and Keith Manley. The Staff at the British Library

(Boston Spa) have given us invaluable assistance. The Brynmor Jones Library at the University of Hull provides the resources for much of the detailed editing work. The editors also wish to express their gratitude to George Slater of the Computer Centre at the University of Hull and to Matthew Palmer for programming assistance, and to Kevin Hall for editorial assistance.

The Annual Bibliography is but one of the historical projects founded by the late Sir Geoffrey Elton. Its form has changed little over the years, showing how well he designed it; its success owes much to his continued support until his death. The general editors, as inheritors of the bibliography, are proud to have been associated with it for seven years.

University of Hull
April 1995

ABBREVIATIONS

Arch.	Archaeological
B.	Bulletin
BAR	British Archaeological Reports
c.	*circa*
CBA	Council for British Archaeology
(comp.)	compiler(s)
(ed.)	editor(s)
edn.	edition
fl.	floruit
HMSO	Her Majesty's Stationery Office
Inst.	Institute
J.	Journal
no.	number
ns	new series
P.	Proceedings
Q.	Quarterly
R.	Review
rev.	revised
ser.	series
Soc.	Society
T.	Transactions
UP	University Press

A. AUXILIARY

(a) *Archives and Bibliography*

1. Aldrich, Richard J. 'Never-never land and wonderland?: British and American policy on intelligence archives', *Contemporary Record* 8 (1994), 133–52.
2. Allison, A.F.; Rogers, D.M. (ed.). *The contemporary printed literature of the English Counter-Reformation between 1558 and 1640, vol. 2: works in English*. Aldershot; Scolar; 1994. Pp xxxv, 250.
3. Anon. *A guide to Cabinet conclusions, 1921–1943*. Belfast; Public Record Office of Northern Ireland; 1993. Pp 333.
4. Anon. *British Library catalogue of additions to the manuscripts, 1981–1985*. London; British Library; 1994. 2 vols. Pp 750.
5. Anon. *British Library catalogue of additions to the manuscripts: the Yelverton collection*. London; British Library; 1994. Pp 690.
6. Anon. *Industrial revolution: a documentary history, series one: the Boulton and Watt archive and the Matthew Boulton papers from Birmingham central library, guide to parts 2–5*. Marlborough; Matthew; 1994. 2 vols.
7. Anon. *Labour history, series 2: minute books and papers of the Royal Arsenal Co-operative Society, 1868–1968; part 1: Political Purposes Committee minutes, 1922–1986: listing and guide to the microfilm edition*. Marlborough; Matthew; 1993. Pp 54.
8. Anon. *Salamanca letters: a catalogue of the Irish college of Salamanca correspondence between 1619–1870, in the library of Maynooth college, Ireland*. Maynooth; St Patrick's college; 1994. Pp 615.
9. Anon. 'A list of essays on Scottish history in books published during the year 1992', *Scottish Historical R.* 72 (1993), 202–10.
10. Anon. 'Bibliography' [medieval queenship], Bc50, 223–52.
11. Anon. 'Business records deposited in 1993', *Business Archives* 68 (1994), 43–67.
12. Anon. 'Major accessions to repositories, 1992', *J. of the Soc. for Army Historical Research* 72 (1994), 180-83.
13. Anon. 'Major accessions to repositories 1992 relating to education', *History of Education Soc. B.* 54 (1994), 44–48.
14. Anon. 'Summary lists of archive surveys and deposits, 1990-1991', *Scottish Industrial History* 14–15 (1992), 53–61.
15. Anon. 'Summary lists of archive surveys and deposits, 1992–1993', *Scottish Industrial History* 17 (1994), 102–10.
16. Anon. 'Transport records deposited in 1991', *J. of Transport History* 3rd ser. 14 (1993), 77–80.
17. Anon. 'Transport records deposited in 1992', *J. of Transport History* 3rd ser. 15 (1994), 78–80.

18. Atherton, Louise. *Never complain, never explain: records of the Foreign Office and State Paper Office, 1500-c.1960*. London; Public Record Office (readers' guide 7); 1994. Pp viii, 189.
19. Atherton, Louise. *Top secret: an interim guide to recent releases of intelligence records at the Public Record Office*. London; Public Record Office; 1993. Pp 32.
20. Ball, Stuart. 'The Conservative party since 1900: a bibliography', Bc141, 727–72.
21. Baxter, Richard. *A subject index to the calendar of the correspondence of Richard Baxter: with another Baxter letter*, ed. N.H. Keeble. London; Dr Williams' Trust; 1994. Pp 20.
22. Beedham, Katharine; English, Barbara; Palmer, J.J.N. (ed.). *Royal Historical Society annual bibliography of British and Irish history: publications of 1993*. Oxford; Oxford UP; 1994. Pp xi, 305.
23. Bennett, John; Matthews, Alison. 'British labour history publications 1993', *Labour History R.* 59/2 (1994), 30-52.
24. Berkhout, Carl T.; Biddle, Martin; Blackburn, Mark A.S.; Foot, Sarah; Keynes, Simon; Rumble, Alexander R. 'Bibliography for 1991', *Anglo-Saxon England* 21 (1992), 265–315.
25. Berkhout, Carl T.; Biddle, Martin; Blackburn, Mark A.S.; Foot, Sarah; Keynes, Simon; Rumble, Alexander R. 'Bibliography for 1992', *Anglo-Saxon England* 22 (1993), 281–325.
26. Bird, Stephen. 'A specialist labour history archive: the National Museum of Labour History archive centre', *History Workshop* 37 (1994), 170-76.
27. Birkett, Dea. *Mary Kingsley, 1862–1900: a biographical bibliography*. Bristol; Bristol UP; 1993. Pp 30.
28. Black, Jeremy. 'Eighteenth-century papers', *British Library J.* 20 (1994), 199–213.
29. Black, Jeremy (comp.). 'The papers of British diplomats, 1689–1793: appendix II', *Archives* 21 (1994), 205–13.
30. Blake, Steven (ed.). 'Recent publications' [on Gloucestershire], *T. of the Bristol & Gloucestershire Arch. Soc.* 111 (1993), 252–56.
31. Bloomberg-Rissman, John. 'Pre-1701 records in the English short-title catalogue: a description and explanation', *Factotum* 38 (1994), 3–5.
32. Bourne, Susan; Chicken, Andrew H. *Records of the medical professions: a practical guide for the family historian*. Privately printed; 1994. Pp 48.
33. Boyns, T. 'Editorial comment: business history through accounting records', *Accounting, Business & Financial History* 3 (1993), 247–52.
34. Britnell, Richard H.; Boulton, Jeremy; Honeyman, Katrina; Collins, Michael (comp.). 'Review of periodical literature published in 1992', *Economic History R.* 2nd ser. 47 (1994), 165–93.
35. Burns, C. 'Pope Clement XI and the British Isles: the inventory of Fondo Albani 163–168', Bc147, 41–85.
36. Cale, Michelle. 'Major accessions to repositories in 1993', *Labour History R.* 59/2 (1994), 23–29.
37. Chamberlain, Eric (ed.). *Catalogue of the Pepys library at Magdalene college, Cambridge, vol. 3; prints and drawings, part 2: portraits*. Woodbridge; Brewer; 1994. Pp xxiv, 250.

38. Clarke, Margaret; Davies, John; Thornton, Loretta (comp.). 'Publications and articles on the history of the east midlands, 1993', *East Midland Historian* 4 (1994), 46–51.
39. Cobb, Henry S. 'Politicians and archives', *J. of the Soc. of Archivists* 15 (1994), 141–149.
40. Cole, Anne E.; Tinley, Ruth (comp.). *Lincolnshire Poor Law index, part 7: index to settlement papers in the Kesteven quarter sessions, 1700-1847*. Lincoln; Lincolnshire Family History Soc.; 1993. Pp 101.
41. Connolly, Sean J. 'The Church of Ireland: a critical bibliography, 1690-1800' [part 4], *Irish Historical Studies* 28 (1993), 362–69.
42. Cowie, Leonard W. (ed.). *Edmund Burke, 1729–1797: bibliography*. London; Greenwood; 1994. Pp 146.
43. Cowley, F.G. (comp.). 'Periodical articles on Welsh religious history, 1989–1991', *J. of Welsh Religious History* 2 (1994), 96–104.
44. Cox, David C. *'This foolish business': Dr Nash and the Worcestershire collections*. Worcester; Worcestershire Historical Soc. (occasional publications 7); 1993. Pp 37.
45. Creaton, Heather (ed.). *Bibliography of printed works on London history to 1939*. London; Library Association; 1994. Pp xxviii, 809.
46. Culling, Joyce. *An introduction to occupations: a preliminary list*. Birmingham; Federation of Family History Societies; 1994. Pp 44.
47. Darwall-Smith, Robin. *The Jowett papers: a summary catalogue of the papers of Benjamin Jowett (1817–1893) at Balliol college, Oxford*. Oxford; Balliol College Library; 1993. Pp xxxii, 205.
48. Davies, K.G. (ed.). *Calendar of state papers, colonial series: America and the West Indies, vol. 45, 1739*. London; HMSO; 1994. Pp xvii, 318.
49. Davison, Stephen (ed.). *Northern Ireland and Canada: a guide to Northern Ireland sources for the study of Canadian history, c.1705–1992*. Belfast; Queen's University of Belfast and Public Record Office of Northern Ireland; 1994. Pp vii, 144.
50. Dixon, Shirley. 'The Royal Army Medical Corps "muniment collection"', *Medical History* 38 (1994), 459–69.
51. Duncan, Robert. 'The papers of John Turner Walton Newbold, 1888–1943: an introductory guide', *B. of the John Rylands University Library of Manchester* 76/2 (1994), 195–203.
52. Dyson, Brian (comp.). *Yorkshire maps and plans in the archives of the university of Hull*. Hull; Centre for Regional and Local History, University of Hull; 1990. Pp iii, 74.
53. Elliot, Matthew. 'A review of Cabinet papers: September-December, 1962', *Contemporary Record* 8 (1994), 153–58.
54. Ellis, Mary (comp.). 'Select digest of manuscript accessions to repositories, 1993, relating to the history of Wales', *Welsh History R.* 17 (1994), 293–99.
55. Eserin, Angela; Hughes, Mike (comp.). *The Sir Frederic Osborn archive: a descriptive catalogue*. Hertford; Hertfordshire Library Service; 1990. Pp v, 137.
56. Eve, David C. 'Archive report number 10, National Monuments Record of Scotland', *Scottish Industrial History* 17 (1994), 90-93.
57. Fehrenbach, Robert J. (ed.). *Private libraries in Renaissance England: a*

collection and catalogue of Tudor and early Stuart book-lists, vol. 3: PLRE 67–86. Marlborough; Matthew; 1993. Pp xxx, 282.
58. Fenning, Hugh. 'The "Udienze" series in the Roman archives of Propaganda Fide, 1750-1820', *Archivium Hibernicum* 48 (1994), 100-106.
59. Fenwick, Gillian. *Women and the 'Dictionary of National Biography': a bibliography of DNB volumes, 1885–1985 and 'missing persons'*. Aldershot; Scolar; 1994. Pp 250.
60. Fonge, C.R. *Tithe awards and maps at the Borthwick institute*. York; Borthwick Inst. (list and index 11); 1994. Pp iii, 70.
61. Ford, Alan. 'The Church of Ireland: a critical bibliography, 1603–1640' [part 2], *Irish Historical Studies* 28 (1993), 352–58.
62. Forster, G.C.F. 'Review of periodical literature and occasional publications [on northern history]', *Northern History* 30 (1994), 208–20.
63. Fowler, Simon. *RAF records in the Public Record Office*. London; Public Record Office; 1994. Pp 138.
64. Franklin, Peter. *Some medieval records for family historians: an introduction to the purposes, contents, and interpretation of pre-1538 records available in print*. Birmingham; Federation of Family History Societies; 1994. Pp 101.
65. Gantt, Jonathan Wes. *A guide to the annual reports of the Port Medical Officer of Health for Swansea*. Swansea; Swansea City Council; 1993. Pp 17.
66. Gibson, Jeremy; Hunter, Judith. *Victuallers' licences: records for family and local historians*. Birmingham; Federation of Family History Societies; 1994. Pp 56.
67. Goose, Nigel R.; Backouche, Isabelle; Raven, Neil. 'Current bibliography of urban history', *Urban History* 21 (1994), 319–81.
68. Graham, T.W. 'A list of articles on Scottish history published during the year 1992', *Scottish Historical R.* 72 (1993), 190-201.
69. Hackett, Ronald Geoffrey. *South African war books: illustrated bibliography of English language publications relating to the Boer war, 1899–1902*. London; De Lotz; 1994. Pp x, 205.
70. Hale, Matthew; Hawkins, Richard; Partridge, Michael Stephen. 'List of publications on the economic and social history of Britain and Ireland published in 1993', *Economic History R.* 2nd ser. 47 (1994), 776–815.
71. Harley, David. 'English archives, local history, and the study of early modern "midwifery"', *Archives* 21 (1994), 145–54.
72. Harper, Peter. *A guide to the manuscript papers of British scientists catalogued by the Contemporary Scientific Archives Centre and the National Cataloguing Unit for the Archives of Contemporary Scientists, 1973–1993*. Bath; NCUACS; 1992. Pp xv, 111.
73. Harris, David John; Shepherd, Jean A. (comp.). *An index of British treaties, vol. 4: 1969–1988*. London; HMSO; 1992. Pp xv, 795.
74. Harris, Mark. 'The finding and use of brief records', *Archives* 21 (1994), 129–44.
75. Haythornthwaite, J.A. *Scotland in the nineteenth century: an analytical bibliography of material relating to Scotland in parliamentary papers, 1800-1900*. Aldershot; Scolar; 1993. Pp viii, 365.
76. Heesom, A.J. '"Agrarian, rural, and bucolic?": local history publications in the early nineties', *Archives* 21 (1994), 73–79.

77. Heilbron, J.L. 'Some uses for catalogues of old scientific instruments', Bc36, 1–16.
78. Hollier, Anita. 'The BP archive', *Contemporary Record* 7 (1993), 654–62.
79. Horden, John. 'A dictionary of Scottish biography', Bc25, 67–84.
80. Horgan, Dominique. 'In our keeping: the archives of the Congregation of the Irish Dominican Sisters, Cabra', *Catholic Archives* 14 (1994), 60-64.
81. Hoyle, Richard W. *Tudor taxation records: a guide for users.* London; Public Record Office (readers' guide 5); 1994. Pp x, 67.
82. Hunter, Isobel. 'The archive of the Physiological Society in the Contemporary Medical Archives Centre', *Medical History* 38 (1994), 328–35.
83. Huws, Daniel. 'Catholic records in the National Library of Wales', *Catholic Archives* 14 (1994), 65–70.4. James, A.R. *A bibliography of world war two HMSO paperbacks: being a bibliography of His Majesty's Stationery Office series of publications in paperback and wrappers dealing with contemporary aspects of world war two and mostly published between 1941 and 1947 and including a price guide.* Southwick; James; 1993. Pp 47.
85. Jebb, Philip. 'The archives of the English Benedictine Congregation kept at St Gregory's, Downside', *Catholic Archives* 14 (1994), 20-36.
86. Kenyon, John R. 'Post-medieval Britain and Ireland in periodic literature in 1992', *Post-Medieval Archaeology* 27 (1993), 297–304.
87. King, Anthony. *British and Irish archaeology: a bibliographical guide.* Manchester; Manchester UP; 1994. Pp xii, 324.
88. Kitching, Christopher. *Archive buildings in the United Kingdom, 1977–1992.* London; HMSO; 1993. Pp vii, 144.
89. Kohler, C.C. *John Milton: a catalogue of the Kohler collection of 550 different editions of the writings of John Milton published between 1641 and 1914.* Dorking; Kohler; 1993. Pp 102.
90. Linton, David; Boston, Ray. *The twentieth-century newspaper press in Britain: an annotated bibliography.* London; Mansell; 1994. Pp 416.
91. Lloyd, Gareth. 'The papers of Dr Thomas Coke: a catalogue', *B. of the John Rylands University Library of Manchester* 76/2 (1994), 205–320.
92. Lowe, Rodney (ed.). *Treasury papers, series 1: papers of the economic section, 1941–1961 (PRO Class T230): listing and guide to part 2 of the microfilm collection.* Marlborough; Matthew; 1994. Pp 56.
93. Macdonald, Catriona. 'Bibliography of works published in 1992 and relevant to Scottish economic and social history', *Scottish Economic & Social History* 14 (1994), 144–58.
94. Mackenzie, Alice. *NEWSPLAN: report of the NEWSPLAN project in Scotland, September 1994.* London; British Library; 1994. Pp ii, 483.
95. Maguire, Martin. 'Select bibliography of writings on Irish economic and social history published in 1993', *Irish Economic & Social History* 21 (1994), 84–111.
96. Manning, Peter. *Contents of the Hussey-Walsh manuscripts covering volumes 1 to 12.* Rainham; Manning; 1993.
97. Marmion, John P. 'The Shrewsbury diocesan archives', *Catholic Archives* 14 (1994), 37–43.

98. McBain, Janet. 'The Scottish Film Archive: 15 years of industrial endeavour', *Scottish Industrial History* 14–15 (1992), 35–37.
99. McGuire, J.I. 'The Church of Ireland: a critical bibliography, 1641–1690' [part 3], *Irish Historical Studies* 28 (1993), 358–62.
100. Medlycott, Mervyn T. *Somerset wills index: printed and manuscript copies.* Weston-super-Mare; Galloway; 1993. Pp xxvi, 165.
101. Merritt, Julia (ed.). *Crown servants, series one: the papers of Thomas Wentworth, 1st earl of Strafford, 1593–1641, from Sheffield city libraries.* Marlborough; Matthew; 1994. Pp 100.
102. Milne, Kenneth. 'The Church of Ireland: a critical bibliography, 1870-1992' [part 6], *Irish Historical Studies* 28 (1993), 376–84.
103. Moat, Janet. 'Preserving British cinema history: the British Film Institute Special Materials Unit', *Contemporary Record* 8 (1994), 400-10.
104. Mollan, Charles. 'The Irish national inventory and one of its "discoveries"', Bc36, 443–53.
105. Morgan, Raine. 'Annual list and brief review of articles on agrarian history, 1992', *Agricultural History R.* 42 (1994), 63–73.
106. Morgan, Raine (comp.). 'Supplement to the bibliography of theses on British agrarian history: 1987–1992', *Agricultural History R.* 42 (1994), 168–85.
107. Morris, V.J.; Orton, D.J. (comp.). 'List of books and pamphlets on agrarian history 1993', *Agricultural History R.* 42 (1994), 163–67.
108. Morriss, Roger; Bursey, Peter (ed.). *A guide to British naval papers in America.* London; Mansell; 1994. Pp 448.
109. Motherway, Michelle. 'The archives of the Little Company of Mary', *Catholic Archives* 14 (1994), 12–19.
110. Murray, James. 'The Church of Ireland: a critical bibliography, 1536–1603' [part 1], *Irish Historical Studies* 28 (1993), 345–52.
111. Mynors, R.A.B.; Thomson, R.M. (comp.). *Catalogue of the manuscripts of Hereford cathedral library.* Woodbridge; Boydell; 1993. Pp xxxiv, 272.
112. Nash, Paul W.; Flavell, A.J. (ed.). *Corvinus press: history and bibliography.* Aldershot; Scolar; 1994. Pp xii, 245.
113. Nockles, Peter B. 'The work of the Methodist archives and research centre, John Rylands university library of Manchester', *Catholic Archives* 14 (1994), 44–52.
114. O'Ferrall, Fergus. 'The Church of Ireland: a critical bibliography, 1800-1870' [part 5], *Irish Historical Studies* 28 (1993), 369–76.
115. Parker, Franklin; Parker, Betty June (comp.). *Education in England and Wales: an annotated bibliography.* London; Garland; 1991. Pp 531.
116. Partridge, Michael Stephen; Partridge, Karen E. (ed.). *Lord Palmerston, 1784–1865: a bibliography.* London; Greenwood; 1994. Pp 328.
117. Patterson, Stephen. *Finding Fife's medical records: a list of the records of the health services in Fife, 1830-1947.* Glasgow; Wellcome Unit for the History of Medicine, University of Glasgow; 1993.
118. Porter, D.S. 'An "unrealized proposal": the Bodleian Library and the Phelps papers', *Bodleian Library Record* 15 (1994), 51–57.
119. Purcell, Mary. 'Dublin diocesan archives: Hamilton papers (5)', *Archivium Hibernicum* 48 (1994), 37–72.

120. Rawsthorne, Lawrence (ed.). *The library of the Bankes family of Soughton Hall and Mynachlog, in Northop, Clwyd: a catalogue*. Mold; Clwyd County Council Library and Information Service; 1994. Pp ix, 86.
121. Raymond, Stuart A. *British genealogical periodicals: a bibliography of their contents; supplement 1: British genealogy in miscellaneous journals*. Exeter; Raymond; 1994. Pp 80.
122. Raymond, Stuart A. *Buckinghamshire: a genealogical bibliography*. Birmingham; Federation of Family History Societies; 1993. Pp 57.
123. Raymond, Stuart A. *Cumberland and Westmorland: a genealogical bibliography*. Birmingham; Federation of Family History Societies; 1993. Pp 68.
124. Raymond, Stuart A. *London and Middlesex: a genealogical bibliography*. Birmingham; Federation of Family History Societies; 1994. Pp 128.
125. Raymond, Stuart A. *Norfolk: a genealogical bibliography*. Birmingham; Federation of Family History Societies; 1993. Pp 90.
126. Raymond, Stuart A. *Oxfordshire: a genealogical bibliography*. Birmingham; Federation of Family History Societies; 1993. Pp 59.
127. Raymond, Stuart A. *Wiltshire: a genealogical bibliography*. Birmingham; Federation of Family History Societies; 1993. Pp 82.
128. Refausse, Raymond. 'Church of Ireland parish registers in the Representative Church Body library, Dublin', *Irish Genealogist* 9 (1994), 73–75.
129. Rickard, Gillian (ed.). *Kent settlement (Poor Law) records: a guide and catalogue, part 2: west Kent (diocese of Rochester)*. Canterbury; Rickard; 1994. Pp 56.
130. Robson, Karen. *MS 175, papers of Chief Rabbi J.H. Hertz, (1872–1946), 1853–1949*. Southampton; Southampton UP; 1993. Pp v, 58.
131. Roffey, Ron A. *Looking back: a brief historical guide to the archives of the Co-operative Wholesale Society's south-east retail group*. Woolwich; Coop South East; 1992. Pp 20.
132. Rowlands, I.W.; O'Leary, Paul (comp.). 'Articles relating to the history of Wales published mainly in 1992', *Welsh History R.* 17 (1994), 300-11.
133. Savage, Nicholas; Beasley, Gerald; Shell, Alison; Coast, John Meriton (ed.). *Early printed books, 1478–1840: catalogue of the British architectural library early imprints collection, vol. 1: A-D*. East Grinstead; Bowker-Saur; 1994. Pp xlviii, 538.
134. Schulze, Rainer. 'Wichtige Archive und Bibliotheken zur Erforschung der Geschichte der Arbeiterbewegung in Grossbritannien: eine Übersicht' [A review of the significant archives and libraries for the study of the history of the labour movement in Great Britain], *Mitteilungsblatt des Instituts zur Erforschung der Europäischen Arbeiterbewegung* 11 (1991), 111–33.
135. Sessions, William K. *Further Irish studies in early printing history*. York; Sessions; 1994. Pp 58.
136. Shackleton, Robert. 'Old books in the university library' [Oxford], Bc25, 121–29.
137. Shepherd, Elizabeth. 'Records management in Britain: a review of some developments in professional principles and practice', *Business Archives* 67 (1994), 13–25.

138. Shipton, Rachel C. *A bibliography of Sir Alexander Fleming, 1881–1955*. London; St Mary's Hospital Medical School; 1993. iv, 26 leaves.
139. Sinclair, Cecil. *Tracing Scottish local history: a guide to local history research in the Scottish Record Office*. Edinburgh; HMSO; 1994. pp viii, 167.
140. Smith, David M. *A guide to the archives of the Company of Merchant Taylors in the city of York*. York; Borthwick Inst. (list and index 12); 1994. Pp 29.
141. Smith, Geoff. 'Theses and dissertations on British and Irish labour history', *Labour History R.* 59/2 (1994), 3–9.
142. Stevenson, Julie. 'A guide to equity suits in the "J" series at the Public Record Office, Chancery Lane', *Business Archives* 68 (1994), 37–42.
143. Storey, Richard. 'The modern records centre' [university of Warwick], *International R. of Social History* 39 (1994), 147–55.
144. Thomas, Garth. *Records of the militia from 1757: including records of the Volunteers, Rifle Volunteers, Yeomanry, Fencibles, Territorials and the Home Guard*. London; Public Record Office (readers' guide 3); 1993. Pp vi, 59.
145. Thompson, Robert Hugh. *Sylloge of coins of the British Isles, vol. 38; the Norweb collection: tokens of the British Isles, 1575–1750, part 2: Dorset, Durham, Essex and Gloucestershire*. London; Spink; 1988. Pp lxxxii, 172.
146. Thompson, Robert Hugh; Dickinson, Michael J. *Sylloge of coins of the British Isles, vol. 38; the Norweb collection: tokens of the British Isles, 1575–1750, part 4: Norfolk to Somerset*. London; Spink; 1993. Pp lvii, 220.
147. Thompson, Robert Hugh; Dickinson, Michael J. *Sylloge of coins of the British Isles, vol. 43; the Norweb collection: tokens of the British Isles, 1575–1750, part 3: Hampshire to Lincolnshire*. London; Spink; 1992. Pp liv, 216.
148. Tite, Colin G.C. *The manuscript library of Sir Robert Cotton*. London; British Library; 1994. Pp 112.
149. Turton, Alison. 'Babcock and Wilcox Ltd: an engineering company's archive (Archive report number 7)', *Scottish Industrial History* 14–15 (1992), 23–26.
150. Walker, David M. 'Listing in Scotland: origins, survey and resurvey', *T. of the Ancient Monuments Soc.* 38 (1994), 31–96.
151. Weinstein, Helen (comp.). *Catalogue of the Pepys library at Magdalene college, Cambridge, vol. 2; ballads, part 2: indexes and lists*. Woodbridge; Brewer; 1994. Pp xxii, 183.
152. Wiles, James L.; Finnegan, Richard B. *A select list of Irish government publications, 1972–1992*. Dublin; Irish Academic; 1994. Pp 80.
153. Williams, Patricia J.; Lewis, D. Geoffrey (ed.). *Saint Francis and the Franciscans: a bibliography of the Bishop John R.H. Moorman Franciscan collection held at St Deinol's library*. Hawarden; St Deinol's Library; 1993. Pp ix, 135.
154. Wilson, John (ed.). *A guide to the South Humberside Area Archive Office*. Grimsby; Humberside County Council Archive Service; 1993. Pp viii, 92.
155. Wilson, John F. 'British business history: a review of the periodical literature for 1992', *Business History* 36/2 (1994), 1–19.

156. Wood, Tom. *An introduction to civil registration*. Birmingham; Federation of Family History Societies; 1994. Pp 55.
157. Woods, C.J. 'Theses on Irish history completed in Irish universities, 1992', *Irish Historical Studies* 28 (1993), 440-41.
158. Wyke, Terry; Rudyard, Nigel (comp.). *Directory of local studies in north-west England*. Manchester; North Western Regional Library System (Bibliography of north-west England, vol. 14); 1993. Pp viii, 272.

(b) *Works of Reference*

1. Adams, J.R.R. (ed.). *Merchants in plenty: Joseph Smyth's Belfast directories of 1807 and 1808*. Belfast; Ulster Historical Foundation; 1992. Pp vi, 88.
2. Adams, Valerie (ed.). *An Irish genealogical source: a guide to church records*. Belfast; Ulster Historical Foundation; 1994. Pp xix, 279.
3. Andrews, J.H. 'The maps of Robert Lythe as a source for Irish place-names', *Nomina* 16 (1992/3), 7–22.
4. Anon. *Abbots Bickington to Zeal Monachorum: a handlist of Devon parish histories compiled on the occasion of the centenary of parish councils*. Exeter; Devon County Council Libraries; 1994. Pp 72.
5. Anon. *Enclosure maps in Derbyshire record office*. Matlock; Derbyshire County Council; 1992. Pp 23.
6. Anon. *Sources for village history, 1: Bratton Fleming*. Barnstaple; North Devon Record Office; 1994. Pp 27.
7. Barclay, Craig. 'The Fountains abbey hoard of civil war silver', *Yorkshire Arch. J.* 66 (1994), 235–37.
8. Beasley, John D. *The origin of names in Peckham and Nunhead*. London; South Riding; 1993. Pp 100.
9. Bird, Enid (comp.). *The organists and organs of the Welsh cathedrals in the 20th century*. Wakefield; Bird; 1992. Pp 44.
10. Bowman, J.H. *Greek printing types in Britain in the nineteenth century: a catalogue*. Oxford; Oxford Bibliographical Soc.; 1992. Pp xiv, 78.
11. Branson, S.J. *A gazetteer of historic Lincolnshire for family and local historians*. Lincoln; Lincolnshire Family History Soc.; 1993. Pp 56.
12. Bryant, Mark; Heneage, Simon (comp.). *A dictionary of British cartoonists and caricaturists, 1730-1980*. Aldershot; Scolar; 1994. Pp xvii, 252.
13. Chapman, Colin R. *An introduction to using newspapers and periodicals*. Birmingham; Federation of Family History Societies; 1993. Pp 30.
14. Cole, Anne E. *An introduction to Poor Law documents before 1834*. Birmingham; Federation of Family History Societies; 1993. Pp 35.
15. Collins, Kenneth E. *Glasgow Jewry: a guide to the history and community of the Jews in Glasgow*. Glasgow; Scottish Jewish Archives Committee; 1993. Pp 32.
16. Cornwall, Julian. *An introduction to reading old title deeds*. Birmingham; Federation of Family History Societies; 1993. Pp 40.
17. Cox, Alan. *Sources for the study of public housing: a London archives guide*. London; Guildhall Library; 1993. Pp 71.
18. Cox, Jane. *An introduction to wills, probate and death duty records: affection defying the power of death*. Birmingham; Federation of Family History Societies; 1993. Pp 44.

19. Cox, Jane. *Never been here before?: a first time guide for family historians at the Public Record Office*. London; Public Record Office; 1993. Pp 108.
20. Cox, Noel; Cox, Alan. *Tokens, checks, metallic tickets, passes and tallies of Wales, 1800-1993: two hundred years of Welsh paranumismatic history*. Privately printed; 1994. Pp 283.
21. Cox, Richard William. *A history of sport: a guide to the literature and sources of information*. Frodsham; British Soc. of Sports History; 1994. Pp x, 101.
22. Daunton, Martin. 'The new Dictionary of National Biography and business history', *Business Archives* 68 (1994), 1–4.
23. Davies, John A. 'A civil war coin hoard from Wortwell, south Norfolk', *Norfolk Archaeology* 42 (1994), 84–89.
24. Deadman, Hugo; Scudder, Elizabeth (ed.). *Introductory guide to the Corporation of London records office*. London; Corporation of London Records Office; 1994. Pp 82.
25. Duffy, Patrick J. *Landscapes of south Ulster: a parish atlas of the diocese of Clogher*. Belfast; Inst. of Irish Studies, Queen's University; 1993. Pp viii, 131.
26. Ebbatson, Linda. 'Context and discourse: Royal Archaeological Institute membership, 1845–1942', Bc85, 22–74.
27. Ellis, Mary. *Using manorial records*. London; Public Record Office (readers' guide 6); 1994. Pp viii, 110.
28. Farrar, Michael. 'Records of nonconformity in Cambridgeshire', Bc38, 150-52.
29. Finnegan, Ruth; Drake, Michael; Eustace, Jaqueline (ed.). *Sources and methods for family and community historians: a handbook*. Cambridge; Cambridge UP; 1994. Pp 322.
30. Flanagan, Deirdre; Flanagan, Laurence. *Irish place-names*. Dublin; Gill & Macmillan; 1994. Pp 224.
31. Fowkes, Dudley Vincent (ed.). *Derbyshire industrial archaeology: a gazetteer of sites, part 3: borough of Amber Valley*. Uttoxeter; Derbyshire Arch. Soc.; 1993. Pp vi, 38.
32. Gandy, Michael (comp.). *Catholic missions and registers, 1700-1800*. London; Gandy; 1993. 6 vol.
33. Gerhold, Dorian. *Courts of equity: a guide to Chancery and other legal records for local and family historians*. Isle of Wight; Pinhorns; 1994. Pp iv, 30.
34. Gibbens, Lillian. *An introduction to church registers*. Birmingham; Federation of Family History Societies; 1994. Pp 44.
35. Grattan-Guiness, Ivor (ed.). *Companion encyclopedia of the history and philosophy of the mathematical sciences*. London; Routledge; 1994. 2 vols. Pp 1806.
36. Green, S.J.D. 'Unestablished versions: voluntary religion in the Victorian north' [review article], *Northern History* 30 (1994), 193–207.
37. Habgood, Wendy. *Chartered accountants in England and Wales: a guide to historical records*. Manchester; Manchester UP; 1994. Pp 241.
38. Hammond, Peter W. 'The Complete Peerage', *Genealogists' Magazine* 24 (1994), 495–500.

39. Hardy, Anne. '"Death is the cure of all diseases": using the General Register Office cause of death statistics for 1837–1920', *Social History of Medicine* 7 (1994), 472–92.
40. Hepper, David J. *British warship losses in the age of sail, 1650-1859.* Rotherfield; Boudriot; 1994. Pp vi, 213.
41. Hilton, John Anthony. *The recusant historian's handbook.* Wigan; North West Catholic History Soc.; 1993. Pp 29.
42. Lockie, John. *Topography of London: facsimile of John Lockie's gazetteer 1813.* London; London Topographical Soc. (publications 148); 1994. Pp 228.
43. Mac Aodha, Breandán S. 'Some commemorative British place-names in Dublin city', *Nomina* 16 (1992/3), 71–77.
44. Marsh, Arthur; Ryan, Victoria; Smethurst, John B. *Historical directory of trade unions, vol. 4: cotton, wool, worsted, linen and jute and other textiles.* Aldershot; Scolar; 1994. Pp 500.
45. Mason, Adair Stuart. 'Tithe commutation maps of Essex', *Essex Archaeology & History* 3rd ser. 25 (1994), 219–25.
46. McDermott, Philip (ed.). *For distinguished conduct in the field: a register of the distinguished conduct medal, 1920-1992.* Colchester; Hayward; 1994. 2 vols. Pp 1072.
47. Millett, Martin; Vyner, Blaise. 'Introduction' [British Archaeological Institute], Bc85, 4–7.
48. Mills, Mary (ed.). *Mapping the past: Wolverhampton, 1577–1986.* Wolverhampton; Borough of Wolverhampton, Library and Information Services Division; 1993. Pp 46.
49. Morris, Albert. *Scotland's paper: the 'Scotsman', 1817–1992.* Edinburgh; The Scotsman; 1992. Pp 64.
50. Nenk, Beverley S.; Margeson, Sue; Hurley, Maurice (comp.). 'Medieval Britain and Ireland in 1992' [excavation reports], *Medieval Archaeology* 37 (1993), 240-313.
51. Owen, Hywel Wyn. *The place-names of east Flintshire.* Cardiff; Wales UP; 1994. Pp xxxvi, 428.
52. Parks, Edwin. 'Long service decorations and medals of the Royal Guernsey militia', *Société Guernesiaise Report & T.* 23 (1994 for 1993), 560-69.
53. Powell, R.F. Peter. *The place-names of Devynock hundred: together with notes on a place-name element peculiar to Breconshire.* Brecon; Powell; 1993.
54. Reid, Andy. *The Union workhouse: a study guide for teachers and local historians.* Chichester; Phillimore; 1994. Pp 112.
55. Saunders, Matthew. 'Conservation in 1993: a review of selected cases', *T. of the Ancient Monuments Soc.* 38 (1994), 169–95.
56. Schichtman, Martin B.; Carley, James P. 'Introduction' [Arthurian legend], Bc100, 4–12.
57. Seaward, Paul; Bedford, Michael J. (ed.). *The Prime Ministers from Walpole to Macmillan.* London; Dod's Parliamentary Companion; 1994. Pp 175.
58. Stanford, Michael. *A companion to the study of history.* Oxford; Blackwell; 1994. Pp vii, 309.
59. Stevenson, Julie. 'The Journal of Transport History, 1953–1992: a cumulative, classified index', *J. of Transport History* 3rd ser. 14 (1993), 181–203.

60. Sturtivant, Ray. *Royal Navy aircraft serials and units, 1911–1919.* Tonbridge; Air Britain; 1992. Pp 490.
61. Sutton, Malcolm (ed.). *Bear in mind these dead: index of deaths from the conflict in Ireland, 1969–1993.* Belfast; Beyond the Pale; 1994. Pp xii, 240.
62. Tyson, Blake. 'Identifying and classifying masons' marks', *Vernacular Architecture* 25 (1994), 4–15.
63. Wade-Martins, Peter (ed.). *An historical atlas of Norfolk.* Norwich; Norfolk Museums Service; 1993. Pp 207.
64. Wallis, Helen (ed.). *Historians' guide to early British maps: a guide to the location of pre-1900 maps of the British Isles preserved in the United Kingdom and Ireland.* London; Royal Historical Soc. (guides and handbooks 18); 1994. Pp 465.
65. Wetherall, David. 'From Canterbury to Winchester: the foundation of the Institute', Bc85, 8–21.
66. Wilbraham, Kevin. 'Harbours, whisky and advertising: report of the Business Archives Council of Scotland surveying officer, 1993–1994', *Scottish Industrial History* 17 (1994), 94–98.
67. Wiseman, William Antony. *G.B. departmental stamps, 1882–1914.* Twyford; Stanley Gibbons; 1994. Pp iii, 67.
68. Wren, Christopher R. *Voided long-cross coinage, 1247–1279: Henry III and Edward I—an illustrated guide to identification.* Herne Bay; Plantagenet; 1993. Pp iv, 80.

(c) *Historiography*

1. Alsop, J.D. 'William Fleetwood and Elizabethan historical scholarship', *Sixteenth Century J.* 25 (1994), 155–76.
2. Anon. 'Haydn Mason: a tribute, with bibliography', Bc134, ix-xii.
3. Ashbee, Paul. 'William Stukeley, the Kit's Coty houses and his coves: a note', *Archaeologia Cantiana* 112 (1994 for 1993), 17–24.
4. Asplin, Peter W. 'Writings of John Durkan, 1931–1994', Bc83, 417–28.
5. Barkan, Elazar. 'Post anti-colonial histories: representing the other in imperial Britain', *J. of British Studies* 33 (1994), 180-203.
6. Barker, Theo C. 'Slow progress: forty years of motoring research', *J. of Transport History* 3rd ser. 14 (1993), 142–65.
7. Bassford, Christopher. *Clausewitz in English: the reception of Clausewitz in Britain and America, 1815–1945.* Oxford; Oxford UP; 1994. Pp x, 293.
8. Beckett, J.V. 'What future for the past in local history?', *East Midland Historian* 4 (1994), 5–15.
9. Berg, Maxine. 'Eileen Power and women's history', *Gender & History* 6 (1994), 265–74.
10. Biller, Peter. 'Heresy and literacy: earlier history of the theme', Bc101, 1–18.
11. Bohstedt, J. 'The moral economy and the discipline of historical context', *J. of Social History* 26 (1993), 265–84.
12. Bradley, James. 'Relational database design and the reconstruction of the British medical profession: constraints and strategies', *History & Computing* 6 (1994), 71–84.
13. Bradley, James; Dupree, Marguerite. 'Interpreting datasets: the experience

of third-party use of a machine-readable "source"', *History & Computing* 5 (1993), 169–78.
14. Brady, Ciaran. 'The dilemma of Ireland's first "new historians"', Bc84, 3–31.
15. Breatnach, Pádraig A. 'Bernhard Bischoff (d. 1991), the Munich school of medieval Latin philology, and Irish medieval studies', *Cambrian Medieval Celtic Studies* 26 (1993), 1–14.
16. Brooke, Christopher J. 'Patrick Collinson', Bc89, xv-xx.
17. Brundage, Anthony. *People's historian: John Richard Green and the writing of history in Victorian England*. London; Greenwood; 1994. Pp 200.
18. Bullock, Alan. 'Has history ceased to be relevant?', *Historian* 43 (1994), 16–20.
19. Coughlan, Anthony. 'Ireland's Marxist historians', Bc84, 288–305.
20. Crafts, N.F.R.; Mills, Terence C. 'The industrial revolution as a macroeconomic epoch: an alternative view', *Economic History R.* 2nd ser. 47 (1994), 769–75.
21. Day, Mildred Leake. 'Valerie M. Lagori: a tribute', Bc100, 1–3.
22. Denley, Peter. 'Models, sources and users: historical database design in the 1990s', *History & Computing* 6 (1994), 33–43.
23. Doyle, David Noel. 'Small differences?: the study of the Irish in the United States and Britain' [review article], *Irish Historical Studies* 29 (1994), 114–19.
24. Farrant, John H. 'The making of Francis Grose's *Antiquities*: evidence from Sussex', *Sussex Arch. Collections* 131 (1993), 152–58.
25. Feeney, Mary; Ross, Seamus. 'Information technology in humanities scholarship: British achievements, prospects and barriers', *Historical Social Research* 19 (1994), 3–59.
26. Field, Geoffrey. 'Signs of crisis, fin-de-siècle doldrums, or middle age?', *International Labor & Working-Class History* 46 (1994), 37–44.
27. Gaimster, David. 'The archaeology of post-medieval society, *c*.1450-1750: material culture studies in Britain since the war', Bc85, 283–312.
28. Gillespie, Raymond. 'Historical revisits: T.W. Moody, *The Londonderry plantation, 1609–1641* (1939)', *Irish Historical Studies* 29 (1994), 109–13.
29. Goose, Nigel R. 'Urban demography in pre-industrial England: what is to be done?', *Urban History* 21 (1994), 273–84.
30. Gourvish, Terry R. 'What kind of railway history did we get?: forty years of research', *J. of Transport History* 3rd ser. 14 (1993), 111–25.
31. Greasley, David; Oxley, Les. 'Rehabilitation sustained: the industrial revolution as a macroeconomic epoch', *Economic History R.* 2nd ser. 47 (1994), 760-68.
32. Haigh, Christopher. 'The recent historiography of the English Reformation', Bc103, 13–32.
33. Harper, Marjory. 'The advantages and drawbacks of Cardbox-Plus as a means of analysing Aberdeen university students in the nineteenth and early twentieth centuries', *History & Computing* 6 (1994), 21–27.
34. Harvey, Charles; Press, Jon. 'Structured query language and historical computing', *History & Computing* 5 (1993), 154–68.
35. Harvey, John H. 'Listing as I knew it in 1949', *T. of the Ancient Monuments Soc.* 38 (1994), 97–104.

36. Hennessy, Peter. 'The pleasures and pains of contemporary history', *History Today* 44/3 (1994), 16–17.
37. Hill, B. 'Women's history: a study in change, continuity or standing still?', *Women's History R.* 2 (1993), 5–22.
38. Hobsbawm, E.J. 'In memoriam: E.P. Thompson, (1924–1993)', *International Labor & Working-Class History* 46 (1994), 2–6.
39. Howells, Robin. '"A la claire Fontaine ...": the Cottret and Ressinger editions of Jacques Fontaine's *Mémoires*', *Huguenot Soc. P.* 26 (1994), 88–93.
40. Humphreys, Robert Arthur. *Elisabeth Humphreys, 1904–1990: a memoir.* Privately printed; 1992. Pp iii, 66.
41. Humphreys, Robin A. 'Alexander Taylor Milne, 1906–1994', *Historical Research* 67 (1994), 227–29.
42. Jenkins, Geraint A. *Theophilus Evans (1693–1767): y dyn, ei deulu, a'i oes* [the man, his family, and his times]. Aberystwyth; Adran Gwasanaethau Diwylliannnol Dyfed; 1993. Pp 40.
43. Jeremy, David J. 'New business history?' [review article], *Historical J.* 37 (1994), 717–28.
44. Jones, Merfyn; Montgomery, David; Rowbotham, Sheila. 'E.P. Thompson' [obituary], *Labour History R.* 59/1 (1994), 2–7.
45. Kroot, Gerard M. 'Historians and economists: study of economic history in Britain, 1920-1950', *History of Political Economy* 25 (1993), 641–75.
46. Kuitenbrouwer, M. 'Drie omwentelingen in de historiografie van het imperialisme: Engeland en Nederland' [Three changes in the historiography of imperialism: England and the Netherlands], *Tijdschrift voor Geschiedenis* 107 (1994), 559–85.
47. Landes, David S. 'What room for accident in history?: explaining big changes by small events', *Economic History R.* 2nd ser. 47 (1994), 637–56.
48. Latham, R.C. 'Pepys and his editors', Bc25, 105–19.
49. Lees, Lynn Hollen. 'The challenge of political change: urban history in the 1990s', *Urban History* 21 (1994), 7–19.
50. Lyth, Peter J. 'The history of commercial air transport: a progress report, 1953–1993', *J. of Transport History* 3rd ser. 14 (1993), 166–80.
51. MacIvor, Arthur. 'Die Forschung zur Geschichte der Arbeiterschaft und der Arbeiterbewegung in England: ein Überblick über die jungsten Entwicklungen' [Research on the history of the workforce and the labour movement in England: an overview of its early development], *Mitteilungsblatt des Instituts zur Erforschung der Europäischen Arbeiterbewegung* 11 (1991), 14–58.
52. Markus, R.A. 'Edward Thompson', *Nottingham Medieval Studies* 38 (1994), 1–2.
53. Mathias, Peter. 'Business history and accounting history: a neighbourly relationship', *Accounting, Business & Financial History* 3 (1993), 253–73.
54. Matthews, R.C.O. 'Political and economic causes of the economic slowdown', *Scottish J. of Political Economy* 40 (1993), 129–42.
55. Merriman, Marcus. 'Gordon Donaldson, 1913–1993: an appreciation', *Scottish Historical R.* 72 (1993), 240-43.
56. Milne, Graeme J.; Paul, Marek. 'Establishing a flexible model for port book

studies: the recent evolution of the Gloucester port book database', *History & Computing* 6 (1994), 106–15.
57. Morton, Graeme. 'Presenting the self: record linkage and referring to ordinary historical persons', *History & Computing* 6 (1994), 12–20.
58. Moss, Michael S. 'University history: navel gazing or a fruitful course of enquiry?', *Scottish Economic & Social History* 14 (1994), 93–99.
59. Nunning, Vera. '"In speech an irony, in fact a fiction": Funktionen englischer Historiographie im 18. Jahrhundert im Spannungsfeld zwischen Anspruch und Wirklichkeit' [The functioning of English historiography in the 18th century, the friction between appearance and reality], *Zeitschrift für Historische Forschung* 21 (1994), 37–63.
60. Ó Muraile, Nollaig. 'The Irish genealogies as an onomastic source', *Nomina* 16 (1992/3), 23–47.
61. Palmer, Bryan D. *Objections and oppositions: histories and politics of E.P. Thompson*. London; Verso; 1994. Pp 224.
62. Phillips, John A. 'Introduction: QUASSHing MPs and electors: new perspectives on parliamentary history', *Parliamentary History* 13 (1994), 1–10.
63. Phythian-Adams, Charles. 'Local history and societal history', *Local Population Studies* 51 (1993), 30-45.
64. Postles, David A. 'Reviewing social networks: using Ucinet', *History & Computing* 6 (1994), 1–11.
65. Powell, David. 'The historical novel: history as fiction and fiction as history', *Historian* 43 (1994), 13–16.
66. Premru, Felicity. 'Oral history at work: London Transport's West Indian workforce', *Business Archives* 67 (1994), 57–71.
67. Price, Gwyn; Gray, Alec. 'Object oriented databases and their application to historical data', *History & Computing* 6 (1994), 44–51.
68. Roberts, Helen E. 'Years of struggle: the life and work of Robin Page Arnot', *Labour History R.* 59/2 (1994), 58–63.
69. Seldon, Anthony. 'Making the *Downing Street years*: an interview with Denys Blakeway', *Contemporary Record* 8 (1994), 84–102.
70. Simon, Joan. 'My life in the history of education', *History of Education Soc. B.* 54 (1994), 29–33.
71. Sisman, Adam. *A.J.P. Taylor: a biography*. London; Sinclair-Stevenson; 1994. Pp xii, 468.
72. Speck, W.A. 'History and computing: some reflections on the past decade', *History & Computing* 6 (1994), 28–32.
73. Steedman, Carolyn. 'Bimbos from hell', *Social History* 19 (1994), 57–67.
74. Swain, Richard M. 'Liddell Hart and the creation of a theory of war, 1919–1933', *Armed Forces & Soc.* 17 (1990), 35–51.
75. Swift, Roger. 'The historiography of the Irish in nineteenth-century Britain: some perspectives', Bc140, 11–18.
76. Thomas, William. 'Recollections of A.J.P. Taylor', *Contemporary European History* 3 (1994), 61–72.
77. Tiller, Kate. 'The VCH: past, present, future', *Historian* 42 (1994), 17–19.
78. Tosh, John. 'What should historians do with masculinity?: reflections on nineteenth-century Britain', *History Workshop* 38 (1994), 179–202.

79. Turk, Stella Maris. 'Cornish and Scillonian marine studies, past and present', Bc136, 76–100.
80. Turner, A.J. 'Interpreting the history of scientific instruments', Bc36, 17–26.
81. Vernon, James. 'Who's afraid of the "linguistic turn"?: the politics of social history and its discontents', *Social History* 19 (1994), 81–97.
82. Wabuda, Susan. 'Bibliography of the published writings of Patrick Collinson, 1957–1992', Bc89, 353–63.
83. Wada, Yoko. 'Bale to Parker on British historical texts in Cambridge college libraries', *T. of the Cambridge Bibliographical Soc.* 4 (1994), 511–19.
84. Watson, George. *Lord Acton's 'History of liberty': a study of his library, with an edited text of his 'History of liberty' notes.* Aldershot; Scolar; 1994. Pp x, 112.
85. Wiener, Martin J. 'The unloved state: twentieth-century politics in the writing of nineteenth-century history', *J. of British Studies* 33 (1994), 283–308.
86. Williams, David M. 'The progress of maritime history, 1953–1993', *J. of Transport History* 3rd ser. 14 (1993), 126–41.
87. Williams, Naomi; Mooney, Graham. 'Infant mortality in an "age of great cities": London and the English provincial cities compared, *c.*1840-1910', *Continuity & Change* 9 (1994), 185–212.
88. Wootton, David. 'Narrative, irony, and faith in Gibbon's *Decline and fall*', *History & Theory* 33 (1994), 77–105.
89. Wrigley, Chris. 'A.J.P. Taylor: a nonconforming radical historian of Europe', *Contemporary European History* 3 (1994), 73–86.

B. GENERAL

(a) *Long Periods—National*

1. Aldcroft, Derek H. 'The European dimension to the modern world', Bc48, 7–36.
2. Anon. *Women advising women, part 2: advice books, manuals, almanacs and journals, c.1625–1837, from the Bodleian Library, Oxford; parts 3–4: 'The Lady's Magazine', 1770-1832. A listing and guide to the microfilm collection.* Marlborough; Matthew; 1994. Pp 56.
3. Anon. 'Tree-ring dates: lists 55–59', *Vernacular Architecture* 25 (1994), 25–44.
4. Anon (ed.). *A history of British deism.* London; Routledge; 1994. Pp 2715.
5. Bacci, Massimo Livi; Reher, David S. 'Other paths to the past: from vital series to population patterns', Bc43, 66–83.
6. Bailey, R.E.; Chambers, M.J. 'Long-term demographic interactions in precensus England', *J. of the Royal Statistical Soc.* ser. A 156 (1993), 339–62.
7. Bailey, Ronald. 'The other side of slavery: black labor, cotton, and textile industrialization in Great Britain and the United States', *Agricultural History* 68 (1994), 35–50.
8. Baker, David. 'Terms of reference' [buildings archaeology], Bc144, 23–35.

9. Barker, Theo C.; Gerhold, Dorian. *The rise and rise of road transport, 1700-1990*. Basingstoke; Macmillan; 1993. Pp 118.
10. Bartine, David. *Reading, criticism, and culture: theory and teaching in the United States and England, 1820-1950*. Columbia (SC); South Carolina UP; 1992. Pp ix, 168.
11. Barty-King, Hugh. *The Baltic story: Baltic coffee house to Baltic Exchange, 1744–1994*. London; Quiller; 1994. Pp xii, 177.
12. Beckett, Ian F.W. 'The amateur military tradition', Bc109, 402–17.
13. Beckett, J.V. 'The decline of the small landowner in England and Wales, 1660-1900', Bc55, 89–112.
14. Belchem, John. 'The Irish in Britain, the United States and Australia: some comparative reflections on labour history', Bc140, 19–28.
15. Bennett, Judith M. 'Woman's history: a study in continuity and change', *Women's History R.* 2 (1993), 173–84.
16. Benson, John. *The rise of consumer society in Britain, 1880-1980*. Harlow; Longman; 1994. Pp 288.
17. Birley, Derek. *Sport and the making of Britain*. Manchester; Manchester UP; 1993. Pp 358.
18. Black, Jeremy. *Convergence or divergence? Britain and the continent*. Basingstoke; Macmillan; 1994. Pp 316.
19. Blum, Carol. 'Of women and the land: legitimizing husbandry', Bc150, 161–69.
20. Bonneuil, Nöel. 'The trend method applied to English data', Bc43, 57–65.
21. Booker, John. *Travellers' money*. Stroud; Sutton; 1994. Pp 160.
22. Bostock, Frances; Jones, Geoffrey. 'Foreign multinationals in British manufacturing', *Business History* 36/1 (1994), 89–126.
23. Brewer, John; Staves, Susan. 'Introduction' [early modern conceptions of property], Bc150, 1–20.
24. Brian, Anthea. 'Lammas meadows', *Landscape History* 15 (1993), 57–69.
25. Brock, Peter (ed.). *Records of conscience: three autobiographical narratives by conscientious objectors, 1665–1865*. York; Sessions; 1993. Pp v, 66.
26. Brodie, Allan. 'Emergency recording: RCHME's rapid response methodology', Bc144, 111–28.
27. Brown, David; Sharman, Frank. 'Enclosure: agreements and acts', *J. of Legal History* 15 (1994), 269–86.
28. Burnett, John. *Idle hands: experience of unemployment, 1790-1990*. London; Routledge; 1994. Pp 384.
29. Byatt, Derrick. *Promises to pay: the first three hundred years of Bank of England notes*. London; Spink; 1994. Pp 246.
30. Carré, Jacques. *The crisis of courtesy: studies in the conduct-book in Britain, 1600-1900*. Leiden; Brill; 1994. Pp 250.
31. Carson, Edward. 'Downing college archives and Dunwich', *Archives* 21 (1994), 83–86.
32. Carter, Harold; Lewis, Roy. 'Towns and villages: social divisions and spatial patterns', Bc116, 143–61.
33. Caunce, Stephen. *Oral history and the local historian*. London; Longman (Approaches to Local History); 1994. Pp 238.
34. Chartres, John A. 'Pre-industrial Britain', Bc56, i-xxxvi.

35. Chown, John F. *A history of money from AD 800*. London; Routledge; 1994. Pp 320.
36. Church, Roy A. 'Enterprise and management', Bc48, 110-55.
37. Clapp, B.W. *An environmental history of Britain since the industrial revolution*. Harlow; Longman; 1994. Pp 288.
38. Clark, Charles E. *The public prints: the newspaper in Anglo-American culture, 1665–1740*. Oxford; Oxford UP; 1994. Pp xiv, 330.
39. Cliff, Andrew; Haggett, Peter; Smallman-Raynor, Matthew. *Measles: a history*. Oxford; Blackwell; 1993. Pp 350.
40. Coad, Jonathan. 'Medieval fortifications and post-medieval artillery defences: developments in post-war research and future trends', Bc85, 215–27.
41. Cooley, Reg. *The unknown fleet: the army's civilian seamen in war and peace*. Stroud; Sutton; 1993. Pp xiv, 233.
42. Cottrell, P.L. 'Investment and finance', Bc48, 250-83.
43. Crafts, N.F.R. 'Managing decline?: 1870-1990', *History Today* 44/6 (1994), 37–42.
44. Crossley, David. 'Early industrial landscapes', Bc85, 244–63.
45. Crowe, Nigel. *The English Heritage book of canals*. London; Batsford/English Heritage; 1994. Pp 128.
46. Daunton, Martin. 'The entrepreneurial state, 1700-1914', *History Today* 44/6 (1994), 11–16.
47. Davey, Peter (ed.). *The archaeology of the clay tobacco pipe, XIII*. Oxford; Tempus Reparatum (BAR British ser. 239); 1994. Pp 153.
48. David, Paul A. 'The reaper and the robot: the adoption of labour-saving machinery in the past and future', Bc55, 275–305.
49. Davies, Glyn. *A history of money: from ancient times to the present day*. Cardiff; Wales UP; 1994. Pp xix, 696.
50. Diamond, Ian; Davies, H. Rhodri; Egger, Peter. 'Some applications of recent developments in event-history analysis for historical demography', Bc43, 223–38.
51. Dietz, Peter. *The British in the Mediterranean*. London; Brassey's; 1994. Pp 250.
52. Digby, Anne. *Making a medical living: doctors and patients in the market for medicine, 1720-1911*. Cambridge; Cambridge UP; 1994. Pp 369.
53. Doel, David C. *Old Chapel and the Unitarian story*. London; Lindsey; 1994. Pp 120.
54. Drake, Michael. 'Introduction' [time, family and community], Bc29, 1–10.
55. Drake, Michael. 'Parliamentary elections and community history', Bc117, 114–30.
56. Duncan, Stephen R.; Scott, Susan; Duncan, Christopher J. 'Smallpox epidemics in cities in Britain', *J. of Interdisciplinary History* 25 (1994), 255–71.
57. Duncan, Stephen R.; Scott, Susan; Duncan, Christopher J. 'The dynamics of smallpox epidemics in Britain, 1550-1800', *Demography* 30 (1993), 405–23.
58. Edwards, J.R.; Newell, E. 'The development of industrial cost and management accounting before 1850', Bc35, 407–33.
59. Emsley, Clive. 'Taking stock of crime', *Historian* 42 (1994), 9–11.

60. Endelman, Todd M. 'Jews, aliens and other outsiders in British history' [review article], *Historical J.* 37 (1994), 959–69.
61. English, Barbara. 'Inheritance and succession in landed families, 1660-1925', *Genealogists' Magazine* 24 (1994), 433–38.
62. Epstein, James A. *Radical expression: political language, ritual and symbol in England, 1790-1850.* Oxford; Oxford UP; 1994. Pp xi, 233.
63. Evans, Eric. 'The enabling state?: welfare and industrial society', *History Today* 44/6 (1994), 17–24.
64. Ewbank, Douglas. 'Coarse and refined methods for studying the fertility transition in historical populations', Bc43, 345–60.
65. Field, Clive D. 'Adam and Eve: gender in the English Free Church constituency', *J. of Ecclesiastical History* 44 (1993), 63–79.
66. Field, Clive D. 'The social composition of English Methodism to 1830: a membership analysis', *B. of the John Rylands University Library of Manchester* 76/1 (1994), 153–78.
67. Finlayson, Geoffrey B.A.M. *The citizen, state and social welfare in Britain, 1830-1990.* Oxford; Oxford UP; 1994. Pp vii, 467.
68. Finnegan, Ruth; Pryce. W.T.R. 'Community and community history', Bc116, 209–17.
69. Fisher, Stephen. 'Introduction' [maritime environment], Bc136, 1–10.
70. Fishman, Robert. 'Megalopolis unbound: decentralization and the end of metropolitan culture', Bc53, 194–203.
71. Fitzgerald, Robert. *Rowntree and the marketing revolution, 1862–1969.* Cambridge; Cambridge UP; 1994. Pp 757.
72. Fitzpatrick, David. 'The Irish in Britain: settlers or transients?', Bc140, 1–10.
73. Foreman-Peck, James. 'Foreign trade and economic growth', Bc48, 216–49.
74. Foreman-Peck, James; Millward, Robert. *Public and private ownership of British industry, 1820-1990.* Oxford; Oxford UP; 1994. Pp xvii, 386.
75. Freer, Wendy. 'Standards of living among canal boat people, 1840-1939', *J. of Transport History* 3rd ser. 13 (1992), 43–58.
76. Gage, T.B. 'The decline of mortality in England and Wales, 1861 to 1964: decomposition by cause of death and component of mortality', *Population Studies* 47 (1993), 47–66.
77. Garrioch, David. 'House names, shop signs and social organisation in western European cities, 1500-1900', *Urban History* 21 (1994), 20-48.
78. Gatrell, V.A.C. *The hanging tree: execution and the English people, 1770-1868.* Oxford; Oxford UP; 1994. Pp xx, 634.
79. Gerhold, Dorian. 'Packhorses and wheeled vehicles in England, 1550-1800', *J. of Transport History* 3rd ser. 14 (1993), 1–26.
80. Gibson, William. *Church, state and society, 1760-1850.* Basingstoke; Macmillan; 1994. Pp 200.
81. Gilley, Sheridan; Sheils, W.J. *A history of religion in Britain: practice and belief from pre-Roman times to the present.* Oxford; Blackwell; 1994. Pp 560.
82. Golby, John. 'Cultural life' [communities and families], Bc117, 194–216.
83. Golby, John. 'Ideas and debates on work', Bc117, 3–10.
84. Golby, John; Finnegan, Ruth. 'Conclusion' [communities and families], Bc117, 217–24.

85. Goldstein, Jacob; Fenster, C. Abraham. 'Anglo-American criteria for resolving child custody disputes from the eighteenth century to the present: reflections on the role of socio-cultural change', *J. of Family History* 19 (1994), 35–56.
86. Goodman, Jordan. *Tobacco in history: the cultures of dependence.* London; Routledge; 1993. Pp 296.
87. Gordon, Robert W. 'Paradoxical property', Bc150, 95–110.
88. Gourvish, Terry R.; Wilson, R.G. *The British brewing industry, 1830-1980.* Cambridge; Cambridge UP; 1994. Pp xxv, 690.
89. Green, Vivian. *The madness of kings: personal trauma and the fate of nations.* Stroud; Sutton; 1993. Pp xiii, 322.
90. Gutmann, Myron P.; Alter, George. 'Family reconstitution as event-history analysis', Bc43, 159–77.
91. Haan, Heiner; Niedhart, Gottfried. *Geschichte Englands vom 16. bis zum 18. Jahrhundert* [The history of England from the 16th to the 18th century]. Munich; Beck; 1993. Pp 291.
92. Haas, J.M.* *Management odyssey: the royal dockyards, 1714–1914.* London; America UP; 1994. Pp 256.
93. Habakkuk, Sir John. *Marriage, debt and the estates system: English landownership, 1650-1950.* Oxford; Oxford UP; 1994. Pp 808.
94. Hammel, E.A.; Mason, Carl. 'My brother's keeper: modelling kinship links in early urbanization', Bc43, 318–42.
95. Hardy, Forsyth. 'The cinema', Bc2, 267–78.
96. Harris, Richard. 'Recording timber-framed buildings', Bc144, 235–48.
97. Heater, Derek. *The idea of European unity.* Leicester; Leicester UP; 1992. Pp ix, 206.
98. Henderson, John; Wall, Richard. 'Introduction' [poor women and children in the European past], Bc151, 1–30.
99. Hendrick, Harry. *Children: new perspectives—child welfare in England, 1872–1989.* London; Routledge; 1994. Pp xv, 354.
100. Hohenberg, Paul M. 'Urban development', Bc48, 284–312.
101. Hook, Robert. 'Survey for strategic planning: RCHME's rapid recording methodology', Bc144, 49–58.
102. Horn, Pamela. *Children's work and welfare, 1780-1880.* London; Macmillan; 1994. Pp 86.
103. Horne, Thomas A. *Property rights and poverty: political argument in Britain, 1605–1834.* Chapel Hill (NC); North Carolina UP; 1990. Pp 296.
104. Hutton, Ronald. *The rise and fall of merry England: the ritual year, 1400-1700.* Oxford; Oxford UP; 1994. Pp 366.
105. Jackson, R.V. 'Inequality of incomes and lifespans in England since 1688', *Economic History R.* 2nd ser. 47 (1994), 508–24.
106. James, David; Jowitt, Tony; Laybourn, Tony. *The centennial history of the Independent Labour party.* Halifax; Ryburn; 1992. Pp 376.
107. Jewell, Helen M. *The north-south divide: the origins of northern consciousness in England.* Manchester; Manchester UP; 1994. Pp 256.
108. Johnson, Paul. 'The employment and retirement of older men in England and Wales, 1881–1981', *Economic History R.* 2nd ser. 47 (1994), 106–28.

109. Jones, R.H. 'Accounting in English local government from the middle ages to *c*.1835', Bc35, 377–403.
110. Katz, David S. *The Jews in the history of England, 1485–1850.* Oxford; Oxford UP; 1994. Pp xv, 447.
111. Kennerley, Alston. 'Education and welfare of merchant seafarers', Bc87, 145–55.
112. Kidd, Colin. 'Gaelic antiquity and national identity in Enlightenment Ireland and Scotland', *English Historical R.* 109 (1994), 1197–1214.
113. Konvitz, Josef W. 'Port functions, innovation and the making of the megalopolis', Bc53, 61–72.
114. Kriedte, P.; Medick, H.; Schlumbohm, J. 'Proto-industrialization revisited: demography, structure and modern domestic industry', *Continuity & Change* 8 (1993), 217–52.
115. Kruft, Hanno Walter; Taylor, Ronald. *A history of architectural theory: from Vitruvius to the present.* London; Zwemmer; 1994. Pp 706.
116. Lacey, Michael J.; Furner, Mary O. 'Social investigation, social knowledge and the state: an introduction', Bc27, 3–62.
117. Laurence, Anne. *Women in England, 1500-1760: a social history.* London; Weidenfeld & Nicolson; 1994. Pp xvi, 301.
118. Lawrence, Christopher. *Medicine in the making of modern Britain, 1700-1920.* London; Routledge; 1994. Pp 112.
119. le Bras, Hervé. 'Simulation of change to validate demographic analysis', Bc43, 259–79.
120. Lee, Ronald D. 'Accidental and systemic change in population history: homeostatis in a stochastic setting', *Explorations in Economic History* 30 (1993), 1–30.
121. Lee, Ronald D. 'Inverse projection and demographic fluctuations: a critical assessment of new methods', Bc43, 7–28.
122. Lerner, Gerda. *The creation of feminist consciousness: from the middle ages to eighteen-seventy.* Oxford; Oxford UP; 1994. Pp xii, 395.
123. Lougheed, Alan. 'Industry and technical change', Bc48, 156–83.
124. Lunt, James. *Jai Sixth!: the Queen Elizabeth's Own 6th Gurkha Rifles, 1817–1994.* London; Cooper; 1994. Pp 240.
125. Mager, W. 'Proto-industrialization and proto-industry: the uses of drawbacks of two concepts', *Continuity & Change* 8 (1993), 181–215.
126. Malcolm, Joyce Lee. *To keep and bear arms: the origins of an Anglo-American right.* London; Harvard UP; 1994. Pp 246.
127. Mann, Michael. *The sources of social power, vol. 2: the rise of classes and nation states, 1760-1914.* Cambridge; Cambridge UP; 1993. Pp x, 826.
128. McCaa, Robert. 'Benchmarks for a new inverse population projection program: England, Sweden and a standard demographic transition', Bc43, 40-56.
129. McFarlane, Anthony. *The British in the Americas, 1480-1815.* Harlow; Longman; 1994. Pp 360.
130. Meeson, Bob. 'Past, present and future: an overview' [buildings archaeology], Bc144, 249–63.
131. Meikle, Maureen M. 'The world of women: recent medieval and early modern publications' [review article], *Innes R.* 45 (1994), 71–77.

132. Meller, Helen. 'Some reflections on the concept of megalopolis and its use by Patrick Geddes and Lewis Mumford', Bc53, 116–29.
133. Messenger, Charles. *For the love of the regiment: a history of British infantry, vol. 1: 1660-1914*. London; Cooper; 1994. Pp 272.
134. Michie, R.C. 'Introduction' [commercial and financial services], Bc62, ix-lxx.
135. Mills, Dennis. 'Community and nation in the past: perception and reality', Bc29, 261–85.
136. Mingay, Gordon. *Land and society in England, 1750-1980*. London; Longman (Themes in British Social History); 1994. Pp 288.
137. Molyneux, Nicholas A.D. 'Targetted recording for limited alterations: secular buildings', Bc144, 155–62.
138. Montin, Kim. 'The seaman's conception of himself', Bc136, 199–209.
139. Morgan, Marjorie. *Manners, morals and class in England, 1774–1858*. Basingstoke; Macmillan; 1994. Pp 256.
140. Morris, Richard. 'Buildings archaeology', Bc144, 13–21.
141. Mountfield, David. *Castles and castle towns of Great Britain*. London; Studio; 1993. Pp 128.
142. Munting, R. 'Social opposition to gambling in Britain: an historical overview', *Interdisciplinary J. of the History of Sport* 10 (1993) 295–312.
143. Nixon, Howard M. 'The literature of English bookbinding', Bc25, 31–46.
144. Nockles, Peter B. *The Oxford Movement in context: Anglican high churchmanship*. Cambridge; Cambridge UP; 1994. Pp 360.
145. Norton, David. *A history of the Bible as literature*, vol. 1. Cambridge; Cambridge UP; 1993. Pp 408.
146. Norton, David. *A history of the Bible as literature*, vol. 2. Cambridge; Cambridge UP; 1993. Pp 505.
147. O'Connor, T.P. 'Pets and pests in medieval and Roman Britain', *Mammal R.* 22 (1992), 107–13.
148. O'Day, Rosemary. *Family and family relationships, 1500-1900: England, France and the United States of America*. London; Macmillan; 1994. Pp 250.
149. Oeppen, Jim. 'Back projection and inverse projection: members of a wider class of constrained projection models', *Population Studies* 47 (1993), 245–67.
150. Oeppen, Jim. 'Generalized inverse projection', Bc43, 29–39.
151. Ogilvie, S.C. 'Proto-industrialization in Europe', *Continuity & Change* 8 (1993), 159–79.
152. Olsen, D.J. 'Cities and culture: an embarrassed silence', Bc53, 166–74.
153. Orlin, Lena Cowen. *Private matters and public culture in post-Reformation England*. Ithaca (NY); Cornell UP; 1994. Pp 336.
154. Palmer, Marilyn; Neaverson, Peter. *Industry in the landscape, 1700-1900*. London; Routledge; 1994. Pp 208.
155. Panayi, Panikos. *Immigration, ethnicity and racism in Britain, 1815–1945*. Manchester; Manchester UP; 1994. Pp 176.
156. Parker, R.H. 'Importing and exporting accounting: the British experience', Bc35, 590-613.
157. Parker, R.H.; Yamey, B.S. 'Introduction' [accounting history], Bc35, 1–10.

158. Parsons, David. 'The Church and its architecture before and after the Reformation', Bc85, 264–82.
159. Parsons, David; Brooke, Christopher J. 'Recording churches and cathedrals', Bc144, 129–54.
160. Pocock, J.G.A. 'Two kingdoms and three histories?: political thought in British contexts', Bc119, 293–312.
161. Porter, Roy. *A social history of London*. London; Hamilton; 1994. Pp 448.
162. Pounds, Norman J.G. *The culture of the English people: iron age to the industrial revolution*. Cambridge; Cambridge UP; 1994. Pp 350.
163. Price, Roger. 'The transformation of agriculture', Bc48, 72–109.
164. Pryce, W.T.R. 'Introduction' [family and community history], Bc116, 1–4.
165. Pryce, W.T.R. 'Towns and their regional settings', Bc116, 121–42.
166. Pryce, W.T.R.; Drake, Michael. 'Studying migration', Bc116, 5–31.
167. Quiney, Anthony. 'Medieval and post-medieval vernacular architecture', Bc85, 228–43.
168. Quinlan, Daniel C.; Shackelford, Jean A. 'Economy and English families, 1500-1850', *J. of Interdisciplinary History* 24 (1994), 431–63.
169. Raven, James. 'Defending conduct and property: the London press and the luxury debate', Bc150, 301–22.
170. Rawson, Claude. *Satire and sentiment, 1660-1830*. Cambridge; Cambridge UP; 1994. Pp xviii, 309.
171. Ross, Stewart. *The Stewart dynasty*. Nairn; Thomas & Lochar; 1993. Pp 328.
172. Ruston, Alan R.; Gilley, Keith. *The 'Inquirer': a history and other reflections*. London; Inquirer; 1993. Pp 56.
173. Sakata, Toshio. *English urban history* [in Japanese]. Tokyo; Sanrei Shobou; 1994. Pp viii, 216.
174. Salmon, Marylynn. 'The cultural significance of breast-feeding and infant care in early modern England and America', *J. of Social History* 27 (1994), 247–69.
175. Savage, Mike; Miles, Andrew. *The remaking of the British working class, 1850-1940*. London; Routledge; 1994. Pp 112.
176. Shapiro, Ian. 'Resources, capacities and ownership: the workmanship ideal and distributive justice', Bc150, 21–42.
177. Shapiro, Laurie Jo. 'Ice-cream in Europe and north America', Bc30, 165–75.
178. Sharpe, Pamela. 'The women's harvest: straw-plaiting and the representation of labouring women's employment, *c*.1793–1885', *Rural History* 5 (1994), 129–42.
179. Sheffield, G.D. *The Redcaps: a history of the Royal Military Police and its antecedents from the middle ages to the gulf war*. London; Brassey's; 1994. Pp 280.
180. Smiles, Sam. *The image of antiquity: ancient Britain and the romantic imagination*. London; Yale UP; 1994. Pp 192.
181. Smith, James E.; Oeppen, Jim. 'Estimating numbers of kin in historical England using demographic microsimulation', Bc43, 280-317.
182. Smith, Russell G. *Medical discipline: the professional conduct jurisdiction of the General Medical Council, 1858–1990*. Oxford; Oxford UP; 1994. Pp xlvi, 397.

183. Sokoll, Thomas. 'The household position of elderly widows in poverty: evidence from two English communities in the late eighteenth and early nineteenth centuries', Bc151, 207–24.
184. Stacey, Robin Chapman. *The road to judgment: from custom to court in medieval Ireland and Wales*. Philadelphia; Philadelphia UP; 1994. Pp xvi, 342.
185. Stammers, Michael. 'Sailing ship seafarers and sea creatures', Bc136, 144–59.
186. Staves, Susan. 'Resentment or resignation? Dividing the spoils among the daughters and younger sons', Bc150, 194–220.
187. Stocker, David. 'Introduction: understanding what we conserve', Bc144, 1–9.
188. Stones, H.R. *British railways in Argentina, 1860-1948*. Bromley; Waters; 1994. Pp 128.
189. Strachan, Hew. 'The British way in warfare', Bc109, 417–35.
190. Sutcliffe, Anthony. 'Introduction: the giant city as a historical phenomenon', Bc53, 1–13.
191. Sutton, Clive. '"Nullius in verba" and "nihil in verbis": public understanding of the role of language in science', *British J. for the History of Science* 27 (1994), 55–64.
192. Szechi, Daniel. *The Jacobites: Britain and Europe, 1688–1788*. Manchester; Manchester UP; 1994. Pp 192.
193. Thackray, David. 'Major conservation projects: recording historic buildings—the National Trust's developing role', Bc144, 165–82.
194. Thornes, Robin. 'Research strategies and priorities' [buildings archaeology], Bc144, 89–96.
195. Tiller, Kate; Thomas, Terry; Collins, Brenda. 'Family, community and religion', Bc117, 155–93.
196. Trussell, James; Guinnane, Timothy W. 'Techniques of event-history analysis', Bc43, 181–205.
197. Tucker, N. 'Boon or burden? Baby love in history', *History Today* 43/9 (1993), 28–35.
198. Turner, Gordon; Turner, Alwyn. *A history of British military bands, vol. 1: cavalry and corps, plus the Parachute Regiment and the Brigade of Gurkhas*. Tunbridge Wells; Spellmount; 1994. Pp 224.
199. Vandereycken, Walter; van Deth, Ron. *From fasting saints to anorexic girls: the history of self-starvation*. London; Athlone; 1994. Pp viii, 296.
200. Ville, Simon P. 'Transport and communications', Bc48, 184–215.
201. Waites, Bernard; Drake, Michael; Finnegan, Ruth. 'Social mobility', Bc117, 89–113.
202. Wall, Richard. 'Some implications of the earnings, income and expenditure patterns of married women in populations in the past', Bc151, 312–35.
203. Ward, K.T. 'Covenants in conveyancing instruments: a note for the vernacular architectural historian', *Vernacular Architecture* 25 (1994), 16–19.
204. Wasson, Ellis Archer. 'The crisis of aristocracy: parliamentary reform, the peerage and the House of Commons, 1750-1914', *Parliamentary History* 13 (1994), 297–311.

205. Wells, John. *The Royal Navy: an illustrated social history, 1870-1982.* Stroud; Sutton; 1994. Pp 306.
206. Westall, Oliver. 'Marketing strategy and the competitive structure of British general insurance', *Business History* 36/2 (1994), 20-46.
207. Williams, David M. 'History of ports and harbours—adaptations to change: conference summary', Bc121, 225–36.
208. Winch, Donald. 'The science of the legislator: the Enlightenment heritage', Bc27, 63–91.
209. Wolffe, John. *God and greater Britain: religion and national life in Britain and Ireland, 1843–1950.* London; Routledge; 1994. Pp 288.
210. Wood, Jason. 'Major conservation projects: recording ancient monuments—LUAU's methodologies and procedures', Bc144, 183–208.
211. Woods, Robert. 'La mortalité infantile en Grande Bretagne: un bilan des connaissances historiques', *Annales de Demographie Historique* (1994), 119–34.
212. Woods, Robert. 'On the historical relationship between infant and adult mortality', *Population Studies* 47 (1993), 195–219.
213. Worsley, Giles. 'The origins of the Gothic revival: a reappraisal', *T. of the Royal Historical Soc.* 6th ser. 3 (1993), 105–50.

(b) *Long Periods—Local*

1. Allcock, Albert. 'The documentary problems in researching the history of Aylsham Baptist church', Bc38, 153–54.
2. Allington, Peter; Greenhill, Basil; Kennerley, Alston. 'Shiphandling and hazards on the Devon coast', Bc87, 14–25.
3. Andrews, David D.; Robey, Tim; Ryan, Pat; Tyers, Ian. 'The granary at Cressing Temple', *Essex Archaeology & History* 3rd ser. 25 (1994), 79–106.
4. Annis, Richard. 'Recording before demolition: vernacular buildings on Stainmore', Bc144, 209–18.
5. Anon. *Mochrum: a parish history, 1794–1994.* Wigtown; GC; 1994. Pp 370.
6. Anon. *The history of mining in the Heanor area.* Heanor; Heanor and District Local History Soc. (publications 6); 1993. Pp 70.
7. Anon. *The survey of London, vol. 43 and 44: Poplar, Blackwall and the Isle of Dogs, the parish of All Saints.* London; Athlone; 1994. 2 vols. Pp 508, 590.
8. Anon. 'Introduction' [Cornish studies], Bc135, 1–2.
9. Atkins, Caroline. 'Recording before dismantling and re-erection: St Edmund's church, Kellington', Bc144, 219–33.
10. Ayers, Brian S. *The English Heritage book of Norwich.* London; Batsford/English Heritage; 1994. Pp 136.
11. Baines, David C. *Two coats colder: chronicles of Offley in Hertfordshire.* Hitchin; Baines; 1994. Pp v, 191.
12. Bairner, Alan. 'Football and the idea of Scotland', Bc3, 9–27.
13. Baker, David. 'The Bedfordshire historic environment record', Bc6, 24–38.
14. Baker-Jones, Leslie. 'Local government, 1815–1974' [Pembrokeshire], Bc96, 272–304; 494–97.
15. Banks, T.M.; Nichol, C.; Bridge, D.G. 'Lead mines in the manor of Kinniside', *T. of the Cumberland & Westmorland Antiquarian & Arch. Soc.* 94 (1994), 215–47.

16. Barge, John (ed.). *A gazetteer of Liverpool breweries.* Manchester; Richardson; 1987. Pp 64.
17. Barker, Theo C. 'London: a unique megalopolis?', Bc53, 43–60.
18. Barnes, Frank O. *Priory demesne to university campus: a topographical history of Nottingham university.* Nottingham; Nottingham UP; 1993. Pp xiv, 513.
19. Barnsby, George J. *A history of the Wolverhampton, Bilston and District Trades Union Council, 1865–1990.* Wolverhampton; Wolverhampton, Bilston and District Trades Union Council; 1994. Pp 112.
20. Beattie, Derek. *Blackburn: the development of a Lancashire cotton town.* Keele; Ryburn; 1993. Pp 192.
21. Bell, Ian. 'Publishing, journalism and broadcasting', Bc2, 385–95.
22. Bellenger, Aidan. 'The vicars apostolic of the western district, 1688–1850', Bc114, 9–22.
23. Berry, Keith Desmond. *John of Gaunt school and the Trowbridge high schools: the first one hundred years.* Bradford-on-Avon; Berry; 1994. Pp 192.
24. Beswick, Molly. *Brickmaking in Sussex: a history and gazetteer.* Midhurst; Middleton; 1993. Pp viii, 242.
25. Birse, Ronald M. *Science at the university of Edinburgh, 1583–1993: an illustrated history to mark the centenary of the faculty of science and engineering, 1893–1993.* Edinburgh; University of Edinburgh, Faculty of Science & Engineering; 1994. Pp 312.
26. Blaikie, Andrew. *Illegitimacy, sex and society: north-east Scotland, 1750-1900.* Oxford; Oxford UP; 1993. Pp xv, 268.
27. Blaikie, Andrew. 'A kind of loving: illegitimacy, grandparents and the rural economy of north-east Scotland, 1750-1900', *Scottish Economic & Social History* 14 (1994), 41–57.
28. Blain, Neil; Boyle, Raymond. 'Battling along the boundaries: the marking of Scottish identity in sports journalism', Bc3, 125–41.
29. Blood, N. Keith; Taylor, Christopher C. 'Cawood: an archiepiscopal landscape', *Yorkshire Arch. J.* 64 (1992), 83–102.
30. Bonnard, Brian. 'Soldiers, smugglers and pirates: some of Alderney's former governors', *B. of the Alderney Soc.* 27 (1992), 69–76.
31. Bowers, Roger. 'Music and worship to 1640' [Lincoln cathedral], Bc5, 47–76.
32. Brading, Rosetta. *(West) Cowes & Northwood, Isle of Wight, 1750-1914.* Newport (IoW); Brading; 1994. Pp 149.
33. Bradney, Joseph Alfred. *A history of Monmouthshire: from the coming of the Normans into Wales down to the present time*, vols. 1–4 [reprint]. Cardiff; Merton Priory; 1992–4. facsimile edn. 4 vols in 9 parts. Pp 1532.
34. Brinkley, Richard. 'Religion, 1815–1974' [Pembrokeshire], Bc96, 369–88; 500-01.
35. Broad, J.P.F. 'The changing face of employment in Buckingham, 1618–1798', *Records of Buckinghamshire* 34 (1994 for 1992), 46–60.
36. Broadie, Alexander. 'A nation of philosophers', Bc2, 61–76.
37. Brock, W.R.; Cooper, P.H.M. *Selwyn college: a history.* Edinburgh; Pentland; 1994. Pp viii, 353.
38. Brooks, Nicholas P. 'Rochester bridge, AD 43–1381', Bc12, 1–40.

39. Brown, A.E.; Taylor, Christopher C. 'Cambridgeshire earthwork surveys, VI', *Cambridge Antiquarian Soc. P.* 82 (1993), 101–11.
40. Brown, David. 'The variety of motives for parliamentary enclosure: the example of the Cannock Chase area, 1773–1887', *Midland History* 19 (1994), 105–27.
41. Brown, T.G.; Sanderson, R.S. *A history of the Sheffield Philatelic Society*. Sheffield; Robinson; 1994. Pp 67.
42. Cameron, Alasdair. 'Theatre in Scotland: 1214 to the present', Bc2, 145–58.
43. Campbell, Ewan. 'Post-medieval pottery in Wales: an archaeological survey', *Post-Medieval Archaeology* 27 (1993), 1–13.
44. Campbell, R.H. 'Too much on the Highlands?: a plea for change', *Scottish Economic & Social History* 14 (1994), 58–76.
45. Carnell, Geoffrey. *The bishops of Peterborough, 1541–1991*. Much Wenlock; Smith; 1993. Pp xiii, 128.
46. Carrington, Peter. *The English Heritage book of Chester*. London; Batsford/English Heritage; 1994. Pp 128.
47. Carter, Jennifer J.; McLaren, C.A. *Crown and gown: an illustrated history of the university of Aberdeen, 1495–1995*. Aberdeen; Aberdeen UP; 1994. Pp 128.
48. Clark, E.A.G. 'The ports of the Exe estuary, 1701–1972', Bc87, 68–78.
49. Clarke, George Anderson. *Deskford parish: loons, lairds, preachers and teachers*. Aberdeen; Aberdeen & North East Scotland Family History Soc.; 1993. Pp viii, 44.
50. Clarke, J. Norman. 'Some notes on the early baptist communities in Horncastle and district', *Lincolnshire Past & Present* 17 (1994), 14–19.
51. Clarke, Patricia A. 'Ickenham manor farm', *London & Middlesex Arch. Soc. T.* 42 (1994 for 1991), 101–13.
52. Cligman, Judith; Crowe, Nigel. *Hertfordshire to Norfolk*. London; English Heritage/HMSO; 1994. Pp x, 123.
53. Coad, Jonathan. 'Architecture and development of Devonport naval base, 1815–1982', Bc87, 167–77.
54. Cookson, Gillian. 'Early textile engineers in Leeds, 1780-1850', Bc112, 40-61.
55. Copsey, Tony. *Book distribution and printing in Suffolk, 1534–1850*. Ipswich; Copsey; 1994. Pp 526.
56. Corder-Birch, Adrian. *A history of Great Yeldham*. Halstead; Halstead and District Local History Soc.; 1994. Pp 132.
57. Coulstock, Patricia H. *The collegiate church of Wimborne minster*. Woodbridge; Boydell; 1993. Pp xii, 279.
58. Coulter, John. *Lewisham: a history and guide*. Stroud; Sutton; 1994. Pp 128.
59. Courtney, Paul. *Medieval and later Usk, vol. 6: report on the excavations at Usk*. Cardiff; Wales UP; 1994. Pp 261.
60. Cox, Barrie. *The place-names of Rutland*. Nottingham; English Place-Name Soc. vols. 67–69; 1994. Pp lxxxvi, 483.
61. Crane, Vera (ed.). *The heart of a London village: the Highgate Literary and Scientific Institution, 1839–1990*. London; Historical Publications; 1991. Pp 80.
62. Craven, Maxwell. *Derby: a history and guide*. Stroud; Sutton; 1994. Pp 128.

63. Cunliffe, Barry; Garratt, Beverley. *Excavations at Portchester castle, vol. 5: post medieval, 1609–1819*. London; Soc. of Antiquaries (report 52); 1994. Pp 250.
64. Currie, Christopher K.; Locock, Martin. 'Excavations at Castle Bromwich Hall gardens, 1989–1991', *Post-Medieval Archaeology* 27 (1993), 111–99.
65. Darvill, Timothy; Gerrard, Christopher. *Cirencester: town and landscape*. Oxford; Oxbow; 1994. Pp 222.
66. Davison, Andrew. 'A genuine and superior article: the last two centuries of brewing in York', *York Historian* 10 (1992), 31–51.
67. Dean, E.B. 'The building of the Methodist church and school hall in Northgate, Oakham, 1865–1935', *Rutland Record* (1993), 95–103.
68. Delair, J.B. 'Pioneer geologists of the Salisbury area', *Wiltshire Arch. & Natural History Magazine* 87 (1994), 127–41.
69. Denning, Royce. *A history of Berkeley Vale*. Parkend; Thornhill; 1994. Pp 160.
70. Devine, Thomas M. *Clanship to crofter's war: the social transformation of the Scottish highlands*. Manchester; Manchester UP; 1994. Pp xv, 258.
71. Dobson, David. *Scottish emigration to colonial America, 1607–1785*. Athens (Ga); Georgia UP; 1994. Pp 216.
72. Dodgshon, R.A. 'West highland and Hebridean settlement prior to crofting and the Clearances: a study in stability or change?', *Soc. of Antiquaries of Scotland P.* 123 (1993), 419–38.
73. Dunlop, A.D. *Hutchesons' Grammar: the history of a Glasgow school*. Glasgow; Hutchesons' Educational Trust; 1992. Pp 160.
74. Dunning, R.W. *Glastonbury: history and guide*. Stroud; Sutton; 1994. Pp 128.
75. Duxbury, Arthur H. 'The decline of the Cumbrian yeoman—Ravenstonedale: a case study', *T. of the Cumberland & Westmorland Antiquarian & Arch. Soc.* 94 (1994), 201–13.
76. Earle, Peter. *A city full of people: men and women of London, 1650-1750*. London; Methuen; 1994. Pp xiii, 320.
77. Ede, Janet; Virgoe, Norma. 'Mapping nonconformity in Norfolk', Bc38, 46–58.
78. Elder, Eileen. 'The Peacocks of north-west Lincolnshire: collectors and recorders of Lincolnshire dialect from *c*.1850 to 1920' [part 2], *Lincolnshire History & Archaeology* 28 (1993), 44–57.
79. Ellson, Peter. 'Stogursey: a monastic fish-pond?', *Somerset & Dorset Notes & Queries* 33 (1994), 323–29.
80. Englander, David. 'Jewish east London, 1850-1950', Bc116, 181–208.
81. Evans, G.G. 'Place-names and field-names of Pennant Melangell', *Montgomeryshire Collections* 82 (1994), 9–22.
82. Evans, John. 'Military aviation' [Pembrokeshire], Bc96, 356–65; 500.
83. Evans, Muriel Bowen. 'The land and its people, 1815–1974' [Pembrokeshire], Bc96, 3–38; 475–78.
84. Farnie, Douglas. 'Cotton waterway: 100 years of Manchester and its ship canal', *History Today* 44/6 (1994), 25–29.
85. Farrant, John H. 'The long man of Wilmington, East Sussex: the documentary evidence reviewed', *Sussex Arch. Collections* 131 (1993), 129–38.

86. Fenton, Alexander. 'Milk products in the everyday diet of Scotland', Bc30, 41–47.
87. Finn, Gerry P.T. 'Faith, hope and bigotry: case studies of anti-Catholic prejudice in Scottish soccer and society', Bc3, 91–112.
88. Fisher, Stephen. 'Introduction' [maritime history of Devon], Bc87, 11–14.
89. Foard, Glenn. 'Ecton: its lost village and landscape park', *Northamptonshire Past & Present* 8 (1993/4), 335–53.
90. Fox, Sir Cyril; Raglan, Lord. *Monmouthshire houses*. Cardiff; Merton Priory/National Museum of Wales. 3 parts. 2nd edn. 1994. Pp 114; 153; 178.
91. Gandy, Michael (ed.). *Catholic parishes in England, Wales and Scotland: an atlas*. London; Gandy; 1993. Pp 32.
92. Geddes, Olive M. *The laird's kitchen: 300 years of food in Scotland*. Edinburgh; HMSO; 1994. Pp x, 110.
93. Gibson, Margaret. *The Liverpool ivories: late antique and medieval ivory and bone carving in Liverpool museum and Walker art gallery*. London; HMSO; 1994. Pp xxvi, 121.
94. Gill, Crispin. 'Ocean liners at Plymouth', Bc87, 226–35.
95. Goddard, Ted. 'Naval activity' [Pembrokeshire], Bc96, 337–45; 498–99.
96. Goodrich, Glyn; Williams, Alan; O'Brien, Colm. 'Excavations at Newcastle quayside: the evolution of the Sandgate', *Archaeologia Aeliana* 5th ser. 22 (1994), 219–33.
97. Grant, Nigel; Humes, Walter. 'Scottish education, 1700-2000', Bc2, 357–72.
98. Gray, Madeleine (ed.). *A history of Monmouthshire: from the coming of the Normans into Wales down to the present time*, vol. 5 [reprint]. Aberystwyth; South Wales Record Soc. no. 8; 1993. Pp xiii, 207.
99. Gray, Vic. 'Introduction' [Cressing Temple: a Templar and Hospitaller manor in Essex] Bc11, 5–10.
100. Greenhill, Basil. 'Epilogue: *The new maritime history of Devon* and the maritime aspects of history', Bc87, 268.
101. Greenhill, Basil; Nix, Michael. 'North Devon shipping, trade and ports, 1786–1939', Bc87, 48–60.
102. Grenville, Jane. 'Research strategies and priorities: an afterthought—the Chester rows', Bc144, 97–107.
103. Griffiths, David. *'A musical place of the first quality': a history of institutional music-making in York, c.1550-1990*. York; York Settlement Trust; 1994. Pp xi, 292.
104. Guillaume, André; Esposito, Marie-Claude. *Londres: histoire d'une place financière*. Paris; P.U.F.; 1993. Pp 305.
105. Hackworth, John. *Sir Henry Fermor school, 1744–1994: a history*. Crowborough; Fermor School; 1994. Pp 230.
106. Haigh, Donald. 'Blue Ball and Baitings, Soyland: a tall story explained', *T. of the Halifax Antiquarian Soc.* ns 2 (1994), 15–27.
107. Hainsworth, Eric. *The doctors of Queensbury: two centuries of general medical practice in a Yorkshire village*. Liskeard; Hainsworth; 1993. Pp vii, 103.

108. Halliday, Robert. 'Wayside graves and crossroad burials', *Norfolk Archaeology* 42 (1994), 80-83.
109. Halliday, S. 'Social mobility, demographic change and the landed elite of County Durham, 1610-1819: an open or shut case?', *Northern History* 30 (1994), 49–63.
110. Harding, Joan Marion; Banks, Joyce F.T. *Newdigate: a Surrey Wealden parish, its history and houses*. Dorking; Banks; 1993. Pp 228.
111. Hargreaves, John D. *Academe and empire: some overseas connections of Aberdeen university, 1860-1970*. Aberdeen; Aberdeen UP; 1994. Pp 128.
112. Harvie, Chris. 'Sport and the Scottish state', Bc3, 43–57.
113. Hatcher, Jane. *Yorkshire to Humberside*. London; HMSO; 1994. Pp x, 123.
114. Hayden, Peter. 'Castle Bromwich Hall gardens: an alternative date', *Post-Medieval Archaeology* 27 (1993), 201–04.
115. Henderson, Hamish. 'The oral tradition', Bc2, 159–72.
116. Henly, Robert. 'The Wiltshire feast', *Genealogists' Magazine* 24 (1994), 393–98.
117. Heslop, D.H.; McCombie, G.; Thomson, C. '"Bessie Surtees house": two merchant houses in Sandhill, Newcastle upon Tyne', *Archaeologia Aeliana* 5th ser. 22 (1994), 1–27.
118. Hilditch, Peter. 'Devon and naval strategy since 1815', Bc87, 155–67.
119. Hilton, John Anthony. *Catholic Lancashire: from Reformation to renewal, 1559–1991*. Chichester; Phillimore; 1994. Pp xi, 146.
120. Hinde, Thomas. *Highgate school: a history*. London; James & James; 1993. Pp 145.
121. Hinde, Thomas. *Paths of progress: a history of Marlborough school*. London; James & James; 1992. Pp 241.
122. Hoare, Peter. *The rectors of St Peter's church, Nottingham, 1241–1991*. Nottingham; St Peter's Parochial Church Council; 1992. Pp 64.
123. Holloway, James. 'Views in Scottish houses, 1630-1770', Bc139, 109–18.
124. Holt, T.G. 'Jesuits in Montgomeryshire, 1670-1873', *J. of Welsh Religious History* 1 (1993), 66–80.
125. Hope, Bryan D. *A curious place: an industrial history of Amlwch, 1550-1950*. Wrexham; Bridge; 1994. Pp 170.
126. Houston, J.F. *Featherbedds and flock bedds: notes on the history of the Worshipful Company of Upholders of the City of London*. Sandy; Three Tents; 1993. Pp 71.
127. Houston, Rab A. 'Elders and deacons: membership of the consistory of the Scots church, Rotterdam (1643–1829) and the kirk session of Tolbooth parish, Edinburgh (1690-1760)', *Tijdschrift voor Sociale Geschiedenis* 20 (1994), 282–308.
128. Howell, David W. 'Farming in Pembrokeshire, 1815–1974', Bc96, 77–110; 479–82.
129. Howell, David W. 'Leisure and recreation, 1815–1974' [Pembrokeshire], Bc96, 418–457; 503–06.
130. Howells, Roscoe. 'The Pembrokeshire islands', Bc96, 208–23; 490-92.

131. Hume, John R. 'Engineering and technology', Bc2, 311–26.
132. Hunt, Donald. *Tunnel: the story of the Channel tunnel, 1802–1994.* Upton-upon-Severn; Images; 1994. Pp 288.
133. Hunter, J.M. 'Littley Park, Great Waltham: historical survey', *Essex Archaeology & History* 3rd ser. 25 (1994), 119–24.
134. Hunter, J.M. 'Medieval and Tudor parks of the middle Chelmer valley', *Essex Archaeology & History* 3rd ser. 25 (1994), 113–18.
135. Hunter, John. 'The historic landscape of Cressing Temple and its environs', Bc11, 25–35.
136. Jarvie, Grant. 'Royal games, sport and the politics of the environment', Bc3, 154–72.
137. Jarvie, Grant; Walker, Graham. 'Ninety-minute patriots?: Scottish sport in the making of the nation', Bc3, 1–8.
138. Jaulmes, Yves. *The French Protestant church of London and the Huguenots: from the church's foundation to the present day.* London; Eglise Protestante Française de Londres; 1993. Pp vi, 59.
139. Jennings, Stuart. *Sermons and stocking frames: the history of Methodism in Ruddington and its links with the framework knitters.* Ruddington; Jennings; 1994. Pp 36.
140. Johnson, Nicholas; Rose, Peter; Bonney, Desmond. *Bodmin Moor: an archaeological survey, vol. 1: the human landscape to c.1800.* London; English Heritage; 1994. Pp 168.
141. Jones, Gareth Elwyn. 'Education, 1815–1974' [Pembrokeshire], Bc96, 389–417; 501–03.
142. Jones, Irene. *The Stalbridge inheritance, 1780-1854.* Dorchester; Larkwood; 1994. Pp xii, 144.
143. Jowitt, J.A. *Mechanization and misery: the Bradford woolcombers' report of 1845.* Halifax; Ryburn; 1991. Pp 44.
144. Kaner, Jennifer. 'Acomb Grange', *York Historian* 10 (1992), 2–17.
145. Kidd, Alan. *Manchester.* Keele; Ryburn; 1993. Pp 256.
146. Kidson, Peter. 'Architectural history' [Lincoln cathedral], Bc5, 14–46.
147. Lambert, Andrew D. 'The impact of naval technology on warship construction and repair at Devonport, 1815–1986', Bc87, 177–88.
148. Lampard, K.J. 'Country banks and economic development in the late eighteenth and early nineteenth centuries: the case of the Margate Bank', *Archaeologia Cantiana* 112 (1994 for 1993), 77–92.
149. Larn, Richard; Larn, Bridget. *Charlestown: the history of a Cornish port.* St Austell; Larn; 1994. Pp xii, 220.
150. le Grice, John. 'A nonconformist descent', Bc38, 155–57.
151. Leach, Peter (ed.). *Ilchester: archaeology, excavations and fieldwork to 1984.* Oxford; Oxbow; 1994. Vol. 2. Pp 261.
152. Leeman, Stephanie Adele. 'Stone walls do not a prison make: the debtors' prison, York', *York Historian* 11 (1994), 23–39.
153. Lenihan, John. 'Mathematics and science', Bc2, 291–310.
154. Lewis, Roy. 'The towns of Pembrokeshire, 1815–1974', Bc96, 39–76; 478–79.
155. Lowerson, John. 'Golf and the making of myths', Bc3, 75–90.
156. Lucas, Robin. 'Ely bricks and roof-tiles and their distribution in Norfolk

and elsewhere in the sixteenth to eighteenth centuries', *Cambridge Antiquarian Soc. P.* 82 (1993), 157–62.

157. Lummis, E.T. 'Officers of the Duke of Norfolk's Regiment and the county of Norfolk', *Norfolk Archaeology* 42 (1994), 97–100.
158. Lyle, Marjorie. *The English Heritage book of Canterbury*. London; Batsford/English Heritage; 1994. Pp 128.
159. Lynch, Michael. 'Scottish culture in its historical perspective', Bc2, 15–46.
160. MacCormick, Neil. 'Law' [Scotland], Bc2, 343–56.
161. MacDougall, Ian (ed.). *'Hard work, ye ken': Midlothian women farmworkers*. Edinburgh; Canongate; 1993. Pp 128.
162. Macmillan, Duncan. 'Scottish art', Bc2, 205–28.
163. Madigan, Tom J. *'The men who wore straw helmets': policing Luton, 1840-1974, including the development and the story of the Luton borough police force, 1876–1947 and the Luton county borough police force, 1964–1966*. Dunstable; Book Castle; 1993. Pp viii, 268.
164. Maguire, W.A. *Belfast*. Keele; Ryburn; 1993. Pp 208.
165. Marks, Lara V. *Model mothers: Jewish mothers and maternity provision in east London, 1870-1939*. Oxford; Oxford UP; 1994. Pp xxi, 320.
166. McAleer, J. Philip. 'Rochester cathedral: the north choir aisle and the space between it and "Gundulf's" tower', *Archaeologia Cantiana* 112 (1994 for 1993), 127–65.
167. McCarra, Kevin. 'Sport in Scotland', Bc2, 279–90.
168. McDonnell, John. 'Antecedents of Border tenant right', *Northern History* 30 (1994), 22–30.
169. McDouall, Andrew, Lord Bankton. *An institute of the laws of Scotland in civil rights with observations upon the agreement or diversity between them and the laws of England in four books after the general method of the viscount of Stair's institutions, vol. 1,* 2. Edinburgh; Stair Soc. vol. 41, 42; 1993, 1994.
170. McGovern, Moira. 'Speech and language therapy education in Edinburgh, 1764–1993', *History of Education Soc. B.* 54 (1994), 34–43.
171. McKay, Ken. 'The port of Milford: oil in the twentieth century', Bc96, 187–96; 488–89.
172. McKay, Ken. 'The port of Milford: the fishing industry', Bc96, 174–88; 487–88.
173. McKean, Charles. 'The Scottishness of Scottish architecture', Bc2, 229–52.
174. Mercer, Peter; Holland, Douglas. *The Hunns mere pit: the story of Woodingdean and Balsdean*. Lewes; Book Guild; 1993. Pp 273.
175. Miles, Dillwyn. 'The tourist industry' [Pembrokeshire], Bc96, 197–207; 489.
176. Mollard, Tom (ed.). *Millgate, Aylsham: a study by the local history research group*. Aylsham; Aylsham Local History Soc.; 1993. Pp vi, 86.
177. Moore, Nicholas J. 'Arlington Row, Bibury, Gloucestershire: early conversion of an industrial building?', *Vernacular Architecture* 25 (1994), 20-24.
178. Moorhouse, H.F. 'From zines like these?: fanzines, tradition and identity in Scottish football', Bc3, 173–94.
179. Moser, Elizabeth. 'The Crescent, Bedford', Bc6, 205–18.

180. Munckton, Thelma. *Somerset paupers: unremembered lives*. Wincanton; Wincanton; 1994. Pp 208.
181. Murray, Athol L.; Burnett, Charles J. 'The seals of the Scottish Court of Exchequer', *Soc. of Antiquaries of Scotland P.* 123 (1993), 439–52.
182. Murray, Catherine. *Oxfordshire to Buckinghamshire*. London; HMSO; 1994. Pp 126.
183. Nelson, Alan H. *Early Cambridge theatres: college, university and town stages, 1464–1720*. Cambridge; Cambridge UP; 1994. Pp 208.
184. Newman, Philip. 'Tinners and tenants on south-west Dartmoor: a case study in landscape history', *Devonshire Association Report & T.* 126 (1994), 199–238.
185. Niblett, Rosalind. 'Excavations at the abbey mill, St Albans', *Hertfordshire Archaeology* 11 (1993), 54–80.
186. Nicolson, Malcolm. 'Medicine' [Scotland], Bc2, 327–42.
187. Noall, Cyril. *The St Ives mining district* vol. 2, ed. Philip Payton and Leonard Truran. Redruth; Dyllansow Truran; 1993. Pp xviii, 158.
188. Northway, Anthony. 'The Devon fishing industry in the eighteenth and nineteenth centuries', Bc87, 126–36.
189. Norton, Ben. *The story of East Dereham*. Chichester; Phillimore; 1994. Pp xiv, 145.
190. Ogden, Mike. *A history of Stockport breweries*. Manchester; Richardson; 1987. Pp 39.
191. Oldham, Arthur A. *The inns and taverns of Wisbech: a survey from the earliest times to 1950*. Wisbech; Spindrift; 1993. Pp 94.
192. Ormrod, David. 'Rochester bridge, 1660-1825', Bc12, 161–219.
193. Pearson, Lynn F. *Building the West Riding: a guide to its architecture and history*. Otley; Smith Settle; 1994. Pp 240.
194. Perrot, Roger. 'The courts of Guernsey', *Guernsey Soc. R.* 48 (1992), 2–8.
195. Pestall, Tim. 'Archaeological investigations into the "lost" village of Eccles-next-the-Sea, Norfolk', *Medieval Settlement Research Group Annual Report* 8 (1993), 17–21.
196. Phillips, C.B.; Smith, J.H. *Lancashire and Cheshire from AD 1540*. London; Longman; 1994. Pp xv, 416.
197. Phillips, Lawrence. 'Pembroke dockyard', Bc96, 152–73; 485–87.
198. Philpin, Audrey. 'Crime and protest, 1815–1974' [Pembrokeshire], Bc96, 305–33; 497–98.
199. Pounds, Norman J.G. 'The Cardiff area: proceedings of the 139th summer meeting of the Royal Archaeological Institute, 1993', *Arch. J.* 150 (1994 for 1993), supplement. Pp. 79.
200. Preston, Harold. *Early East Anglian banks and bankers*. Thetford; Preston; 1994. Pp 168.
201. Preston, Joseph H. *Arundel: a history of the town and the castle*. Selinsgrove (Pe); Susquehanna UP; 1993. Pp 292.
202. Price, M.R. Connop. 'The Pembrokeshire coal industry', Bc96, 111–37; 482–84.
203. Priestley, Ursula. 'The Norwich textile industry: the London connection, 1570-1760', *London J.* 19 (1994), 108–18.
204. Purser, John. 'Music' [Scotland], Bc2, 173–90.

205. Quarmby, Ernest. *Banknotes and banking in the Isle of Man: a guide for historians and collectors*. London; Spink; 1994. Pp 248.
206. Rackham, Oliver. 'Woodland management and timber economy as evidenced by the buildings at Cressing Temple', Bc11, 85–92.
207. Railton, Margaret. *Early medical services: Berkshire and south Oxfordshire from 1740*. Reading; Polmood; 1994. Pp xii, 243.
208. Richards, John; Richards, Margaret. *Timber frame houses in the Scottish countryside*. Edinburgh; HMSO; 1994. Pp v, 65.
209. Richards, Maurice. *A history of Newtown*. Welshpool; Powysland Club; 1993. Pp v, 94.
210. Richards, Stewart. *Wye College and its world: a centenary history*. Wye; Wye College; 1994. Pp 336.
211. Riches, Anne. 'The architect's house and the search for a national style', Bc139, 325–36.
212. Roberts, Brian K. 'Dating villages: theory and practice', *Landscape History* 14 (1992), 19–30.
213. Roberts, Brian K. 'Some relict landscapes in Westmorland: a reconsideration', *Arch. J.* 150 (1994 for 1993), 433–55.
214. Roberts, Dafydd. 'The Pembrokeshire slate industry', Bc96, 138–51; 484–85.
215. Roberts, Martin. *The English Heritage book of Durham*. London; Batsford/English Heritage; 1994. Pp 128.
216. Robey, Tim. 'The archaeology of Cressing Temple', Bc11, 37–50.
217. Robinson, P.W. 'The delights of the Hammam: Turkish baths in Calderdale, *c.*1862–*c.*1940', *T. of the Halifax Antiquarian Soc.* ns 2 (1994), 79–85.
218. Royle, Edward (ed.). *A history of the non-conformist churches of York*. York; Borthwick Inst. (Borthwick texts & calendars 18); 1993. Pp x, 179.
219. Ruderman, Arthur. *A history of Ashford*. Chichester; Phillimore; 1993. Pp x, 150.
220. Rudling, David; Barber, Luke. 'Excavations at the Phoenix Brewery site, Hastings, 1988', *Sussex Arch. Collections* 131 (1993), 73–113.
221. Ryan, Pat. 'The history of Cressing Temple from the documentary sources', Bc11, 11–24.
222. Ryan, Pat; Andrews, David D. 'A brick typology for Cressing Temple', Bc11, 93–103.
223. Ryan, Pat; Andrews, David D. 'The walled garden', Bc11, 105–16.
224. Sampson, Mark. *Brighton: history and guide*. Stroud; Sutton; 1994. Pp 128.
225. Scott, Paul H. 'Introduction' [Scotland: a concise cultural history], Bc2, 9–14.
226. Scott, Vernon. 'Pembrokeshire soldiery', Bc96, 346–55, 499–500.
227. Sewell, Andrew. 'St Michael's church, Aldbourne', *Wiltshire Arch. & Natural History Magazine* 87 (1994), 102–15.
228. Shepherd, Robbie. 'Dance', Bc2, 195–204.
229. Skyrme, J.E. *A Casterbridge ironmonger: Thurmans of Dorchester, 1863–1966*. Weymouth; Rowbotham; 1993. Pp ii, 53.
230. Smith, Donald. 'Culture and religion' [Scotland], Bc2, 47–60.

231. Smith, John T. *Hertfordshire houses: a selective inventory*. London; Royal Commission on the Historical Monuments of England; 1993. Pp xviii, 248.
232. Smith, John T. 'Nine hundred years of St Albans: architecture and social history', *Hertfordshire Archaeology* 11 (1993), 1–22.
233. Smith, Leonard. *Religion and the rise of labour: nonconformity and politics in Lancashire and the West Riding, 1880-1914*. Keele; Ryburn; 1994. Pp 224.
234. Smith, M.A. *Religion in industrial society: Oldham and Saddleworth, 1740-1865*. Oxford; Oxford UP; 1994. Pp xi, 311.
235. Smith, Stuart. *The cinemas and theatres of Exeter*. Wakefield; Mercia Cinema Soc.; 1994. Pp 74.
236. Stanier, Peter H. 'Dorset limekilns: a first survey', *Dorset Natural History & Arch. Soc. P.* 115 (1994 for 1993), 33–49.
237. Starkey, David J. 'Devon's shipbuilding industry, 1786–1970', Bc87, 78–91.
238. Starkey, David J. 'The ports, seaborne trade and shipping industry of south Devon', Bc87, 32–48.
239. Stell, Christopher. *Nonconformist chapels and meeting-houses in the north of England*. London; HMSO (Royal Commission on the Historical Monuments of England); 1994. Pp 380.
240. Stell, Christopher. 'Nonconformist chapels in East Anglia', Bc38, 11–20.
241. Stenning, Dave F. 'The barley barn, the wheat barn and the early development of barns in south-east England', Bc11, 51–75.
242. Stenning, Dave F. 'The Red Lion hotel, Colchester', *Essex Archaeology & History* 3rd ser. 25 (1994), 134–59.
243. Stevenson, Sara. 'Photography', Bc2, 253–66.
244. Stewart, F.J. 'The New Abbey Curling Club: a history', *Dumfriesshire & Galloway Natural History & Antiquarian Soc. T.* 3rd ser. 66 (1991), 79–87.
245. Stewart, James A. *Settlements of western Perthshire: land and society north of the highland line, 1480-1851*. Edinburgh; Pentland; 1990. Pp xi, 303.
246. Storm, Alan. 'Seasonality of births and marriages in a seafaring community before the age of steam', *Local Population Studies* 52 (1994), 43–47.
247. Stratton, Michael. *Ironbridge and the electric revolution: the history of electricity generation at Ironbridge A and B power stations*. London; Murray; 1994. Pp 120.
248. Stringer, Stan. 'Wilmington parochial charities', *Dartford Historical & Antiquarian Soc. Newsletter* 30 (1993), 4–12.
249. Taylor, John Hilton. *The King's school, Tynemouth: a history*. Tynemouth; King's School; 1991. Pp viii, 122.
250. Thistlethwaite, Nicholas. 'Music and worship, 1660-1980' [Lincoln cathedral], Bc5, 77–111.
251. Thomas, W.S.K. *Georgian and Victorian Brecon: portrait of a Welsh county town*. Llandysul; Gomer; 1993. Pp xxiii, 230.
252. Thompson, David M. 'Historical survey, 1750-1949' [Lincoln cathedral], Bc5, 210-318.
253. Thomson, Derick. 'Gaelic literature', Bc2, 127–44.
254. Thorne, Ronald. 'Pembrokeshire and national politics, 1815–1974', Bc96, 227–71; 492–94.
255. Till, Barry. *York against Durham: the guardianship of the spiritualities in*

the diocese of Durham 'sede vacante'. York; Borthwick Inst. (Borthwick paper 84); 1993. Pp vi, 31.

256. Tolman, Susan. *Newbury: a history and guide*. Stroud; Sutton; 1994. Pp 128.
257. Tranter, Neil L. 'Population, migration and labour supply', Bc48, 37–71.
258. Travis, John. 'The rise of the Devon seaside resorts, 1750-1900', Bc87, 136–145.
259. Travis, John F. 'Devon seaside tourism and the coastal environment, 1750-1900', Bc136, 160-79.
260. Tyack, Geoffrey. *Warwickshire country houses*. Chichester; Phillimore; 1994. Pp xxviii, 284.
261. Tyers, Ian. 'Tree-ring dating at Cressing Temple and the Essex curve', Bc11, 77–83.
262. Verey, Anthony. *Berkshire yeomanry*. Stroud; Sutton; 1994. Pp 128.
263. Vincent, E.A. *Geology and mineralogy at Oxford, 1860-1986: history and reminiscence*. Oxford; Vincent; 1994. Pp vii, 245.
264. Wake, Roy. *St Mary's school, Ascot, and its antecedents*. London; Haggerston; 1994. Pp 232.
265. Walker, Graham. 'Nancy Riach and the Motherwell swimming phenomenon', Bc3, 142–53.
266. Walker, John. 'A lodging range at Newland Hall, Roxwell, Essex', *Essex Archaeology & History* 3rd ser. 25 (1994), 160-73.
267. Watkin, Brenda. 'The buildings of Littley Park', *Essex Archaeology & History* 3rd ser. 25 (1994), 125–33.
268. Watson, Roderick. 'Dialects of "voice" and "place": literature in Scots and English from 1700', Bc2, 99–126.
269. Weaver, Cora; Osborne, Bruce E. *Aquae Malvernensis: a history and topography of the springs, spouts, fountains and wells of the Malverns and the development of a public water supply*. Malvern; Weaver; 1994. Pp iv, 248.
270. Webb, Adrian James. *Exchequer special commissions of inquiry: Somerset and Dorset, 1560-1846*. Weston-super-Mare; Galloway; 1993. Pp iv, 24.
271. Welch, Edwin. 'The geography of dissent in Bedfordshire', Bc6, 54–60.
272. Wells, James A. *Signals to danger: railway accidents at Newcastle upon Tyne and in Northumberland, 1851–1992*. Morpeth; Northumberland County Library; 1992. Pp xii, 140.
273. Werbner, Pnina. 'Avoiding the ghetto: Pakistani migrants and settlement shifts in Manchester', Bc29, 240-60.
274. White, Andrew. *A history of Lancaster, 1193–1993*. Keele; Ryburn; 1993. Pp 256.
275. White, Elizabeth (ed.). *Keynsham and Saltford: life and work in times past, 1539–1945*. Keynsham; Keynsham & Saltford Local History Soc.; 1990. Pp 118.
276. Whitworth, Alan. 'Yorkshire dovecots and pigeon lofts: a preliminary survey', *Yorkshire Arch. J.* 65 (1993), 75–89.
277. Williams, K.J. *A history of Boldmere*. Sutton Coldfield; Westwood; 1994. Pp 112.

278. Williamson, Tom. 'The Norfolk nonconformist chapels survey: some preliminary results', Bc38, 59–72.
279. Wilson, Bob; Edwards, Peter. 'Butchery of horse and dog at Witney palace, Oxfordshire, and the knackering and feeding of meat to hounds during the post-medieval period', *Post-Medieval Archaeology* 27 (1993), 43–56.
280. Wilson, John (ed.). *An index of the freemen's roll book of Grimsby, 1780-1980*. Beverley; Humberside County Council, County Archives Service; 1994. Pp iv, 110.
281. Woodall, Joy. *Gin, ale and poultices ... lasers and scanners: Solihull workhouse and hospital, 1742–1993*. Solihull; Woodall; 1994. Pp 96.
282. Woodford, Doreen E. *Methodism in Much Wenlock, 1800-1960: a story of the unremarked*. Newport; GILDERtext; 1994. Pp 100.
283. Woodruff, William. *Billy boy: the story of a Lancashire weaver's son*. Keele; Ryburn; 1993. Pp 288.
284. Worssam, Bernard C.; Tatton-Brown, Tim. 'Kentish Rag and other Kent building stones', *Archaeologia Cantiana* 112 (1994 for 1993), 93–125.
285. Wurtzburg, Rosemary. 'The sub-letting of smallholdings on a Devon estate, 1788–1834', *Devonshire Association Report & T.* 126 (1994), 85–105.
286. Yates, Nigel. *Religion and society in Kent, 1640-1914*. Woodbridge; Boydell (Kent history project 2); 1994. Pp xii, 244.

(c) *Collective Volumes*

1. Groenewegen, Peter (ed.). *Feminism and political economy in Victorian England*. Aldershot; Elgar; 1994. Pp viii, 197.
2. Scott, Paul H. (ed.). *Scotland: a concise cultural history*. Edinburgh; Mainstream; 1993. Pp 416.
3. Jarvie, Grant; Walker, Graham (ed.). *Scottish sport in the making of the nation: ninety-minute patriots?* London; Leicester UP; 1994. Pp ix, 200.
4. Anon. *A miscellany*. Norwich; Norfolk Record Soc vol. 56; 1993. Pp 160.
5. Owen, Dorothy M. (ed.). *A history of Lincoln minster*. Cambridge; Cambridge UP; 1994. Pp xviii, 334.
6. Anon. *Bedfordshire historical miscellany: essays in honour of Patricia Bell*. Bedford; Bedfordshire Historical Record Soc. vol. 72; 1993. Pp xii, 235.
7. O'Sullivan, Patrick (ed.). *The Irish world wide, vol. 3: the creative migrant*. London; Pinter; 1994. Pp 255.
8. Harrison, Brian (ed.). *The history of the university of Oxford, vol. 8: the twentieth century*. Oxford; Oxford UP; 1994. Pp xxix, 872.
9. Matheson, Lister M. (ed.). *The popular and practical science of medieval England*. East Lansing (Mi); Colleagues; 1994. Pp xiii, 425.
10. Hunter, Michael (ed.). *Robert Boyle reconsidered*. Cambridge; Cambridge UP; 1994. Pp xviii, 231.
11. Andrews, David D. (ed.). *Cressing Temple: a Templar and Hospitaller manor in Essex*. Chelmsford; Essex County Council Planning; 1993. Pp 122.
12. Yates, Nigel; Gibson, James M. (ed.). *Traffic and politics: the construction and management of Rochester bridge, AD 43–1993*. Woodbridge; Boydell; 1994. Pp xviii, 408.

13. Hills, John; Ditch, John; Glennerster, Howard (ed.). *Beveridge and social security: an international retrospective*. Oxford; Oxford UP; 1994. Pp xii, 251.
14. Vann, J. Don; VanArsdel, Rosemary T. (ed.). *Victorian periodicals and Victorian society*. Aldershot; Scolar; 1994. Pp ix, 370.
15. Downie, J.A.; Corns, Thomas N. (ed.). *Telling people what to think: early eighteenth-century periodicals from 'The Review' to 'The Rambler'*. London; Cass; 1993. Pp 131.
16. Clayton, Dorothy J.; Davies, Richard G.; McNiven, Peter (ed.). *Trade, devotion and governance: papers in later medieval history*. Stroud; Sutton; 1994. Pp xiv, 208.
17. Sharpe, Kevin; Lake, Peter (ed.). *Culture and politics in early Stuart England*. London; Macmillan; 1994. Pp x, 382.
18. Jones, Charles W. *Bede, the schools and the 'Computus'*, ed. Wesley M. Stevens [reprints not itemised]. London; Variorum; 1994. Pp x, 356.
19. Phillips, John A. (ed.). *Computing parliamentary history: George III to Victoria* [reprints not itemised]. Edinburgh; Edinburgh UP; 1994. Pp v, 146.
20. Pedersen, Susan; Mandler, Peter (ed.). *After the Victorians: private conscience and public duty in modern Britain, essays in memory of John Clive*. London; Routledge; 1994. Pp x, 265.
21. Minnis, A.J. (ed.). *Late-medieval religious texts and their transmission: essays in honour of A.I. Doyle*. Woodbridge; Boydell; 1994. Pp x, 208.
22. Garnett, George; Hudson, John (ed.). *Law and government in medieval England and Normandy: essays in honour of Sir James Holt*. Cambridge; Cambridge UP; 1994. Pp xviii, 387.
23. Batey, Colleen E.; Jesch, Judith; Morris, Christopher D. (ed.). *The viking age in Caithness, Orkney and the north Atlantic*. Edinburgh; Edinburgh UP; 1993. Pp 554.
24. Davies, J.L.; Kirby, D.P. (ed.). *Cardiganshire county history, vol. 1: from the earliest times to the coming of the Normans*. Cardiff; Wales UP; 1994. Pp xx, 445.
25. Horden, John (ed.). *Bibliographia: lectures, 1975–1988, by recipients of the Marc Fitch prize for bibliography*. Oxford; Leopard's Head; 1992. Pp xvi, 160.
26. Reynolds, David (ed.). *Allies at war: the Soviet, American and British experience, 1939–1945*. London; Macmillan; 1994. Pp xxiv, 456.
27. Lacey, Michael J.; Furner, Mary O. (ed.). *The state and social investigation in Britain and the United States*. Cambridge; Cambridge UP; 1993. Pp xii, 352.
28. Munro, John H. (ed.). *Textiles, towns and trade: essays in the economic history of late medieval England and the Low Countries* [reprints not itemised]. London; Variorum; 1994. Pp 344.
29. Drake, Michael (ed.). *Time, family and community: perspectives on family and community history*. Oxford; Blackwell; 1994. Pp ix, 302.
30. Lysaght, Patricia (ed.). *Milk and milk products: from medieval to modern times*. Edinburgh; Canongate; 1994. Pp ix, 229.
31. Pheby, John (ed.). *J.A. Hobson after fifty years: freethinker of the social sciences*. Basingstoke; Macmillan; 1994. Pp xiii, 283.

32. Parkinson, G.H.R. (ed.). *Routledge history of philosophy, vol. 4: the Renaissance and seventeenth-century rationalism.* London; Routledge; 1993. Pp xxix, 444.
33. Wiggershaus, Norbert; Foerster, Roland G. (ed.). *The western security community, 1948–1950: common problems and conflicting national interests during the foundation phase of the north Atlantic alliance.* Oxford; Berg; 1993. Pp xi, 461.
34. Reynolds, David (ed.). *The origins of the cold war in Europe: international perspectives.* London; Yale UP; 1994. Pp xii, 271.
35. Parker, R.H.; Yamey, B.S. (ed.). *Accounting history: some British contributions.* Oxford; Oxford UP; 1994. Pp ix, 661.
36. Anderson, R.G.W.; Bennett, J.A.; Ryan, W.F. (ed.). *Making instruments count: essays on historical scientific instruments presented to Gerard l'Estrange Turner.* Aldershot; Variorum; 1993. Pp xix, 492.
37. Gorodetsky, Gabriel (ed.). *Soviet foreign policy, 1917–1991: a retrospective.* London; Cass; 1994. Pp 227.
38. Virgoe, Norma; Williamson, Tom (ed.). *Religious dissent in East Anglia.* Norwich; Norfolk Arch. & Historical Research Group/Centre of East Anglian Studies; 1993. Pp 170.
39. Schürer, Kevin; Arkell, Tom (ed.). *Surveying the people: the interpretation and use of document sources for the study of population in the later seventeenth century.* Oxford; Leopard's Head; 1992. Pp xv, 308.
40. Buckton, David; Heslop, T.A. (ed.). *Studies in medieval art and architecture presented to Peter Lasko.* Stroud; Sutton; 1994. Pp iv, 220.
41. Copley, Stephen; Garside, Peter (ed.). *The politics of the picturesque: literature, landscape and aesthetics since 1770.* Cambridge; Cambridge UP; 1994. Pp xiv, 304.
42. Holland, Robert (ed.). *Emergencies and disorder in the European empires after 1945.* London; Cass; 1994. Pp xi, 256.
43. Reher, David S.; Schofield, Roger (ed.). *Old and new methods in historical demography.* Oxford; Oxford UP; 1993. Pp viii, 426.
44. Bickers, Robert A. (ed.). *Ritual and diplomacy: the Macartney mission to China, 1792–1794. Papers presented at the 1992 conference of the British Association for Chinese Studies.* London; Wellsweep for British Association for Chinese Studies; 1993. Pp 93.
45. Tarling, Nicholas (ed.). *The Cambridge history of south-east Asia, vol. 2: the nineteenth and twentieth centuries.* Cambridge; Cambridge UP; 1992. Pp 706.
46. Beckles, Hilary; Shepherd, Verene (ed.). *Caribbean freedom: society and economy from emancipation to the present* [reprints not itemised]. London; Currey; 1993. Pp xiv, 581.
47. Fox, Robert; Guagnini, Anna (ed.). *Education, technology and industrial performance in Europe, 1850-1939.* Cambridge; Cambridge UP; 1993. Pp xi, 302.
48. Aldcroft, Derek H.; Ville, Simon P. (ed.). *The European economy, 1750-1914: a thematic approach.* Manchester; Manchester UP; 1994. Pp x, 323.
49. Field, J.V.; James, Frank A.J.L. (ed.). *Renaissance and revolution: humanists, scholars, craftsmen and natural philosophers in early modern Europe.* Cambridge; Cambridge UP; 1993. Pp xv, 291.

50. Parsons, John Carmi (ed.). *Medieval queenship*. Stroud; Sutton; 1993. Pp 264.
51. Wright, D.F. (ed.). *Martin Bucer: reforming church and community*. Cambridge; Cambridge UP; 1994. Pp xiv, 195.
52. Gillingham, John (ed.). *Richard III: a medieval kingship*. London; Collins & Brown. 1993. Pp 154.
53. Barker, Theo C.; Sutcliffe, Anthony (ed.). *Megalopolis: the giant city in history*. New York; St Martin's; 1993. Pp xi, 213.
54. Rumble, Alexander R. (ed.). *The reign of Cnut: king of England, Denmark and Norway*. London; Leicester UP; 1994. Pp xvii, 341.
55. Thompson, F.M.L. (ed.). *Landowners, capitalists and entrepreneurs: essays for Sir John Habakkuk*. Oxford; Oxford UP; 1994. Pp xiii, 312.
56. Chartres, John A. (ed.). *The industrial revolutions, vol. 1: pre-industrial Britain* [mostly reprints]. Oxford; Blackwell for the Economic History Soc.; 1994. Pp 498.
57. Hoppit, Julian; Wrigley, E.A. (ed.). *The industrial revolutions, vols. 2–3: the industrial revolution in Britain* [mostly reprints]. Oxford; Blackwell for the Economic History Soc.; 1994. Pp 522; 436.
58. O'Brien, P.K. (ed.). *The industrial revolutions, vols. 4–5: the industrial revolution in Europe* [mostly reprints]. Oxford; Blackwell for the Economic History Soc.; 1994.
59. Jenkins, D.T. (ed.). *The industrial revolutions, vol. 8: the textile industries* [mostly reprints]. Oxford; Blackwell for the Economic History Soc.; 1994. Pp 438.
60. Pollard, Sidney (ed.). *The industrial revolutions, vol. 9: the metal fabrication and engineering industries* [mostly reprints]. Oxford; Blackwell for the Economic History Soc.; 1994. Pp 490.
61. Church, Roy A. (ed.). *The industrial revolutions, vol. 10: the coal and iron industries* [mostly reprints]. Oxford; Blackwell for the Economic History Soc.; 1994. Pp 514.
62. Michie, R.C. (ed.). *The industrial revolutions, vol. 11: commercial and financial services* [mostly reprints]. Oxford; Blackwell for the Economic History Soc.; 1994. Pp 436.
63. Garrigan, Kristine Ottesen (ed.). *Victorian scandals: representations of gender and class*. Athens (Oh); Ohio UP; 1992. Pp xi, 330.
64. Gilchrist, Roberta Lynn; Mytum, Harold C. (ed.). *Advances in monastic archaeology*. Oxford; Tempus Reparatum (BAR British ser. 227); 1993. Pp ii, 148.
65. James, Winston; Harris, Clive (ed.). *Inside Babylon: the Caribbean diaspora in Britain*. London; Verso; 1993. Pp ix, 317.
66. Barron, Caroline M.; Sutton, Anne F. (ed.). *Medieval London widows, 1300-1500*. London; Hambledon; 1994. Pp xxxiv, 271.
67. Gillingham, John. *Richard Coeur de Lion: kingship, chivalry and war in the twelfth century* [reprints not itemised]. London; Hambledon; 1993. Pp xix, 266.
68. Horrox, Rosemary (ed.). *Fifteenth-century attitudes: perceptions of society in late medieval England*. Cambridge; Cambridge UP; 1994. Pp xii, 244.

69. Chibnall, Marjorie (ed.). *Anglo-Norman studies, XVI*. Woodbridge; Boydell (Proceedings of the Battle conference); 1993. Pp xii, 289.
70. King, Edmund (ed.). *The anarchy of King Stephen's reign*. Oxford; Oxford UP; 1994. Pp xiii, 332.
71. Taylor, A.J.P. *From Napoleon to the second International: essays on nineteenth-century Europe*, ed. Chris Wrigley [reprints not itemised]. London; Hamilton; 1003. Pp xxiii, 426.
72. Shorter, Alfred H. *Studies on the history of papermaking in Britain*, ed. R.L. Hills [reprints not itemised]. Aldershot; Variorum; 1993. Pp 336.
73. Bloomfield, Arthur I. (ed.). *Essays in the history of international trade theory* [reprints not itemised]. Aldershot; Elgar; 1994. Pp 288.
74. Dyer, Christopher C. (ed.). *Everyday life in medieval England* [reprints not itemised]. London; Hambledon; 1994. Pp 340.
75. Sherborne, James. *War, politics and culture in fourteenth-century England*, ed. Anthony Tuck [reprints not itemised]. London; Hambledon; 1994. Pp xvi, 200.
76. Turner, Ralph V. (ed.). *Judges, administrators and the common law in Angevin England* [reprints not itemised]. London; Hambledon; 1994. Pp 336.
77. Bates, David; Curry, Anne E. (ed.). *England and Normandy in the middle ages*. London; Hambledon; 1994. Pp xiv, 336.
78. Curry, Anne E.; Hughes, Michael (ed.). *Arms, armies and fortifications in the hundred years war*. Woodbridge; Boydell; 1994. Pp xvii, 221.
79. Benson, John (ed.). *The miners of Staffordshire, 1840-1914*. Keele; Centre for Local History, Keele University; 1993. Pp 117.
80. Meek, Donald E. (ed.). *Essays in appreciation of the Rev. Christopher Anderson (1782–1852)*. Edinburgh; The Scottish Baptist History Project; 1992. Pp 42.
81. Bawcutt, N.W. (ed.). *The Malone society collections*, vol. 15. Oxford; Oxford UP; 1994. Pp 205.
82. Lascombes, André (ed.). *Spectacle and image in Renaissance Europe*. Leiden; Brill; 1993. Pp viii, 368.
83. MacDonald, A.A.; Lynch, Michael; Cowan, Ian B. (ed.). *The Renaissance in Scotland: studies in literature, religion, history and culture offered to John Durkan*. Leiden; Brill; 1994. Pp xxiv, 428.
84. Brady, Ciaran (ed.). *Interpreting Irish history: the debate on historical revisionism, 1938–1994* [mainly reprints]. Dublin; Irish Academic; 1994. Pp x, 348.
85. Vyner, Blaise (ed.). *Building on the past: papers celebrating 150 years of the Royal Archaeological Institute*. London; Royal Archaeological Institute; 1994. Pp viii, 319.
86. Cooper, Janet (ed.). *A history of the county of Essex, vol. 9: the borough of Colchester*. Oxford; Oxford UP (Victoria History of the Counties of England); 1994. Pp xxiii, 439.
87. Fisher, Stephen (ed.). *A new maritime history of Devon, vol. 2: from the late eighteenth century to the present day*. London; Conway; 1994. Pp 272.
88. Razzell, Peter. *Essays in English population history*. London; Caliban; 1994. Pp v, 229.

89. Fletcher, Anthony; Roberts, Peter (ed.). *Religion, culture and society in early modern Britain: essays in honour of Patrick Collinson*. Cambridge; Cambridge UP; 1994. Pp xx, 372.
90. MacGregor, Arthur (ed.). *Sir Hans Sloane: collector, scientist, antiquary—founding father of the British Museum*. London; British Museum; 1994. Pp 308.
91. Noll, Mark A.; Bebbington, David W.; Rawlyk, George A. (ed.). *Evangelicalism: comparative studies of popular Protestantism in North America, the British Isles and beyond, 1700-1990*. Oxford; Oxford UP; 1994. Pp xv, 430.
92. Russell, G.A. (ed.). *The 'arabick' interest of the natural philosophers in seventeenth-century England*. Leiden; Brill; 1994. Pp xi, 320.
93. Anon (ed.). *Camden miscellany, XXXII*. London; Royal Historical Soc. (Camden 5th ser. vol. 3); 1994. Pp 513.
94. Lilley, J.M.; Stroud, G.; Brothwell, D.R.; Williamson, M.H. *The archaeology of York, vol 12: medieval cemeteries, fasc. 3: the Jewish burial ground at Jewbury* [reprints not itemised]. York; CBA for York Arch. Trust; 1994. Pp 572.
95. Gray, Margaret. *The Trinitarian order in England: excavations at Thelsford priory* [reprints not itemised]. Oxford; Tempus Reparatum (BAR British ser. 226); 1993. Pp vii, 144.
96. Howell, David W. (ed.). *Pembrokeshire county history, vol. IV: modern Pembrokeshire, 1815–1974*. Haverfordwest; Pembrokeshire Historical Soc.; 1993. Pp xxii, 521.
97. Rogers, Nicholas (ed.). *England in the fifteenth century*. Stamford; Watkins (Proceedings of the 1992 Harlaxton Symposium); 1994. Pp xii, 252.
98. Saul, Nigel (ed.). *England in Europe*. London; Collins & Brown; 1994. Pp 192.
99. McEvoy, James J. (ed.). *Robert Grosseteste, exegete and philosopher* [reprints not itemised]. Aldershot; Variorum; 1994. Pp 320.
100. Schichtman, Martin B.; Carley, James P. *Culture and the king*. New York; State of New York UP; 1994. Pp vi, 324.
101. Biller, Peter; Hudson, Anne (ed.). *Heresy and literacy, 1000-1530*. Cambridge; Cambridge UP; 1994. Pp xxv, 313.
102. Fryde, E.B. (ed.). *Essays and poems presented to Daniel Huws*. Aberystwyth; National Library of Wales; 1994. Pp 420.
103. Todd, Margo (ed.). *Reformation to revolution*. London; Routledge; 1994. Pp 302.
104. Beckett, John V.; Polak, J.P.; Riden, Dorothy M.; Kiernan, David T. (ed.). *A seventeenth-century Scarsdale miscellany*. Chesterfield; Derbyshire Record Soc. vol. 20; 1993. Pp iv, 192.
105. Kermode, Jennifer I.; Walker, Garthine (ed.). *Women, crime and the courts in early modern England*. London; UCL; 1994. Pp vii, 216.
106. Erickson, Charlotte (ed.). *Leaving England: essays on British emigration in the nineteenth century*. Ithaca (NY); Cornell UP; 1994. Pp xiv, 272.
107. Meirion-Jones, Gwyn; Jones, Michael C.E. (ed.). *Manorial domestic buildings in England and northern France*. London; Soc. of Antiquaries (occasional paper 15); 1993. Pp xvii, 206.

108. Rollason, David W.; Harvey, Margaret; Prestwich, Michael (ed.). *Anglo-Norman Durham, 1093–1193*. Woodbridge; Boydell; 1994. Pp xxix, 506.
109. Chandler, David; Beckett, Ian F.W. (ed.). *The Oxford illustrated history of the British army*. Oxford; Oxford UP; 1994. Pp xviii, 494.
110. Malloy, Robin Paul; Evensky, Jerry (ed.). *Adam Smith and the philosophy of law and economics*. Dordrecht; Kluwer; 1994. Pp x, 226.
111. Ritchie, Anna (ed.). *Govan and its early medieval sculpture*. Stroud; Sutton; 1994. Pp xiv, 168.
112. Anon (ed.). *Publications of the Thoresby society: miscellany*. Leeds; Thoresby Soc.; 1993. Pp 108.
113. Anon (ed.). *Collections for a history of Staffordshire*. Stafford; Staffordshire Record Soc.; 1994. Pp 253.
114. Bellenger, Aidan (ed.). *Fathers in faith: the western district, 1688–1988—tercentenary essays*. Bath; Downside; 1991. Pp v, 82.
115. Chappell, Vere. *The Cambridge companion to Locke*. Cambridge; Cambridge UP; 1994. Pp x, 330.
116. Pryce, W.T.R. (ed.). *From family to community history*. Cambridge; Cambridge UP; 1994. Pp xvi, 240.
117. Golby, John (ed.). *Communities and families*. Cambridge; Cambridge UP; 1994. Pp xiii, 242.
118. Greengrass, Mark; Leslie, Michael; Raylor, Timothy (ed.). *Samuel Hartlib and the universal reformation: studies in intellectual communication*. Cambridge; Cambridge UP; 1994. Pp xix, 372.
119. Mason, Roger A. *Scots and Britons: Scottish political thought and the union of 1603*. Cambridge; Cambridge UP; 1994. Pp xiv, 323.
120. Ville, Simon P. (ed.). *Shipbuilding in the United Kingdom in the nineteenth century*. St John's (Newfoundland); International Maritime Economic History Association; 1993. Pp 195.
121. Edwards, John; Holm, Poul (ed.). *North sea ports and harbours: adaptations to change*. Esbjerg; Fiskeri og Søfartsmuseet (Proceedings of the second North Sea history conference); 1992. Pp 257.
122. Jablonsky, David (ed.). *Churchill and Hitler: selected essays on the political-military direction of total war* [mostly reprints]. London; Cass; 1994. Pp 323.
123. Aston, Michael; Lewis, Carenza (ed.). *The medieval landscape of Wessex*. Oxford; Oxbow (monograph 46); 1994. Pp 284.
124. Saul, Nigel. *An historical atlas of Britain: prehistoric and medieval*. Stroud; Sutton/National Trust; 1994. Pp x, 214.
125. Anon (ed.). *Bede and his world: the Jarrow lectures, 1958–1993* [reprints not itemised]. Aldershot; Variorum; 1994. Pp 1024.
126. Barber, Malcolm (ed.). *The military orders: fighting for the faith and caring for the sick*. Aldershot; Variorum; 1994. Pp 428.
127. Friedland, Klaus (ed.). *Maritime food transport*. Cologne; Bohlau (Quellen und Darstellungen zur Hansischen Geschichte, neue folge 40); 1994.
128. Migliazza, Alessandro; Decleva, Enrico (ed.). *Diplomazia e storia delle relazioni internazionali: studi in onore di Enrico Serra* [Diplomacy and the history of international relations: studies in honour of Enrico Serra]. Milan; Guiffre; 1991.

129. Anon. *Naissances et fonctionnement des réseaux monastiques et canoniaux: actes du 1er Colloque International du C.E.R.C.O.M.* St Etienne; C.E.R.C.O.M.; 1991.
130. Paravacini, Werner (ed.). *Mare Balticum.* Sigmaringen; Thorbecke; 1992.
131. Binneveld, Hanns; Decker, Rudolf (ed.). *Curing and insuring: essays on illness in past times.* Hilversum; Verloren; 1993. Pp xiv, 169.
132. Puhle, Hans-Jurgen (ed.). *Bürger in der Gesellschaft* [townspeople in society]. Göttingen; Vandenhoeck & Ruprecht; 1992.
133. Higgitt, John (ed.). *Medieval art and architecture in the diocese of St Andrews.* London; British Arch. Association; 1994. Pp viii, 231.
134. Hemming, T.D.; Freeman, E.; Meakin, David. (ed.). *The secular city: studies in the Enlightenment.* Exeter; Exeter UP; 1994. Pp 256.
135. Payton, Philip (ed.). *Cornish studies*, vol. 2. Exeter; Exeter UP; 1994. Pp 179.
136. Fisher, Stephen (ed.). *Man and the maritime environment.* Exeter; Exeter UP; 1994. Pp 264.
137. Lowrey, John (ed.). *Robert Adam* [reprints not itemised]. Edinburgh; Edinburgh UP; 1994. Pp x, 126.
138. Rogers, G.A.J. (ed.). *Locke's philosophy: content and context.* Oxford; Oxford UP; 1994. Pp 280.
139. Gow, Ian; Rowan, Alistair (ed.). *Scottish country houses, 1600-1914.* Edinburgh; Edinburgh UP; 1994. Pp xi, 353.
140. Buckland, Patrick; Belchem, John (ed.). *The Irish in British labour history.* Liverpool; Institute of Irish Studies (Conference Proceedings in Irish Studies 1); 1993. Pp 114.
141. Seldon, Anthony; Ball, Stuart (ed.). *Conservative century: the Conservative party since 1900.* Oxford; Oxford UP; 1994. Pp xix, 842.
142. Forey, Alan (ed.). *Military orders and crusades* [reprints not itemised]. Aldershot; Variorum; 1994. Pp viii, 318.
143. Hall, A. Rupert (ed.). *Science and society: historical essays on the relations of science, technology and medicine* [reprints not itemised]. Aldershot; Variorum; 1994. Pp x, 324.
144. Wood, Jason (ed.). *Buildings archaeology: applications in practice.* Oxford; Oxbow (monograph 43); 1994. Pp 263.
145. Robbins, Keith. *Politicians, diplomacy and war in modern British history.* London; Hambledon; 1994. Pp 320.
146. Fassl, P.; Liebhart, W.; Wust, W. (ed.). *Aus Schwaben und Altbayern* [from Swabia and Old Bavaria]. Sigmaringen; Thorbecke; 1991.
147. Henkel, W. (ed.). *Ecclesiae memoria: miscellanea in onore del R.P. Josef Metzler.* Rome; Herder; 1991.
148. Johnson, Paul (ed.). *Twentieth-century Britain.* London; Longman; 1994. Pp 528.
149. Frantzen, Allen J.; Moffat, Douglas. *The world of work: servitude, slavery and labor in medieval England.* Glasgow; Cruithne; 1994. Pp 232.
150. Brewer, John; Staves, Susan (ed.). *Early modern conceptions of property.* London; Routledge; 1994. Pp 620.
151. Henderson, John; Wall, Richard (ed.). *Poor women and children in the European past.* London; Routledge; 1994. Pp 360.

(d) *Genealogy and Heraldry*

1. Allen, Brigid. *Morrells of Oxford: the family and their brewery, 1743–1993.* Oxford; Oxfordshire Books; 1994. Pp xxi, 169.
2. Anon. *Chirk parish registers, vol. 1: baptisms, 1666–1706; marriages, 1612–1706; burials, 1611–1706.* Wrexham; Clwyd Family History Soc.; 1994. Pp xiv, iv, 31.
3. Anon. *Chirk parish registers, vol. 4: baptisms, burials and marriages, 1803–1812.* Wrexham; Clwyd Family History Soc.; 1993. Pp v, 28.
4. Anon. *Cwm parish registers, 1727–1837.* Wrexham; Clwyd Family History Soc.; 1993. Pp iii, 16, 22.
5. Anon. *Denbigh parish registers, vol. 8: baptisms, 1790-1800; burials, 1790-1800.* Wrexham; Clwyd Family History Soc.; 1994. Pp vii, 41.
6. Anon. *Derwen parish registers, 1632–1812,* vols. 1–2. Wrexham; Clwyd Family History Soc.; 1994. 2 vols.
7. Anon. *Dyserth parish registers, 1602–1812.* Wrexham; Clwyd Family History Soc.; 1992. 3 vols.
8. Anon. *Erbistock parish registers, 1679–1812.* Wrexham; Clwyd Family History Soc.; 1993. 2 vols.
9. Anon. *Flint parish registers, vol. 1 part 1: baptisms, 1598–1685; marriages, 1598–1687.* Wrexham; Clwyd Family History Soc.; 1994.
10. Anon. *Flint parish registers, vol. 1 part 2: burials, 1598–1685 and bishop's transcripts [of] baptisms, 1685–1709; marriages, 1684–1708; burials, 1684–1709.* Wrexham; Clwyd Family History Soc.; 1994.
11. Anon. *Flint parish registers, vol. 3: baptisms, 1737–1764; marriages, 1727–1754; burials, 1727–1764.* Wrexham; Clwyd Family History Soc.; 1993. Pp viii, 67.
12. Anon. *Halkyn parish registers: baptisms, 1706–1719; marriages, 1698–1754; burials, 1698–1764.* Wrexham; Clwyd Family History Soc.; 1993.
13. Anon. *Halkyn parish registers, baptisms, 1803–1812; burials, 1803–1812.* Wrexham; Clwyd Family History Soc.; 1994.
14. Anon. *Halkyn parish registers: marriages, 1754–1812.* Wrexham; Clwyd Family History Soc.; 1993.
15. Anon. *Halkyn parish registers: marriages, 1813–1837.* Wrexham; Clwyd Family History Soc.; 1993. Pp v, 29.
16. Anon. *Hanmer parish registers, vol. 1, part 1: baptisms, 1563–1660.* Wrexham; Clwyd Family History Soc.; 1994. Pp 43.
17. Anon. *Hanmer parish registers, vol. 1, part 2: marriages, 1563–1600; burials, 1563–1600.* Wrexham; Clwyd Family History Soc.; 1994. Pp 36.
18. Anon. *Hanmer parish registers, vol. 2, part 1: baptisms, 1653–1708; marriages, 1655–1708.* Wrexham; Clwyd Family History Soc.; 1993.
19. Anon. *Hanmer parish registers, vol. 2, part 3: baptisms and marriages, 1709–1748.* Wrexham; Clwyd Family History Soc.; 1993.
20. Anon. *Hanmer parish registers, vol. 3, part 3: marriages, 1749–1787; (burials, 1782 only).* Wrexham; Clwyd Family History Soc.; 1993. Pp 41.
21. Anon. *Hawarden parish registers, vol. 11: baptisms and burials, 1755–1770.* Wrexham; Clwyd Family History Soc.; 1994. Pp xvii, 58.
22. Anon. *Hawarden parish registers, vol. 13: marriages & burials, 1771–1790.* Wrexham; Clwyd Family History Soc.; 1994.

23. Anon. *Hawarden parish registers, vol. 15: baptisms, 1801–1810*. Wrexham; Clwyd Family History Soc.; 1994.
24. Anon. *Hawarden parish registers, vol. 5: marriages, 1668–1696; burials: 1666–1698*. Wrexham; Clwyd Family History Soc.; 1994. Pp xvii, 65.
25. Anon. *Hawarden parish registers, vol. 6 part 1: baptisms, 1697–1717*. Wrexham; Clwyd Family History Soc.; 1994.
26. Anon. *Hawarden parish registers, vol. 6 part 2: marriages, 1697–1717; burials, 1699–1716*. Wrexham; Clwyd Family History Soc.; 1994.
27. Anon. *Hawarden parish registers, vol. 7: baptisms and marriages, 1718–1726; burials, 1717–1726*. Wrexham; Clwyd Family History Soc.; 1994.
28. Anon. *Hawarden parish registers, vol. 8: baptisms, 1727–1741; marriages, 1727–1741; burials, 1726–1741*. Wrexham; Clwyd Family History Soc.; 1994. Pp xv, 71.
29. Anon. *Henllan parish registers, vol. 1: baptisms, 1684–1746*. Wrexham; Clwyd Family History Soc.; 1993. Pp xxv, 34.
30. Anon. *Henllan parish registers, vol. 4: baptisms, 1746–1787*. Wrexham; Clwyd Family History Soc.; 1994.
31. Anon. *Henllan parish registers, vol. 5: burials, 1746–1800*. Wrexham; Clwyd Family History Soc.; 1994.
32. Anon. *Holt parish registers, vol. 4: baptisms and burials, 1740-1772*. Wrexham; Clwyd Family History Soc.; 1991. Pp vi, 26.
33. Anon. *Holywell parish registers, vol. 1, part 1: baptisms, 1714–1741*. Wrexham; Clwyd Family History Soc.; 1994. Pp xii, 45.
34. Anon. *Holywell parish registers, vol. 2, part 2: marriages, 1714–1740; vol. 4, part 1: baptisms, 1773–1793*. Wrexham; Clwyd Family History Soc.; 1994. 2 vols.
35. Anon. *Holywell parish registers, vol. 3, part 3: marriages, 1714–1740*. Wrexham; Clwyd Family History Soc.; 1994. Pp viii, 37.
36. Anon. *Holywell parish registers, vol. 4, part 2: marriages, 1772–1787*. Wrexham; Clwyd Family History Soc.; 1994. Pp x, 45.
37. Anon. *Llanarmon Dyffryn Ceiriog parish registers, vol. 1: 1624–1769*. Wrexham; Clwyd Family History Soc.; 1994. Pp xxiv, 54.
38. Anon. *Llanarmon Dyffryn Ceiriog parish registers, vol. 2: baptisms, 1769–1815; marriages and burials, 1769–1813*. Wrexham; Clwyd Family History Soc.; 1993. Pp v, 43.
39. Anon. *Llanasa parish records*, vols. 1, 2, 4. Wrexham; Clwyd Family History Soc.; 1992. 3 vol.
40. Anon. *Llandegla parish registers. vol. 1: baptisms, 1710-1743; marriages, 1710-1738; burials, 1710-1741*. Wrexham; Clwyd Family History Soc.; 1994. Pp vii, 29.
41. Anon. *Llanelidan parish registers: baptisms, 1686–1754; burials, 1695–1754*. Wrexham; Clwyd Family History Soc.; 1992. Pp x, 53.
42. Anon. *Llanferres parish registers, vol. 1: baptisms and burials, 1587–1730; marriages, 1588–1730*. Wrexham; Clwyd Family History Soc.; 1994. Pp xxii, 53.
43. Anon. *Llanferres parish registers, vol. 2: baptisms and burials, 1731–1790; marriages, 1733–1754*. Wrexham; Clwyd Family History Soc.; 1994.
44. Anon. *Llangedwyn parish registers, 1672–1812*. Wrexham; Clwyd Family History Soc.; 1993. Pp 63, 24.

45. Anon. *Llannefydd parish registers, vol. 2: baptisms, 1754–1812 reconstituted from bishop's transcripts*. Wrexham; Clwyd Family History Soc.; 1994. Pp xix, 28.
46. Anon. *Marchwiel parish registers, vol. 1: baptisms, 1652–1700; marriages, 1665–1700; burials, 1622–1700*. Wrexham; Clwyd Family History Soc.; 1993. Pp xii, 20.
47. Anon. *Meliden parish registers, 1602–1812*. Wrexham; Clwyd Family History Soc.; 1992–3. 3 vols.
48. Anon. *Mold parish registers, vol. 3: baptisms, 1650-1675; marriages, 1653–1673; burials, 1652–1675*. Wrexham; Clwyd Family History Soc.; 1991. Pp xxix, 58.
49. Anon. *Mold parish registers, vol. 7: marriages, 1722–1754; vol. 11: burials, 1773–1812*. Wrexham; Clwyd Family History Soc.; 1992. 2 vol.
50. Anon. *Nannerch parish registers, vol. 1: baptisms and burials, 1664–1770; marriages, 1664–1754*. Wrexham; Clwyd Family History Soc.; 1993. Pp xviii, 65.
51. Anon. *Nerquis parish registers: baptisms, 1665–1731; marriages, 1669–1754; burials, 1669–1731*. Wrexham; Clwyd Family History Soc.; 1993.
52. Anon. *Overton parish registers, vol. 1: baptisms, 1602–1639; marriages, 1602–1644; burials, 1602–1650*. Wrexham; Clwyd Family History Soc.; 1994. Pp xx, 33.
53. Anon. *Registers of the church of Holy Trinity, Coventry, Warwickshire, part 3: baptisms, 1653–1745*. Birmingham; Birmingham & Midland Soc. for Genealogy & Heraldry; 1993. Pp 174.
54. Anon. *Rhuddlan parish registers, vol. 1: baptisms, 1681–1734*. Wrexham; Clwyd Family History Soc.; 1994. Pp xviii, 49.
55. Anon. *Rhuddlan parish registers, vol. 2: marriages, 1681–1734; burials, 1681–1734*. Wrexham; Clwyd Family History Soc.; 1994. Pp xiv, 52.
56. Anon. *Rhuddlan parish registers, vol. 4: baptisms, 1772–1797*. Wrexham; Clwyd Family History Soc.; 1994.
57. Anon. *Rhuddlan parish registers, vol. 5: marriages, 1773–1795; burials, 1773–1796*. Wrexham; Clwyd Family History Soc.; 1994.
58. Anon. *Ruthin parish registers, vol. 1: baptisms, 1592–1744*. Wrexham; Clwyd Family History Soc.; 1994. Pp lx, 95.
59. Anon. *St Asaph parish registers, vol. 1: baptisms, 1593–1669; marriages, 1606–1689; burials, 1594–1669*. Wrexham; Clwyd Family History Soc.; 1994.
60. Anon. *St Asaph parish registers, vol. 7, part 2: marriages, 1750-1780; burials, 1750-1780*. Wrexham; Clwyd Family History Soc.; 1994. 2 vols.
61. Anon. *St George parish registers, vol. 1: 1694–1742*. Wrexham; Clwyd Family History Soc.; 1992. Pp xi, 19.
62. Anon. *St George parish registers, vol. 2: baptisms, marriages and burials, 1741–1812*. Wrexham; Clwyd Family History Soc.; 1993. Pp xx, 45.
63. Anon. *Sutton-on-Hull: monumental inscriptions*. Cottingham; East Yorkshire Family History Soc.; 1993. Pp xxxii, 74.
64. Anon. *Trelawnyd parish registers, vol. 1: baptisms, 1700-1812*. Wrexham; Clwyd Family History Soc.; 1993. Pp xxiv, 42.
65. Anon. *Trelawnyd parish registers, vol. 2: marriages, 1700-1836; burials, 1699–1812*. Wrexham; Clwyd Family History Soc.; 1993. Pp xviii, 66.

66. Anon. *Tremeirchion parish registers, vol. 1: baptisms, 1599–1694; marriages, 1600-1689; burials, 1599–1692*. Wrexham; Clwyd Family History Soc.; 1994. Pp xxvi, 61.
67. Anon. *Treuddyn parish registers, 1622–1812*. Wrexham; Clwyd Family History Soc.; 1993. 3 vol.
68. Anon. 'The Clayton Family of Preston, Lancashire', *Catholic Ancestor* 5 (1994), 104–107.
69. Beckett, J.V. *The rise and fall of the Grenvilles: the dukes of Buckingham and Chandos, 1710 to 1921*. Manchester; Manchester UP; 1994. Pp xi, 308.
70. Blake, Philip H. 'The early Derings', *Archaeologia Cantiana* 112 (1994 for 1993), 299–308.
71. Campbell, Mary V. (ed.). *The registers of the church of St Andrew, Clifton, part 1: baptisms, 1813–1837; part 2, burials and marriages, 1813–1837*. Bristol; Bristol Record Office; 1994. 2 vols. Pp 285, 325.
72. Chapman, Colin R. 'Pre-1841 census and population listings in Bedfordshire', Bc6, 3–23.
73. Charlesworth, Andrew. 'The spatial diffusion of riots: popular disturbances in England and Wales, 1750-1850', *Rural History* 5 (1994), 1–22.
74. Cirket, Alan F. (ed.). *An index of Bedfordshire probate records, 1484–1858, part 2: Kimpton-Z*. London; British Record Soc.; 1994. Pp 408.
75. Clifford, John G.; Clifford, Francine (ed.). *Eyam parish register, 1630-1700*. Chesterfield; Derbyshire Record Soc. vol. 21; 1993. Pp xii, 206.
76. Cole, Anne E.; Miller, V.E. (ed.). *Marriage index of Lincolnshire, vol. 4: Graffoe deanery, 1754–1812*. Lincoln; Lincolnshire Family History Soc.; 1994. 2 parts. Pp 4, 73.
77. Cole, Jean A. (ed.). *Wiltshire militia orders, 1759–1770: an index to names, courts martial, promotions, demotions, discharges, etc., with the movement of the Wiltshire regiment of militia and some other reginments*. Trowbridge; Wiltshire Family History Soc.; 1994. Pp 48.
78. Elton, Margaret. *The Eltons of Somerset: rise and fall of a Bristol merchant family*. Stroud; Sutton; 1994. Pp 288.
79. Emmison, Frederick G. (ed.). *Essex wills, vol. 8: the commissary court, 1558–1569*. Chelmsford; Essex Record Office; 1993. Pp xxv, 274.
80. Everard, Richard H.A.J. 'The family of Everard, part 5: Everard of Randlestown, co. Meath', *Irish Genealogist* 9 (1994), 43–72.
81. Green, Peter (ed.). *Norfolk archdeaconry marriage licence bonds, 1813–1837*. Norwich; Norfolk & Norwich Genealogical Soc. vol. 25; 1993. Pp vi, 170.
82. Hanks, Patrick. 'The present-day distribution of surnames in the British Isles', *Nomina* 16 (1992/3), 79–98.
83. Hawgood, David. *An introduction to using computers for genealogy*. Birmingham; Federation of Family History Societies; 1994. Pp 51.
84. Hebden, J.R. (ed.). *Masham peculier: an indexed calendar of wills and administrations*. Ripon; Ripon Historical Soc.; 1994. 2 vols. Pp 180.
85. Hill, John (ed.). *Hertfordshire militia lists: Aspenden*. Ware; Hertfordshire Family & Population History Soc.; 1992. Pp iii,26.
86. Hill, John (ed.). *Hertfordshire militia lists: Baldock*. Ware; Hertfordshire Family & Population History Soc.; 1994. Pp 55.

87. Hill, John (ed.). *Hertfordshire militia lists: Hitchin.* Ware; Hertfordshire Family & Population History Soc.; 1994. 3 vols. Pp 49; 49; 53.
88. Hill, John (ed.). *Hertfordshire militia lists: Reed and Buckland; Datchworth; Westmill; Bygrave, Clothall and Wallingon; Barley.* Ware; Hertfordshire Family & Population History Soc.; 1993. 5 vol. Pp 27; 24; 31; 39; 29.
89. Hill, John (ed.). *Hertfordshire militia lists: St Pauls Walden and Stagenhoe.* Ware; Hertfordshire Family & Population History Soc.; 1992. Pp iii, 44.
90. Hill, John (ed.). *Hertfordshire militia lists: the parish of Bushey, covering the years 1758–1786.* Ware; Hertfordshire Family & Population History Soc.; 1994. Pp 38.
91. Hodgetts, Michael. 'The Yates of Harvington, 1631–1696', *Recusant History* 22 (1994), 152–81.
92. Hollingsbee, Kathleen; Webster, Martyn C. 'St Mary's, Dover, monumental inscriptions', *Archaeologia Cantiana* 112 (1994 for 1993), 309–16.
93. Hurley, Beryl (ed.). *An index of Wiltshire registration districts and parishes from 1837.* Devizes; Wiltshire Family History Soc.; 1992. Pp 17.
94. Hurley, Beryl (ed.). *Local censuses in Wiltshire: miscellaneous censuses, 1695–1887; part 1–2; surviving north Wiltshire 1695 tax censuses, part 1–2.* Trowbridge; Wiltshire Family History Soc.; 1994. 4 vols.
95. Hurley, Beryl (ed.). *Miscellaneous censuses, 1695–1887* [Wiltshire]. Devizes; Wiltshire Family History Soc.; 1994. 2 vol. Pp 52, 52.
96. Hurley, Beryl (ed.). *Surviving north Wiltshire 1695 tax censuses.* Devizes; Wiltshire Family History Soc.; 1994. 2 vol. Pp 54, 54.
97. Jones-Baker, Doris. '"The rest left no posterity here ..." in Plymouth colony: the "lost" descendants of the Mayflower pilgrims of 1620 in England and Holland', *Genealogists' Magazine* 24 (1994), 489–95.
98. Kerswell, R.E. *Bennets of Tesillian.* Edinburgh; Pentland; 1994. Pp xx, 362.
99. Mark, Gordon St George. 'The Joyces of Merview, part 2', *Irish Genealogist* 9 (1994), 89–113.
100. McInnes, Ian. *Annuity Meritorious Service Medal, 1847–1953.* Haslemere; Jade; 1994. Pp xvi, 208.
101. Mitchell, Alison M.; Mitchell, Angus (ed.). *Pre-1855 gravestone inscriptions: an index for the Stewartry of Kirkcudbright, vol. 6.* Edinburgh; Scottish Genealogy Soc.; 1993. Pp 51.
102. Moir, George (comp.). *Peterhead parish, 1801: census material.* Aberdeen; Aberdeen & North-East Scotland Family History Soc.; 1993. Pp iv, 46.
103. Morris, H.F. 'Faulkner's Dublin Journal 1766: births, marriages and deaths', *Irish Genealogist* 9 (1994), 14–42.
104. Musty, John. 'A group of Salisbury poll books', *Wiltshire Arch. & Natural History Magazine* 87 (1994), 143–47.
105. Newsome, Eileen Betty. *An index to the archbishop of York's marriage bonds and allegations, 1765–1779.* London; Garland; 1994. Pp xiii, 915.
106. Newsome, Eileen Betty. (comp.). *An index to the archbishop of York's marriage bonds and allegations, 1750-1764.* York; Borthwick Inst. (list and index 13); 1994. Pp iii, 331.
107. Murchadha, Diarmuid. 'Memorial inscriptions of Irish interest from Somerset', *Irish Genealogist* 9 (1994), 5–13.

108. Ó Murchadha, Diarmuid. 'Nationality names in the Irish Annal', *Nomina* 16 (1992/3), 49–70.
109. Pace, Norman Bedford; Whitehead, Barbara (ed.). *The parish register of St Olave, York, 1650-1785*. Leeds; Yorkshire Arch. Soc. (Parish Register Section); 1993. Pp 364.
110. Raymond, Stuart A. *Londoners' occupations: a genealogical guide*. Birmingham; Federation of Family History Societies; 1994. Pp 48.
111. Ruston, Alan R. *My ancestors were English Presbyterians/Unitarians: how can I find out more about them?* London; Soc. of Genealogists; 1993. Pp 56.
112. Streeter, P.T. (ed.). *The Heath Mount school register: a biographical register of pupils at the school from 1865 to 1992*. Watton-at-Stone; Old Heath Mount School Association; 1992. Pp 136.
113. Thurley, Clifford A.; Thurley, Dorothea (comp.). *An index of the probate records of the consistory court of Ely, 1449–1858, part 1: A-E*, ed. Elisabeth Leedham-Green and Rosemary Rodd. London; British Record Soc. (Index Library vol. 103); 1994. Pp lxii, 376.
114. Upton, Anthony A. *Foleshill probate wills and inventories, 1535–1599*. Ligthorne; Upton; 1993. Pp 82.
115. Webb, Adrian James. *The roll of the 1st and 2nd battalions of the Somerset militia, 1758*. Weston-super-Mare; Galloway; 1993. Pp v, 42.
116. Webb, Adrian James; Skeggs, James. *List of commissioned officers in the local militia for the county of Somerset*. Weston-super-Mare; Galloway; 1993. Pp ix, 28.
117. Willson, F.M.G. *A strong supporting cast: the Shaw Lefèvres, 1789–1936*. London; Athlone; 1993. Pp ix, 409.

C. ROMAN BRITAIN

See also Ba147,b38; Gi67

(a) *Archaeology*

1. Austen, Paul S. 'Recent excavations on Hadrian's wall at Burgh-by-Sands', *T. of the Cumberland & Westmorland Antiquarian & Arch. Soc.* 94 (1994), 35–54.
2. Betts, Ian M.; Foot, Robert. 'A newly identified late Roman tile group from southern England', *Britannia* 25 (1994), 21–34.
3. Bewley, Robert Howard. *Prehistoric and Romano-British settlement in the Solway plain, Cumbria*. Oxford; Oxbow (monograph 36); 1994. Pp 100.
4. Bidwell, Paul T.; Speak, Stephen. *Excavations at South Shields Roman fort*, vol. 1. Newcastle upon Tyne; Soc. of Antiquaries of Newcastle upon Tyne; 1994. Pp 285.
5. Bishop, M.C. *Corstopitum: an Edwardian excavation*. Oxford; Oxbow; 1994. Pp 48.
6. Black, E.W. 'Villa-owners: Romano-British gentlemen and officers', *Britannia* 25 (1994), 99–110.

7. Blockley, Kevin; Ashmore, Fiona; Ashmore, Patrick J. 'Excavations on the Roman fort at Abergavenny, Orchard site, 1972–1973', *Arch. J.* 150 (1994 for 1993), 169–242 + microfiche.
8. Booth, Paul. 'A Roman burial near Welford-on-Avon, Warwickshire', *Birmingham & Warwickshire Arch. Soc. T.* 98 (1994), 37–50.
9. Booth, Paul; Boyle, Angela; Keevil, Graham D. 'A Romano-British kiln site at Lower Farm, Nuneham Courtenay, and other sites on the Didcot to Oxford and Wooton to Abingdon water mains, Oxfordshire', *Oxoniensia* 58 (1993), 87–217.
10. Bowman, Alan K. *Life and letters on the Roman frontier: Vindolanda and its people.* London; British Museum; 1994. Pp 160.
11. Bowman, Alan K.; Thomas, J. David (ed.). *Vindolanda writing tablets: tabulae Vindulandenses, II.* London; British Museum; 1994. Pp 440.
12. Caruana, D.; Shotter, David C.A.; Pirie, E.J.E. 'Roman and medieval coins found during sewer renewal in Carlisle, 1983–1987', *T. of the Cumberland & Westmorland Antiquarian & Arch. Soc.* 94 (1994), 65–76.
13. Casey, P. John; Davies, J.L.; Evans, Jeremy. *Excavations at Segontium (Caernarfon) Roman fort, 1975–1979.* London; CBA (research report 90); 1993. Pp xviii, 346.
14. Crummy, Philip. 'Late iron-age and Roman Colchester', Bc86, 2–18.
15. Dalwood, C.H.; Buteux, V.A.; Darlington, John. 'Excavations at Farrier Street and other sites north of the city wall, Worcester 1988–1992', *T. of the Worcestershire Arch. Soc.* 14 (1994), 75–114.
16. Darlington, John; Evans, June. 'Roman Sidbury, Worcester: excavations 1959–1989', *T. of the Worcestershire Arch. Soc.* 13 (1992), 5–104.
17. Dawson, Michael. *A late-Roman cemetery at Bletsoe.* Oxford; Oxbow; 1994. Pp 68.
18. Ellis, Peter; Evans, Jane; Hannaford, Hugh R.; Hughes, Gwilym; Jones, Alex. 'Excavations in the Wroxeter hinterland, 1988–1990', *Shropshire History & Archaeology* 69 (1994), 1–119.
19. Evans, Edith. 'Military architects and building design in Roman Britain', *Britannia* 25 (1994), 143–64.
20. Ford, S.D. 'The Silchester church: a dimensional analysis and a new reconstruction', *Britannia* 25 (1994), 119–26.
21. Frere, S.S.; Tomlin, R.S.O.; Hassall, M.W.C. (ed.). *The Roman inscriptions of Britain*, vol. 2, fasc. 6. Stroud; Sutton; 1994. Pp xiv, 102.
22. Fulford, Michael G.; Allen, J.L.R.; Rippon, S.J. 'The settlement and drainage of the Wentlooge Level, Gwent: excavation and survey at Rumney Great Wharf', *Britannia* 25 (1994), 175–211.
23. Greep, Stephen. 'Antler roundel pendants from Britain and the north-western Roman provinces', *Britannia* 25 (1994), 79–97.
24. Hands, Anthony Richard. *The Romano-British roadside settlement at Wilcote, Oxfordshire, 1: excavations, 1990-1992.* Oxford; Tempus Reparatum (BAR British ser. 232); 1993. Pp 213.
25. Huskinson, Janet. *Roman sculpture from eastern England.* Oxford; Oxford UP; 1994. Pp 46.
26. Johns, Catherine; Bland, Roger. 'The Hoxne late Roman treasure', *Britannia* 25 (1994), 165–73.

27. Jones, Alex. 'Archaeological excavations at Little Paxton, Diddington, Cambridgeshire, 1992–1993. First interim report: the Romano-British period', *P. of the Cambridge Antiquarian Soc.* 82 (1993), 55–66.
28. Maltby, Mark. 'The animal bones from a Romano-British well at Oakridge II, Basingstoke', *P. of the Hampshire Field Club & Arch. Soc.* 49 (1994), 47–76.
29. Manning, W.H. *Report on the excavations at Usk, 1965–1976: the Roman pottery*. Cardiff; Wales UP; 1993. Pp xviii, 461.
30. Marsden, Peter. *Ships of the port of London: first to eleventh centuries AD*. London; English Heritage; 1994. Pp 237.
31. Meadows, I.D. 'Three Roman sites in Northamptonshire: excavations by E. Greenfield at Bozeat, Higham Ferrers and Great Oakley between 1961 and 1966', *Northamptonshire Archaeology* 24 (1992), 77–94.
32. Medlycott, Maria. 'Iron age and Roman material from Birchanger, near Bishop's Stortford; excavations at Woodside industrial park, 1992', *Essex Archaeology & History* 3rd ser. 25 (1994), 28–45.
33. Medlycott, Maria. 'The Othona community site, Bradwell-on-Sea, Essex: the extra-mural settlement', *Essex Archaeology & History* 3rd ser. 25 (1994), 60-71.
34. O'Connell, Martin G.; Bird, Joanna. 'The Roman temple at Wanborough: excavations, 1985–1986', *Surrey Arch. Collections* 82 (1994), 1–168 + microfiche.
35. Rawlings, Mick. 'Romano-British sites observed along the Codford-Ilchester water pipeline', *Somerset Archaeology & Natural History* 136 (1992), 29–60.
36. Shimmin, D. 'Excavations at Osborne Street, Colchester', *Essex Archaeology & History* 3rd ser. 25 (1994), 46–59.
37. Snape, Margaret E. *Roman brooches from north Britain: a classification and a catalogue of brooches from sites on the Stanegate*. Oxford; Tempus Reparatum (BAR British ser. 235); 1993. Pp v, 123.
38. Snape, Margaret E. 'An excavation of the Roman cemetery at South Shields', *Archaeologia Aeliana* 5th ser. 22 (1994), 43–66.
39. Swan, Vivien G.; Monaghan, Jason. 'Head-pots: a north African tradition in Roman York', *Yorkshire Arch. J.* 65 (1993), 21–38.
40. Symonds, R.P.; Tomber, R.S. 'Late Roman London: an assessment of the ceramic evidence from the city of London', *London & Middlesex Arch. Soc. T.* 42 (1994 for 1991), 59–99.
41. Thorpe, Reuben; Sharman, Josephine. 'An iron age Romano-British enclosure system at Normanton le Heath', *Leicestershire Arch. & Historical Soc. T.* 68 (1994), 1–63.
42. Whitworth, Alan. 'Recording the Roman wall', *Archaeologia Aeliana* 5th ser. 22 (1994), 67–77.
43. Witts, Pat. 'Interpreting the Brading "Abraxas" mosaic', *Britannia* 25 (1994), 111–17.
44. Woodward, Ann; Leach, Peter. *The Uley shrines: excavation of a ritual complex on West Hill, Uley, Gloucestershire, 1977–1979*. London; English Heritage (arch. report 17); 1993. Pp xv, 360.
45. Woolliscroft, D.J. 'Signalling and the design of the Cumberland coast

system', *T. of the Cumberland & Westmorland Antiquarian & Arch. Soc.* 94 (1994), 55–64.

46. Yeoman, P.A.; Stewart, I.J. 'A Romano-British villa estate at Mantles Green, Amersham, Buckinghamshire', *Records of Buckinghamshire* 34 (1994 for 1992), 107–82.
47. Zeepvat, R.J. 'A Roman coin manufacturing hoard from *Magiovinium*, Fenny Stratford, Bucks', *Britannia* 25 (1994), 1–19.

(b) *History*

1. Casey, P. John. *Carausius and Allectus: the British usurpers.* London; Batsford; 1994. Pp 213.
2. Davies, J.L. 'The Roman period' [Cardiganshire], Bc24, 275–317.
3. Dixon, Philip. 'Roman and early medieval Britain', Bc124, 55–112.
4. Fishwick, Duncan. '*Numinibus Aug(ustorum)*', *Britannia* 25 (1994), 127–41.
5. Hanson, W.S. 'Dealing with barbarians: the Romanization of Britain', Bc85, 149–63.
6. Higham, N.J. 'Gildas and "Agitius": a comment on *De excidio*, XX, 1', *B. of the Board of Celtic Studies* 40 (1993), 123–34.
7. Jarrett, Michael G. *Early Roman campaigns in Wales.* Cardiff; National Museum of Wales (Seventh annual Caerleon lecture); 1994. Pp 36.
8. Jarrett, Michael G. 'Non-legionary troops in Roman Britain: part one, the units', *Britannia* 25 (1994), 35–77.
9. Shotter, David C.A. 'Rome and the Brigantes: early hostilities', *T. of the Cumberland & Westmorland Antiquarian & Arch. Soc.* 94 (1994), 21–34.

D. ENGLAND 450-1066

See also Aa24–25; Bb38,c54,124–125; Ca12,30,b3; Eg83,j10–11,25; Ja3,7

(a) *General*

1. Banham, Debby. 'Anglo-Saxon attitudes: in search of the origins of English racism', *European R. of History* 1 (1994), 143–56.
2. Blair, John. *Anglo-Saxon Oxfordshire.* Stroud; Sutton; 1994. Pp xxv, 230.
3. Cessford, Craig. 'The death of Æthelfrith of Lloegr', *Northern History* 30 (1994), 179–83.
4. Dark, Kenneth R. *Civitas to kingdom: British political continuity, 300-800.* London; Leicester UP; 1994. Pp xv, 322.
5. Fairless, Peter. *Northumbria's golden age: the kingdom of Northumbria, AD 547–735.* York; Sessions; 1994. Pp vi, 114.
6. Hall, Richard. *The English Heritage book of viking age York.* London; Batsford/English Heritage; 1994. Pp 128.
7. Higham, N.J. *The English conquest: Gildas and Britain in the fifth century.* Manchester; Manchester UP; 1994. Pp viii, 220.
8. Lund, Niels. *De haerger og de braender: Danmark og England i Vikingetiden*

[They harried and they burned: Denmark and England in the viking age] Copenhagen; Gyldendal; 1993. Pp 212.

9. Mason, Emma. 'Nationalism and national cults in England and on the continent between the tenth and the twelfth centuries', *The Historian* 44 (1994), 9–11.
10. Rumble, Alexander R. 'Introduction' [Cnut in context], Bc54, 1–9.
11. Sarnowsky, Jurgen. 'England und der Kontinent im 10. Jahrhundert' [England and the continent in the tenth century], *Historisches Jahrbuch* 114 (1994), 47–75.

(b) *Politics and Institutions*

1. Bailey, K.A. 'The hidation of Buckinghamshire, part 2: before Domesday', *Records of Buckinghamshire* 34 (1994 for 1992), 87–96.
2. Baines, Arnold H.J. 'Ealdorman Byrhtnoth and the Brayfield charter of 967', *Records of Buckinghamshire* 34 (1994 for 1992), 30–45.
3. Balkwill, C.J. 'Old English *wic* and the origin of the hundred', *Landscape History* 15 (1993), 5–12.
4. Brooks, Sister Beda. 'Archbishop Wulfred (805–832) and the lordship of Minster in Thanet in the early ninth century', *Downside R.* 3 (1993), 217–27.
5. Clarke, Peter A. *The English nobility under Edward the Confessor*. Oxford; Oxford UP; 1994. Pp xi, 386.
6. Clarkson, T.J. 'Richmond and Catraeth', *Cambrian Medieval Celtic Studies* 26 (1993), 15–20.
7. Gretsch, Mechthild. 'The language of the "Fonthill letter"', *Anglo-Saxon England* 23 (1994), 57–102.
8. Hooper, Nicholas. 'Military developments in the reign of Cnut', Bc54, 89–100.
9. Hough, Carole A. 'The early Kentish "divorce laws": a reconsideration of Æthelberht, chs. 79 and 80', *Anglo-Saxon England* 23 (1994), 19–34.
10. Hunt, John. 'Land tenure and lordship in tenth and eleventh-century Staffordshire', *Staffordshire Studies* 4 (1991/2), 1–20.
11. Keynes, Simon. 'Cnut's earls', Bc54, 43–88.
12. Keynes, Simon. 'The "Dunstan B" charters', *Anglo-Saxon England* 23 (1994), 165–93.
13. Keynes, Simon. 'The west Saxon charters of King Æthelwulf and his sons', *English Historical R.* 109 (1994), 1109–49.
14. Klingelhöfer, Eric. *Manor, vill, and hundred: the development of rural institutions in early medieval Hampshire*. Toronto; Pontifical Inst. of Medieval Studies; 1992. Pp 213.
15. Lawson, M.K. 'Archbishop Wulfstan and the homiletic element in the laws of Æthelred II and Cnut' [reprint], Bc4, 141–64.
16. Nelson, Janet L. 'England and the continent in the Anglo-Saxon period', Bc98, 21–35.
17. Taviani-Carozzi, Huguette. 'De l'histoire au mythe: la genéalogie royale anglo-saxonne', *Cahiers de Civilisation Médiévale* 4 (1993), 355–73.
18. Williamson, Tom. *The origins of Norfolk*. Manchester; Manchester UP; 1993. Pp xi, 208.

(c) *Religion*

1. Charles-Edwards, T.M. 'The new edition of Adomnan's *Life of Columba*' [review article], *Cambrian Medieval Celtic Studies* 26 (1993), 65–73.
2. Clayton, Mary; Magennis, Hugh. *Old English lives of St Margaret.* Cambridge; Cambridge UP; 1994. Pp 239.
3. Fell, Christine. 'Saint Æþelðryþ: a historical-hagiographical dichotomy revisited', *Nottingham Medieval Studies* 38 (1994), 18–34.
4. Holder, Arthur G. (ed.). *On the tabernacle* [Bede]. Liverpool; Liverpool UP; 1994. Pp xxvi, 192.
5. Marsden, Richard. 'Old Latin interventions in the Old English *Heptateuch*', *Anglo-Saxon England* 23 (1994), 229–64.
6. McCarthy, Daniel. 'The origin of the *latercus* Paschal cycle of the insular Celtic churches', *Cambrian Medieval Celtic Studies* 28 (1994), 25–49.
7. Meens, Rob. 'A background to Augustine's mission to Anglo-Saxon England', *Anglo-Saxon England* 23 (1994), 5–17.
8. Murphy, Elinor. 'The nunnery that Alfred built at Shaftesbury', *Hatcher R.* 37 (1994), 40-53.
9. O'Loughlin, T.; Conrad-O'Briain, H. 'The "baptism of tears" in early Anglo-Saxon sources', *Anglo-Saxon England* 22 (1993), 65–83.
10. Orchard, A.P. McD. 'Crying wolf: oral style and the *Sermones lupi*', *Anglo-Saxon England* 21 (1992), 239–64.
11. Orchard, Nicholas. 'An eleventh-century Anglo-Saxon missal fragment', *Anglo-Saxon England* 23 (1994), 283–89.
12. Rumble, Alexander R.; Morris, Rosemary. 'Osbern's account of the translation of St Ælfheah's relics from London to Canterbury, 8–11 June 1023', Bc54, 283–315.
13. Schaferdiek, Knut. 'Fragen der frühen angelsächsischen Festlandmission' [problems of the early Anglo-Saxon mission to the continent], *Frühmittelalterliche Studien* 28 (1994), 172–95.
14. Warin, Anne. *Wilfrid, AD 634 to 709.* York; Sessions; 1992. Pp xii, 91.
15. Wood, Ian N. 'The mission of Augustine of Canterbury to the English', *Speculum* 69 (1994), 1–17.

(d) *Economic Affairs and Numismatics*

1. Blackburn, Mark A.S.; Rogerson, A. 'Two viking-age silver ingots from Ditchingham and Hindringham, Norfolk: the first East Anglian ingot finds', *Medieval Archaeology* 37 (1993), 222–24.
2. Hill, David. 'An urban policy for Cnut?', Bc54, 101–05.
3. Jonsson, Kenneth. 'The coinage of Cnut', Bc54, 193–230.
4. O'Hara, Michael Dennis. 'An iron reverse die of the reign of Cnut', Bc54, 231–82.

(e) *Intellectual and Cultural*

1. Adams, J.N.; Deegan, Marilyn. 'Bald's *Leechbook* and the *Physica Plinii*', *Anglo-Saxon England* 21 (1992), 87–114.

2. Carley, James P. 'More pre-Conquest manuscripts from Glastonbury abbey', *Anglo-Saxon England* 23 (1994), 265–81.
3. Clayton, Mary. 'Ælfric's *Judith*: manipulative or manipulated?', *Anglo-Saxon England* 23 (1994), 215–27.
4. Crook, John. '"A worthy antiquity": the movement of King Cnut's bones in Winchester cathedral', Bc54, 165–92.
5. Crook, John. 'King Edgar's reliquary of St Swithun', *Anglo-Saxon England* 21 (1992), 177–202.
6. Cross, J.E. 'A newly identified manuscript of Wulfstan's "commonplace book", Rouen Bibliothèque Municipale, MS 1382 (U.109), fols. 173v-198v', *J. of Medieval Latin* 2 (1992), 63–83.
7. Cross, J.E. 'Atto of Vercelli, *De pressuris ecclesiasticis*, Archbishop Wulfstan, and Wulfstan's "commonplace book"', *Traditio* 48 (1993), 238–46.
8. Davis, Craig R. 'Cultural assimilation in the Anglo-Saxon royal genealogies', *Anglo-Saxon England* 21 (1992), 23–36.
9. Dumville, David N. 'English square miniscule: the mid-century phases', *Anglo-Saxon England* 23 (1994), 133–64.
10. Frank, Roberta. 'King Cnut in the verse of his skalds', Bc54, 106–24.
11. Gameson, Richard. 'English manuscript art in the mid-eleventh century: the decorative tradition', *Antiquaries J.* 71 (1993 for 1991), 64–122.
12. Gameson, Richard. 'The decoration of the Tanner Bede', *Anglo-Saxon England* 21 (1992), 115–59.
13. Graham, Timothy. 'A Parkerian transcript of the list of Bishop Leofric's procurements for Exeter cathedral: Matthew Parker, the Exeter Book, and Cambridge university library MS Ii.2.11', *T. of the Cambridge Bibliographical Soc.* 4 (1994), 421–60.
14. Heslop, T.A. 'A dated "late Anglo-Saxon" illuminated psalter', *Antiquaries J.* 72 (1994 for 1992), 171–74.
15. Hill, Joyce. 'Ælfric and Smaragdus', *Anglo-Saxon England* 21 (1992), 203–37.
16. King, Vernon. 'An unreported early use of Bede's *De natura rerum*', *Anglo-Saxon England* 22 (1993), 85–91.
17. Marshall, Anne; Marshall, Garry. 'Differentiation, change and continuity in Anglo-Saxon buildings', *Arch. J.* 150 (1994 for 1993), 366–402.
18. Milovanovic-Barham, Celica. 'Aldhelm's *Enigmata* and Byzantine riddles', *Anglo-Saxon England* 22 (1993), 51–64.
19. Moffat, Douglas. 'Anglo-Saxon scribes and Old English verse', *Speculum* 67 (1992), 805–27.
20. Muir, Bernard (ed.). *Exeter anthology of Old English poetry: an edition of Exeter Dean and Chapter MS 3501*. Exeter; Exeter UP; 1994. 2 vols. Pp 396; 441.
21. Nees, Lawrence. 'Ultan the scribe', *Anglo-Saxon England* 22 (1993), 127–46.
22. Ohlgren, Thomas H (ed.). *Anglo-Saxon textual illustration.* Toronto; Pontifical Inst. of Medieval Studies; 1992. Pp 576.
23. Powell, Timothy E. 'The "three orders" of society in Anglo-Saxon England', *Anglo-Saxon England* 23 (1994), 103–32.
24. Rusche, Philip G. 'Dry-point glosses to Aldhelm's *De laudibus virginitatis* in Beinecke 401', *Anglo-Saxon England* 23 (1994), 195–213.

25. Schipper, W. 'Dry-point compilation notes in the Benedictional of St Æthelwold', *British Library J.* 20 (1994), 17–34.
26. Sims-Williams, Patrick. 'Byrhtferth's Ogam signature', Bc102, 283–91.
27. Spurrell, Mark. 'The architectural interest of the *Regularis Concordia*', *Anglo-Saxon England* 21 (1992), 161–76.
28. Williams, Peter. *The organ in western culture, 750-1250.* Cambridge; Cambridge UP; 1994. Pp 416.
29. Winkler, John Frederick. 'Roman law in Anglo-Saxon England', *J. of Legal History* 13 (1992), 101–27.
30. Wollmann, Alfred. 'Early Latin loan-words in Old English', *Anglo-Saxon England* 22 (1993), 1–26.

(f) *Society and Archaeology*

1. Alcock, N.W.; Walsh, D. 'Architecture at Cowdery's Down: a reconsideration', *Arch. J.* 150 (1994 for 1993), 403–09.
2. Anderson, Trevor. 'An Anglo-Saxon case of hyperostosis frontalis interna', *Archaeologia Cantiana* 112 (1994 for 1993), 253–59.
3. Aston, Michael. 'Medieval settlement studies in Somerset', Bc123, 219–38.
4. Aston, Michael; Lewis, Carenza. 'Introduction' [landscape of Wessex], Bc123, 1–12.
5. Bayley, Justine; Bryant, Richard; Heighway, Carolyn. 'A tenth-century bell-pit and bell-mould from St Oswald's priory, Gloucester', *Medieval Archaeology* 37 (1993), 224–36.
6. Bond, James. 'Forests, chases, warrens and parks in medieval Wessex', Bc123, 115–58.
7. Brownsword, Roger; Hines, John. 'The alloys of a sample of Anglo-Saxon great square-headed brooches', *Antiquaries J.* 73 (1994 for 1993), 1–10.
8. Costen, Michael. 'Settlement in Wessex in the tenth century: the charter evidence', Bc123, 97–113.
9. Cox, Barrie. 'The pattern of Old English *burh* in early Lindsey', *Anglo-Saxon England* 23 (1994), 35–56.
10. Croft, Robert A. 'Protecting medieval settlement sites in Somerset', Bc123, 255–69.
11. Dickinson, Tania M.; Farley, Michael E. 'An Anglo-Saxon saucer brooch from Hedsor, Buckinghamshire', *Records of Buckinghamshire* 34 (1994 for 1992), 97–100.
12. Eagles, Bruce. 'The archaeological evidence for settlement in the fifth to seventh centuries AD', Bc123, 13–32.
13. Eagles, Bruce; Mortimer, Catherine. 'Early Anglo-Saxon artefacts from Hod Hill, Dorset', *Antiquaries J.* 73 (1994 for 1993), 132–40.
14. Evans, Angela Care. *The Sutton Hoo ship burial.* London; British Museum; 2nd edn. 1994. Pp 128.
15. Evison, Vera. *An Anglo-Saxon cemetery at Great Chesterford, Essex.* Oxford; Oxbow; 1994. Pp 234.
16. Farley, Michael E. 'The Cop round barrow at Bledlow, Buckinghamshire: prehistoric or Anglo-Saxon?', *Records of Buckinghamshire* 34 (1994 for 1992), 11–13.

17. Fellows-Jensen, Gillian. 'Danish place-names and personal names in England: the influence of Cnut? ', Bc54, 125–40.
18. Field, Naomi; Leahy, Kevin. 'Prehistoric and Anglo-Saxon remains at Nettleton Top, Nettleton', *Lincolnshire History & Archaeology* 28 (1993), 9–38.
19. Gardiner, Mark. 'The excavation of a late Anglo-Saxon settlement at Market Field, Steyning, 1988–1989', *Sussex Arch. Collections* 131 (1993), 21–67.
20. Girsch, Elizabeth Stevens. 'Metaphorical usage, sexual exploitation, and divergence in the Old English terminology for male and female slaves', Bc149, 30-54.
21. Goodburn, Damian. 'Fragments of a 10th-century timber arcade from Vintner's Place on the London waterfront', *Medieval Archaeology* 37 (1993), 78–92.
22. Grinsell, L.V. 'Barrows in the Anglo-Saxon land charters', *Antiquaries J.* 71 (1993 for 1991), 46–63.
23. Hamerow, Helena. 'Migration theory and the migration period', Bc85, 164–77.
24. Hare, John. 'Agriculture and rural settlement in the chalklands of Wiltshire and Hampshire from *c.*1200-*c.*1500', Bc123, 159–70.
25. Hase, Patrick H. 'The Church in the Wessex heartlands', Bc123, 47–81.
26. Hills, Catherine; Penn, Kenneth; Rickett, Robert. *The Anglo-Saxon cemetery at Spong Hill, North Elmham, part V: catalogue of cremations.* Norwich; East Anglian Archaeology (report 67); 1994. Pp 252.
27. Hines, John. *Clasps, Hektespenner, Agraffen: Anglo-Scandinavian clasps of classes A-C of the third to sixth centuries AD. Typology, diffusion and function.* Stockholm; Almquist & Wiksell; 1993. Pp 126.
28. Hines, John. 'A gold bracteate from Kingston Bagpuize, Oxfordshire', *Medieval Archaeology* 37 (1993), 219–22.
29. Hinton, David A. 'The archaeology of eighth to eleventh-century Wessex', Bc123, 33–46.
30. Hooke, Della. *Pre-Conquest charter-bounds of Devon and Cornwall.* Woodbridge; Boydell; 1994. Pp x, 265.
31. Hooke, Della. 'The administrative and settlement framework of early medieval Wessex', Bc123, 83–95.
32. Hughes, Michael. 'Towns and villages in medieval Hampshire', Bc123, 195–212.
33. Karras, Ruth Mazo. 'Desire, descendants and dominance: slavery, the exchange of women, and masculine power', Bc149, 16–29.
34. Lewis, Carenza. 'Patterns and processes in the medieval settlement of Wiltshire', Bc123, 171–94.
35. Liddle, Peter; Middleton, Samantha. 'An Anglo-Saxon cemetery at Wigston Magna, Leicestershire', *Leicestershire Arch. & Historical Soc. T.* 68 (1994), 64–86.
36. Lowden, John. 'On the purpose of the Sutton Hoo ship burials', Bc40, 91–101.
37. McDougall, Ian. 'Serious entertainments: an examination of a peculiar type of viking atrocity', *Anglo-Saxon England* 22 (1993), 201–25.

38. McKinley, Jaqueline I. *Spong Hill, part VIII: the cremations*. Oxford; Oxbow; 1994. Pp 148.
39. Okasha, Elisabeth. 'A second supplement to *Hand-list of Anglo-Saxon non-Runic inscriptions*', *Anglo-Saxon England* 21 (1992), 37–85.
40. Oosthuizen, Susan. 'Saxon commons in south Cambridgeshire', *Cambridge Antiquarian Soc. P.* 82 (1993), 93–100.
41. Ovitt, George, Jr. 'The cultural context of western technology: early Christian attitudes toward manual labour', Bc149, 71–94.
42. Rippon, Stephen. 'Medieval wetland reclamation', Bc123, 239–54.
43. Ruffing, John. 'The labour structure of Ælfric's colloquy', Bc149, 55–70.
44. Samson, Ross. 'The end of medieval slavery', Bc149, 95–124.
45. Taylor, Christopher C. 'The regular village plan: Dorset revisited and revised', Bc123, 213–18.
46. Timby, Jane. 'Sancton I Anglo-Saxon cemetery: excavations carried out between 1976 and 1980', *Arch. J.* 150 (1994 for 1993), 243–365.

E. ENGLAND 1066–1500

See also Aa64,b50,62,68; Ba40,47,104,141,147,b38,79,99,134,166,183,185,195,255, c9,11,16,18,21–22,28,40,50,52,64,66–70,74–78,94–95,97,99,101,108,123–124,126, 133,142,149; Ca12,30; Da9,e28,f3–4,6,10,24,31–32,34,42,45; Fe24,g61,k36; Gi22

(a) *General*

1. Barrow, G.W.S. 'The kings of Scotland and Durham', Bc108, 311–23.
2. Barrow, G.W.S. 'The Scots and the north of England' [during the Anarchy], Bc70, 231–53.
3. Bates, David. 'The forged charters of William the Conqueror and Bishop William of St Calais', Bc108, 111–24.
4. Bates, David. 'The rise and fall of Normandy, *c.*911–1204', Bc77, 19–35.
5. Bevan, Bryan. *Henry IV*. London; Rubicon; 1994. Pp ix, 166.
6. Biggs, Douglas. '"A wrong whom conscience and kindred bid me to right": a reassessment of Edmund Langley, duke of York, and the usurpation of Henry IV', *Albion* 26 (1994), 253–72.
7. Bouet, Pierre. 'Henri Ier Beauclerc, roi d'Angleterre et duc de Normandie vu par Orderic Vital', *Les Conférences d'Histoire Locale du Lycée de Domfront* 12 (1992), 4–20.
8. Carley, James P. 'A grave event: Henry V, Glastonbury abbey, and Joseph of Arimathaea's bones', Bc100, 129–48.
9. Carpenter, Christine. 'Who ruled the midlands in the later middle ages?', *Midland History* 19 (1994), 1–20.
10. Carpenter, David. 'King Henry III and the Tower of London', *London J.* 19 (1994), 95–107.
11. Chamberlain, David. 'Marie de France's Arthurian *Lai*: subtle and political', Bc100, 15–34.

12. Chaplais, Pierre. *Piers Gaveston: Edward II's adoptive brother*. Oxford; Oxford UP; 1994. Pp xiv, 150.
13. Chibnall, Marjorie. 'Normandy' [during the Anarchy], Bc70, 93–115.
14. Chibnall, Marjorie. 'The charters of the Empress Matilda', Bc22, 276–98.
15. Contamine, Phillipe. 'The Norman "nation" and the French "nation" in the fourteenth and fifteenth centuries', Bc77, 215–234.
16. Crawford, Anne (ed.). *Letters of the queens of England, 1100-1547*. Stroud; Sutton; 1994. Pp 250.
17. Crook, David. '"Thieves and plunderers": an anti-ministerial protest of 1296', *Historical Research* 67 (1994), 327–36.
18. Crouch, David. 'The hidden history of the twelfth century', *Haskins Soc. J.* 3 (1993), 111–30.
19. Crouch, David. 'The March and the Welsh kings', Bc70, 255–89.
20. Curry, Anne E. 'Lancastrian Normandy: the jewel in the crown?', Bc77, 235–52.
21. Dalton, Paul. *Conquest, anarchy and lordship: Yorkshire, 1066–1154*. Cambridge; Cambridge UP; 1994. Pp xxii, 345.
22. Dalton, Paul. 'Scottish influence on Durham, 1066–1214', Bc108, 339–52.
23. Davies, R.R. 'The failure of the first British empire? England's relations with Ireland, Scotland and Wales, 1066–1500', Bc98, 121–32.
24. Dunbabin, Jean. 'Geoffrey of Chaumont, Thibaud of Blois and William the Conqueror', Bc69, 101–16.
25. Gillingham, John. 'Introduction: interpreting Richard III', Bc52, 1–19.
26. Golding, Brian. *Conquest and colonization: the Normans in Britain, 1066–1100*. London; Macmillan; 1994. Pp xiv, 227.
27. Halliday, Robert. 'Robert de Vere, ninth earl of Oxford', *Medieval History* 3 (1993), 71–85.
28. Hammond, Peter W. 'The reputation of Richard III', Bc52, 133–49.
29. Harriss, G.L. 'The king and his subjects' [in the fifteenth century], Bc68, 13–28.
30. Helmerichs, Robert. 'King Stephen's Norman itinerary', *Haskins Soc. J.* 3 (1993), 89–97.
31. Hicks, Michael. 'Richard, duke of Gloucester: the formative years', Bc52, 21–37.
32. Horrox, Rosemary. 'Fifteenth-century attitudes', Bc68, 1–12.
33. Horrox, Rosemary. 'The government of Richard III', Bc52, 75–74.
34. Hubert, Madeleine. *A la reconquête d'un trône: la succession du Conquérant au XIIè siècle avec la complicité de l'Emperesse Mathilde et de Robert de Caen*. Condé-sur-Noireau; Corlet; 1992. Pp 185.
35. Justice, Steven. *Writing and rebellion: England in 1381*. Berkeley (Ca); California UP; 1994. Pp xiv, 289.
36. King, Edmund. 'Introduction' [the Anarchy], Bc70, 1–36.
37. Lund, Niels. 'Knut des Heiligen beabsichtiger Zug nach England im Jahre 1085' [The intended expedition of Cnut the holy to England in 1085], Bc130, 101–10.
38. Maddicott, J.R. *Simon de Montfort*. Cambridge; Cambridge UP; 1994. Pp xxv, 404.
39. Maddicott, J.R. 'Follower, leader, pilgrim, saint: Robert de Vere, earl of

Oxford, at the shrine of Simon de Montfort', *English Historical R.* 109 (1994), 641–53.
40. Matthew, D.J.A. 'Durham and the Anglo-Norman world', Bc108, 1–22.
41. Matthew, D.J.A. 'The English cultivation of Norman history', Bc77, 1–18.
42. McLaren, Mary-Rose. 'The aims and interests of the London chroniclers of the fifteenth century', Bc16, 158–76.
43. Michael, Michael. 'The little land of England is preferred before the great kingdom of France: the quartering of the royal arms by Edward III', Bc40, 113–26.
44. Mortimer, Richard. *Angevin England, 1154–1272.* Oxford; Blackwell; 1994. Pp 266.
45. Moss, Vincent. 'Normandy and England in 1180: the pipe roll evidence', Bc77, 185–95.
46. Ormrod, W.M. 'St Maurus' wind: the weather in medieval England', *Medieval History* 3 (1993), 59–63.
47. Ormrod, W.M. 'The domestic response to the hundred years war', Bc78, 83–101.
48. Powell, Edward. 'After "After McFarlane": the poverty of patronage and the case for constitutional history', Bc16, 1–16.
49. Strickland, Matthew. 'Against the Lord's anointed: aspects of warfare and baronial rebellion in England and Normandy, 1075–1265', Bc22, 56–79.
50. Thompson, Kathleen. 'Orderic Vitalis and Robert de Bellême', *J. of Medieval History* 20 (1994), 133–41.
51. Turner, Ralph V. *King John.* London; Longman; 1994. Pp xii, 306.
52. Walker, David. *The Normans in Britain.* Oxford; Blackwell; 1994. Pp viii, 178.
53. Wall, Valerie. 'Malcolm III and the foundation of Durham cathedral', Bc108, 325–37.
54. Watkiss, Leslie; Chibnall, Marjorie (ed.). *The Waltham chronicles.* Oxford; Oxford UP; 1994. Pp 152.
55. Watts, J.L. 'When did Henry VI's minority end?', Bc16, 116–39.
56. White, Graeme. 'Continuity in government' [during the Anarchy], Bc70, 117–43.
57. Wickson, A.R.D. 'England in 1092', *Medieval History* 3 (1993), 118–27.
58. Williams, Daniel. 'Richard III and his overmighty subjects: in defence of a king', Bc97, 56–71.
59. Young, Alan. 'The bishopric of Durham in Stephen's reign', Bc108, 353–68.

(b) *Politics*

1. Arthurson, Ian. *The Perkin Warbeck conspiracy, 1491–1499.* Stroud; Sutton; 1994. Pp xii, 244.
2. Cron, B.M. 'The duke of Suffolk, the Angevin marriage and the ceding of Maine, 1445', *J. of Medieval History* 20 (1994), 75–99.
3. Crouch, David. 'A Norman "conventio" and bonds of lordship in the middle ages', Bc22, 299–324.
4. Given-Wilson, Chris. 'Richard II, Edward II and the Lancastrian inheritance', *English Historical R.* 109 (1994), 553–71.

5. Hollister, C. Warren. 'The magnates of Stephen's reign: reluctant anarchists', *Haskins Soc. J.* 3 (1993), 77–87.
6. Holt, J.C. '1153: the treaty of Winchester', Bc70, 291–316.
7. Leland, John L. 'The abjuration of 1388', *Medieval Prosopography* 15 (1994), 115–39.
8. McNiven, Peter. 'Rebellion, sedition and the legend of Richard II's survival in the reigns of Henry IV and Henry V', *B. of the John Rylands University Library of Manchester* 76/1 (1994), 93–117.
9. Richmond, Colin F. '1483: the year of decision (or taking the throne)', Bc52, 39–55.
10. Thompson, Kathleen. 'William Talvas, count of Ponthieu, and the politics of the Anglo-Norman realm', Bc77, 169–84.

(c) *Constitution, Administration and Law*

1. Allen, Martin. 'The Durham mint before Boldon book', Bc108, 381–98.
2. Barlow, Frank. 'English kings and the Church (1066–1154)', *Medieval History* 3 (1993), 171–77.
3. Barton, J.L. 'On *The teaching of Roman law in England around 1200*' [review article], *J. of Legal History* 14 (1993), 53–58.
4. Blackburn, Mark A.S. 'Coinage and currency' [during the Anarchy], Bc70, 145–205.
5. Boorman, Julia. 'The sheriffs of Henry II and the significance of 1170', Bc22, 255–75.
6. Brand, Paul. '"Time out of mind": the knowledge and use of the eleventh- and twelfth-century past in thirteenth-century litigation', Bc69, 37–54.
7. Cheney, Mary. '"Possessio/proprietas" in ecclesiastical courts in mid twelfth-century England', Bc22, 245–54.
8. Crook, David. 'The archbishopric of York and the extent of the forest in Nottinghamshire in the twelfth century', Bc22, 325–40.
9. Harriss, G.L. 'The medieval parliament' [review article], *Parliamentary History* 13 (1994), 206–26.
10. Hudson, John. *Land, law and lordship in Anglo-Norman England*. Oxford; Oxford UP; 1994. Pp ix, 320.
11. Hudson, John. 'Anglo-Norman land law and the origins of property', Bc22, 198–222.
12. Ibbetson, David. 'From property to contract: the transformation of sale in the middle ages', *J. of Legal History* 13 (1992), 1–22.
13. Jefferson, Lisa. 'MS Arundel 48 and the earliest statutes of the Order of the Garter', *English Historical R.* 109 (1994), 356–85.
14. Kemp, Brian R. 'Archdeacons and parish churches in England in the twelfth century', Bc22, 341–64.
15. Kiralfy, Albert. 'Taking the will for the deed: the mediaeval criminal attempt', *J. of Legal History* 13 (1992), 95–100.
16. Maxfield, David K. 'A fifteenth-century lawsuit: the case of St Anthony's hospital', *J. of Ecclesiastical History* 44 (1993), 199–223.
17. McDonald, Peter. 'St Augustine's abbey and the see of Canterbury: the struggle with Archbishop Mepham', *Southern History* 15 (1993), 1–17.

18. Meaney, Audrey L. 'Gazetteer of hundred and wapentake meeting-places of the Cambridge region', *Cambridge Antiquarian Soc. P.* 82 (1993), 67–92.
19. Milsom, S.F.C. 'The origin of prerogative wardship', Bc22, 223–44.
20. Neville, Cynthia J. 'Keeping the peace on the northern marches in the later middle ages', *English Historical R.* 109 (1994), 1–25.
21. Payling, Simon J. 'Arbitration, perpetual entails and collateral warranties in late-medieval England', *J. of Legal History* 13 (1992), 32–62.
22. Powell, Edward. 'Law and justice' [in the fifteenth century], Bc68, 29–41.
23. Robinson, O.F. 'Bracton and Hengham: the wages of virtue', *J. of Legal History* 15 (1994), 73–76.
24. Round, J. Horace; Powell, W.R. 'The liberties of the borough of Colchester', *Essex Archaeology & History* 3rd ser. 25 (1994), 72–78.
25. Seipp, David J. 'The concept of property in the early common law', *Law & History R.* 12 (1994), 29–91.
26. Sharpe, Richard. 'The prefaces of *Quadripartitus*', Bc22, 148–72.
27. Stein, Peter. 'The Vacarian school', *J. of Legal History* 13 (1992), 23–31.
28. Summerson, Henry R.T. 'Old and new bishoprics: Durham and Carlisle', Bc108, 369–80.
29. Summerson, Henry R.T. 'The enforcement of the statute of Winchester, 1285–1327', *J. of Legal History* 13 (1992), 232–50.
30. Vincent, Nicholas. 'The origins of the Winchester pipe rolls', *Archives* 21 (1994), 25–42.
31. Virgoe, Roger. 'Inheritance and litigation in the fifteenth century: the Buckenham disputes', *J. of Legal History* 15 (1994), 23–40.
32. Wormald, Patrick. '*Quadripartitus*', Bc22, 111–47.

(d) *External Affairs*

1. Arnold, Benjamin. 'Germany and England, 1066–1453', Bc98, 76–87.
2. Fauchon, Jacques. 'Avranches, point d'appui de la diplomatie anglaise en Bretagne à la fin de la guerre de cent ans', *Revue de l'Avranchin et du Pays de Granville* 69 (1992), 45–51.
3. Goodman, Anthony. 'Before the Armada: Iberia and England in the middle ages', Bc98, 108–20.
4. Grant, Alexander. 'Foreign affairs under Richard III', Bc52, 113–32.
5. O'Connor, Stephen. 'Finance, diplomacy and politics: royal service by two London merchants in the reign of Edward III', *Historical Research* 67 (1994), 18–39.
6. Ormrod, W.M. 'England, Normandy and the beginnings of the hundred years war, 1259–1360', Bc77, 197–213.
7. Saul, Nigel. 'England and Europe: problems and possibilities', Bc98, 9–20.
8. Vale, Malcolm. 'The end and aftermath of the hundred years war', Bc98, 151–62.

(e) *Religion*

1. Allmand, Christopher. 'The English and the Church in Lancastrian Normandy', Bc77, 287–97.

2. Barrow, Julia. 'English cathedral communities and reform in the late tenth and eleventh centuries', Bc108, 25–39.
3. Bird, Nicholas du Q. 'A medieval inventory from Glastonbury abbey', *Somerset & Dorset Notes & Queries* 33 (1994), 310-20.
4. Burgess, Clive; Kümin, Beat. 'Penitential bequests and parish regimes in late medieval England', *J. of Ecclesiastical History* 44 (1993), 610-30.
5. Burton, Janet E. *Monastic and religious orders in Britain, 1000-1300.* Cambridge; Cambridge UP; 1994. Pp xi, 354.
6. Burton, Janet E. 'The monastic revival in Yorkshire: Whitby and St Mary's, York', Bc108, 41–51.
7. Carnwath, Julia. 'The churchwardens' accounts of Thame, Oxfordshire, *c.*1443–1524', Bc16, 177–97.
8. Chibnall, Marjorie. 'Le problème des réseaux monastiques en Angleterre', Bc129, 341–52.
9. Chibnall, Marjorie. 'Monastic foundations in England and Normandy, 1066–1189', Bc77, 37–49.
10. Connolly, Margaret. 'Public revisions or private responses?: the oddities of BL Arundel MS 197, with special reference to *Contemplations of the dread and love of God*', *British Library J.* 20 (1994), 55–64.
11. Crosby, Everett U. *Bishop and chapter in twelfth-century England: a study of the 'mensa episcopalis'.* Cambridge; Cambridge UP; 1994. Pp xiv, 450.
12. Davril, Anselme (ed.). *The Winchcombe sacramentary: Orléans, Bibliothèque Municipale 127 (105).* London; Henry Bradshaw Soc. vol. 109; 1994. Pp viii, 280.
13. Doyle, A.I. 'Introduction' [late-medieval religious texts], Bc21, 1–7.
14. Duncan, Thomas G. 'Two Middle English penitential lyrics: sound and scansion', Bc21, 55–65.
15. Eckert, Thomas. 'Nichthäretische Papstkritik in England vom Beginn des 14. bis zur zweiten Hälfte des 15. Jahrhunderts' [Non-heretical criticism of the papacy in England from the beginning of the fourteenth century to the second half of the fifteenth century], *Annuarium Historiae Conciliorum* 23 (1991), 116–359.
16. Edwards, A.S.G. 'The transmission and audience of Osbern Bokenham's *Legendys of hooly wummen*', Bc21, 157–67.
17. Erler, Mary C. 'Three fifteenth-century vowesses', Bc66, 165–83.
18. Fletcher, Alan J. 'A hive of industry or a hornets' nest?: MS Sidney Sussex 74 and its scribes', Bc21, 131–55.
19. Foreville, Raymonde. 'Naissance d'un ordre double: l'ordre de Sempringham', Bc129, 163–74.
20. Fröhlich, Walter (ed.). *The letters of St Anselm of Canterbury*, vol. 2. Kalamazoo (Mi); Cistercian Publications; 1993. Pp 352.
21. Gilchrist, Roberta Lynn; Oliva, Marilyn. *Religious women in medieval East Anglia, c.1100-1540.* Norwich; Centre for East Anglian Studies (Studies in East Anglian History 1); 1993. Pp 116.
22. Gillespie, Vincent. 'Thy will be done: *Piers Plowman* and the *Pater noster*', Bc21, 95–119.
23. Hanna, Ralph, III. '"Meddling with makings" and Will's work', Bc21, 85–94.

24. Hargreaves, Harry. 'The Wycliffite *Glossed Gospels* as source: further evidence', *Traditio* 48 (1993), 247–51.
25. Hillaby, Joe. 'Beth miqdash me'at [Sanctuary in minature]: the synagogues of medieval England', *J. of Ecclesiastical History* 44 (1993), 182–98.
26. Holdsworth, Christopher. 'The Church' [during the Anarchy], Bc70, 207–29.
27. Hudson, Anne. 'Aspects of the "publication" of Wyclif's Latin sermons', Bc21, 121–29.
28. Hudson, Anne. '*Laicus litteratus*: the paradox of Lollardy', Bc101, 222–36.
29. Kemp, Brian R. 'Towards admission and institution: English episcopal formulae for the appointment of parochial incumbents in the twelfth century', Bc69, 155–76.
30. Long, R.J. 'Richard Fishacre's *Super S. Augustini librum de haeresibus adnotationes*: an edition and commentary', *Archives d'Histoire Doctrinale et Litteraire du Moyen Age* 60 (1993), 207–79.
31. Loyn, Henry. 'Abbots of English monasteries in the period following the Norman Conquest', Bc77, 95–103.
32. Marx, C.W. 'British Library Harley MS 1740 and popular devotion', Bc97, 207–22.
33. Millett, Bella. '*Mouvance* and the medieval author: re-editing *Ancrene wisse*', Bc21, 9–20.
34. Moore, R.I. 'Literacy and the making of heresy, *c*.1000-*c*.1150', Bc101, 19–37.
35. Morenzoni, F. (ed.). *Thomas de Chobham: Sermones*. Turnhout; Brepols; 1993 (Corpus Christianorum continuatio medievalis 82A) . Pp xxvii, 322.
36. Morgan, Nigel. 'The coronation of the Virgin by the Trinity and other texts and images of the glorification of Mary in fifteenth-century England', Bc97, 223–41.
37. Nichols, Ann Eljenholm. *Seeable signs: the iconography of the seven sacraments, 1350-1544*. Woodbridge; Boydell; 1994. Pp xx, 412.
38. Orme, Nicholas. 'The clergy of Clyst Gabriel, Devon, 1312–1508', *Devonshire Association Report & T*. 126 (1994), 107–21.
39. Owen, Dorothy M. 'Historical survey, 1091–1450' [Lincoln cathedral], Bc5, 112–63.
40. Owen, Dorothy M. 'Introduction: the English Church in eastern England, 1066–1110', Bc5, 1–13.
41. Philpott, Mark. 'The *De iniusta uexacione Willelmi episcopi primi* and canon law in Anglo-Norman Durham', Bc108, 125–37.
42. Pickering, O.S. 'The outspoken *South English legendary* poet', Bc21, 21–37.
43. Poulle, Béatrice. 'Savigny and England', Bc77, 159–168.
44. Powell, Sue. 'The transmission and circulation of *The lay folk's catechism*', Bc21, 67–84.
45. Richmond, Colin F. 'Margins and marginality: English devotion in the later middle ages', Bc97, 242–52.
46. Richmond, Colin F. 'Religion' [in the fifteenth century], Bc68, 183–201.
47. Robson, Michael. 'Franciscan lectors at Christ Church cathedral priory, Canterbury, 1275–1314', *Archaeologia Cantiana* 112 (1994 for 1993), 261–81.

48. Ruthven, Suzanne. 'The survival of witchcraft in medieval England', *Medieval History* 3 (1993), 166–70.
49. Spencer, H. Leith. *English preaching in the late middle ages*. Oxford; Oxford UP; 1993. Pp xvi, 542.
50. Swanson, R.N. 'Literacy, heresy, history and orthodoxy: perspectives and permutations for the later middle ages', Bc101, 279–93.
51. Thompson, Benjamin. 'Free alms tenure in the twelfth century', Bc69, 221–43.
52. Thompson, Benjamin. 'The laity, the alien priories and the redistribution of ecclesiastical property', Bc97, 19–41.
53. Thompson, John J. 'Another look at the religious texts in Lincoln cathedral library, MS 91', Bc21, 169–87.
54. Tillotson, John. 'Visitation and reform of the Yorkshire nunneries in the fourteenth century', *Northern History* 30 (1994), 1–21.
55. Timmins, T.C.B. (ed.). *The register of John Waltham, bishop of Salisbury, 1388–1395*. Woodbridge; Boydell (Canterbury & York Soc. 80); 1994. Pp xxiv, 331.
56. Tudor, Victoria. 'Durham priory and its hermits in the twelfth century', Bc108, 67–78.
57. Vaughn, Sally N. 'Anselm in Italy, 1097–1100', Bc69, 245–70.
58. Vincent, Nicholas (ed.). *English episcopal acta, vol. 9: Winchester, 1205–1238*. Oxford; Oxford UP for British Academy; 1994. Pp xxvi, 263.
59. Walker, John. 'Crusaders and patrons: the influence of the crusades on the patronage of the Order of St Lazarus in England', Bc126, 327–32.
60. Watson, Nicholas. 'The composition of Julian of Norwich's *Revelation of love*', *Speculum* 68 (1993), 637–83.
61. Windeatt, Barry A. (ed.). *English mystics of the middle ages*. Cambridge; Cambridge UP; 1994. Pp 311.
62. Wogan-Browne, Jocelyn. 'The apple's message: some post-Conquest hagiographic accounts of textual transmission', Bc21, 39–53.

(f) *Economic Affairs*

1. Adams, Mary. 'History of the demesne farm at Appledore from contemporary building records', *Archaeologia Cantiana* 112 (1994 for 1993), 283–98.
2. Adams, Terence R. 'Aliens, agriculturalists and entrepreneurs: identifying the market-makers in a Norfolk port from the water-bailiffs' accounts, 1400-1460', Bc16, 140-57.
3. Atkin, M.A. 'Land use and management in the upland demesne of the de Lacy estate of Blackburnshire, *c.*1300', *Agricultural History R.* 42 (1994), 1–19.
4. Baxter, W.T. 'Early accounting: the tally and the checker-board' [reprint], Bc35, 197–235.
5. Bradley, Helen. 'The Datini factors in London, 1380-1410', Bc16, 55–79.
6. Campbell, Bruce M.S.; Galloway, James A.; Keene, Derek; Murphy, Margaret. *A medieval capital and its grain supply: agrarian production and distribution in the London region, c.1300.* Cheltenham; Inst. of British Geographers (research paper 30); 1993. Pp 235.

7. Clark, Elaine. 'Mothers at risk of poverty in the medieval English countryside', Bc151, 139–59.
8. Clegg, Nancy W.; Reed, Clyde G. 'The economic decline of the Church in medieval England', *Explorations in Economic History* 31 (1994), 261–80.
9. Cooper, Janet. 'Medieval Colchester', Bc86, 19–66.
10. Cromarty, D.; Cromarty, R. (ed.). *The wealth of Shrewsbury in the early fourteenth century: six local subsidy rolls—text and commentary.* Shrewsbury; Shropshire Arch. & Historical Soc.; 1993. Pp 125.
11. Faith, Rosamund. 'Demesne resources and labour rent on the manors of St Paul's cathedral, 1066–1222', *Economic History R.* 2nd ser. 47 (1994), 657–78.
12. Gervers, Michael. '*Pro defensione Terre Sancte*: the development and exploitation of the Hospitallers' landed estate in Essex', Bc126, 3–20.
13. Harvey, P.D.A. 'Boldon Book and the wards between Tyne and Tees', Bc108, 399–405.
14. Harvey, P.D.A. 'Manorial accounts' [reprint], Bc35, 91–115.
15. Jenkins, H.J.K. 'Medieval barge traffic and the building of Peterborough cathedral', *Northamptonshire Past & Present* 8 (1993/4), 255–62.
16. Kermode, Jennifer I. 'Medieval indebtedness: the regions *versus* London', Bc97, 72–88.
17. Langdon, John. 'Lordship and peasant consumerism in the milling industry of early fourteenth-century England', *Past & Present* 145 (1994), 3–46.
18. Lewis, Elisabeth A. *The Southampton port and brokage books, 1448–1449.* Southampton; Southampton UP (Southampton Records ser. vol. 36); 1993. Pp xliv, 276.
19. Nobes, C.W. 'The Gallerani account book of 1305–1308' [reprint], Bc35, 239–49.
20. Noke, C. 'Agency and the *excessus* balance in manorial accounts' [reprint], Bc35, 139–59.
21. Owen, A.E.B. 'Beyond the sea bank: sheep on the Huttoft outmarsh in the early thirteenth century', *Lincolnshire History & Archaeology* 28 (1993), 39–41.
22. Postles, David A. 'Some differences between seigneurial demesnes in medieval Oxfordshire', *Oxoniensia* 58 (1993), 219–32.
23. Postles, David A. 'The perception of profit before the leasing of demesnes' [reprint], Bc35, 116–38.
24. Richmond, Colin F. 'The transition from feudalism to capitalism in the archives of Magdalen college, Oxford: a note', *History Workshop* 37 (1994), 165–69.
25. Runyan, Timothy J. 'Wine and war: the Anglo-Gascon wine trade in the later middle ages', Bc127, 245–55.
26. Stopford, J. 'Modes of production among medieval tilers', *Medieval Archaeology* 37 (1993), 93–108.
27. Stuart, D.G. 'A rental of the borough of Burton, 1319', Bc113, 1–51; 227–32.
28. Titow, Jan. 'Lost rents, vacant holdings and the contraction of peasant cultivation after the Black Death', *Agricultural History R.* 42 (1994), 97–114.
29. Tringham, Nigel J. 'The early thirteenth-century survey of the earl of Chester's fee of Leek', *Staffordshire Studies* 5 (1993), 1–12.

30. Wade, J.F. 'The overseas trade of Newcastle upon Tyne in the late middle ages', *Northern History* 30 (1994), 31–48.

(g) *Social Structure and Population*

1. Aers, David. 'Justice and wage-labor after the Black Death: some perplexities for William Langland', Bc149, 169–90.
2. Aird, W.M. 'An absent friend: the career of Bishop William of St Calais', Bc108, 283–97.
3. Almond, Richard. 'Medieval hunting: ruling classes and the community', *Medieval History* 3 (1993), 147–55.
4. Angers, Denise. 'La guerre et le pluralisme linguistique: aspects de la guerre de cent ans', *Annales de Normandie* 43 (1993), 124–39.
5. Aston, Margaret. 'Corpus Christi and corpus regni: heresy and the Peasants' revolt', *Past & Present* 143 (1994), 3–47.
6. Aston, Margaret. 'Death' [in the fifteenth century], Bc68, 202–28.
7. Bailey, Mark. 'Rural society' [in the fifteenth century], Bc68, 150-68.
8. Barron, Caroline M. 'Johanna Hill (d. 1441) and Johanna Sturdy (d. *c.*1460), bell-founders', Bc66, 99–111.
9. Barron, Caroline M. 'The widows' world in later medieval London', Bc66, xiii-xxxiv.
10. Birrell, Jean. 'A great thirteenth-century hunter: John Giffard of Brimpsfield', *Medieval Prosopography* 15 (1994), 37–66.
11. Bishop, Louise M. 'Hearing God's voice: kind wit's call to labor in *Piers Plownam*', Bc149, 191–205.
12. Bradley, Helen. 'Lucia Visconti, countess of Kent (d. 1424)', Bc66, 77–84.
13. Brault, Gerard J.; Hines, John. 'A French source of the Lord Marshal's roll (1295–1296)', *Antiquaries J.* 73 (1994 for 1993), 27–36.
14. Britnell, Richard H. 'The Black Death in English towns', *Urban History* 21 (1994), 195–210.
15. Brown, A.L. 'The House of Commons, 1386–1421: a commentary', *Parliaments, Estates & Representations* 14 (1994), 1–18.
16. Brown, S.D.B. 'Leavetaking: lordship and mobility in England and Normandy in the twelfth century', *History* 79 (1994), 199–215.
17. Brown, Vivien (ed.). *Eye priory cartulary and charters*, part 2. Woodbridge; Boydell (Suffolk charters 13); 1994. Pp xviii, 193.
18. Burgess, Clive. *The parish church and the laity in late medieval Bristol.* Bristol; Historical Association, Bristol Branch; 1992. Pp 22.
19. Carpenter, Christine. 'Gentry and community in medieval England', *J. of British Studies* 33 (1994), 340-80.
20. Castor, Helen. '"Walter Blount was gone to serve traytours": the sack of Elvaston and the politics of the north midlands in 1454', *Midland History* 19 (1994), 21–39.
21. Clark, Elaine. 'Social welfare and mutual aid in the medieval countryside', *J. of British Studies* 33 (1994), 381–406.
22. Cownie, Emma. 'Gloucester abbey, 1066–1135: an illustration of religious patronage in Anglo-Norman England', Bc77, 143–57.

23. Crouch, David. 'Normans and Anglo-Normans: a divided aristocracy?', Bc77, 51–67.
24. Cullum, Patricia H. '"For pore people harberles": what was the function of the maisonsdieu?', Bc16, 36–54.
25. Dane, Joseph A. 'Who is buried in Chaucer's tomb?—prolegomena', *Huntington Library Q.* 57 (1994), 99–123.
26. Davies, Matthew. 'Thomasyne Percyvale, "the maid of Week" (d. 1512)', Bc66, 186–207.
27. Düll, Siegrid; Luttrell, Anthony; Keen, Maurice. 'Faithful unto death: the tomb slab of Sir William Neville and Sir John Clanvowe, Constantinople 1391', *Antiquaries J.* 71 (1993 for 1991), 174–90.
28. Dyer, Christopher C. 'The English medieval village community and its decline', *J. of British Studies* 33 (1994), 407–29.
29. Finch, Andrew John. 'Sexual morality and canon law: the evidence of the Rochester consistory court', *J. of Medieval History* 20 (1994), 261–75.
30. Foulds, Trevor (ed.). *The Thurgarton cartulary.* Stamford; Watkins; 1994. Pp ccxxi, 814.
31. Frantzen, Allen J. 'The work of work: servitude, slavery and labor in medieval England', Bc149, 1–15.
32. Galloway, Andrew. 'The making of a social ethic in late-medieval England: from *Gratitudo* to "kyndnesse"', *J. of the History of Ideas* 55 (1994), 365–83.
33. Gillingham, John. '1066 and the introduction of chivalry into England', Bc22, 31–55.
34. Goldberg, P. Jeremy P. 'Women in later medieval English archives', *J. of the Soc. of Archivists* 15 (1994), 59–71.
35. Goldberg, P. Jeremy P. 'Women' [in the fifteenth century], Bc68, 112–31.
36. Hanawalt, Barbara A. *Growing up in medieval London: the experience of childhood in history.* Oxford; Oxford UP; 1993. Pp xvii, 300.
37. Harper-Bill, Christopher (ed.). *Charters of the medieval hospitals of Bury St Edmunds.* Woodbridge; Boydell (Suffolk charters 14); 1994. Pp xvi, 272.
38. Hatcher, John. 'England in the aftermath of the Black Death', *Past & Present* 144 (1994), 3–35.
39. Hayes, Rosemary. 'The "private life" of a late medieval bishop: William Alnwick, bishop of Norwich and Lincoln', Bc97, 1–18.
40. Hettinger, Madonna J. 'Defining the servant: legal and extra-legal terms of employment in fifteenth-century England', Bc149, 206–28.
41. Hollister, C. Warren. 'The aristocracy' [during the Anarchy], Bc70, 37–66.
42. Horrox, Rosemary. 'Service' [in the fifteeenth century], Bc68, 61–78.
43. Humphery-Smith, Cecil R. *Hugh Revel: master of the Hospital of St John of Jerusalem, 1258–1277.* Chichester; Phillimore; 1994. Pp xv, 136.
44. Huneycutt, Lois L. 'Female succession and the language of power in the writings of twelfth-century churchmen', Bc50, 189–201; 220-21.
45. Illsley, John Sherwood. 'The medieval gentry and the English constitution from the Conquest to the reign of Edward I', *Medieval History* 3 (1993), 3–20.
46. James, Susan E. 'Sir William Parr of Kendal, part 2: 1471–1483', *T. of the Cumberland & Westmorland Antiquarian & Arch. Soc.* 94 (1994), 105–20.

47. Jefferies, P.J. 'Profitable fourteenth-century legal practice and landed investment: the case of Judge Stonor, *c.*1281–1354', *Southern History* 15 (1993), 18–35.
48. Jones, Michael C.E.; Walker, Simon (ed.). 'Private indentures for life service in peace and war, 1278–1476', Bc93, 1–190.
49. Keefe, Thomas K. 'Proffers for heirs and heiresses in the Pipe rolls: some observations on indebtedness in the years before Magna Carta', *Haskins Soc. J.* 3 (1993), 99–109.
50. Keene, Derek. 'Tanners' widows, 1300-1350', Bc66, 1–27.
51. Kirkby, J.L. *The Hungerford cartulary: a calendar of the earl of Radnor's cartulary of the Hungerford family*. Devizes; Wiltshire Record Soc. vol. 49; 1994. Pp xix, 300.
52. Lacey, Kay. 'Margaret Croke (d. 1491)', Bc66, 143–64.
53. Lachaud, Frederique. 'An aristocratic wardrobe of the late thirteenth century: the confiscation of the goods of Osbert de Spaldington in 1298', *Historical Research* 67 (1994), 91–100.
54. Laughton, Jane. 'Women in court: some evidence from fifteenth-century Chester', Bc97, 89–99.
55. Lewis, Chris P. 'The Domesday jurors', *Haskins Soc. J.* 3 (1993), 17–44.
56. Lucas, Peter J. 'John Capgrave, friar of Lynn', *The Historian* 44 (1994), 23–24.
57. Maddern, Phillipa. '"Best trusted friends": concepts and practices of friendship among fifteenth-century Norfolk gentry', Bc97, 100-17.
58. Mason, Emma. 'The purposeful patronage of Henry of Blois', *Medieval History* 3 (1993), 30-50.
59. McRee, Ben R. 'Peacemaking and its limits in later medieval Norwich', *English Historical R.* 109 (1994), 831–66.
60. Mertes, Kate. 'Aristocracy' [in the fifteenth century], Bc68, 42–60.
61. Moffat, Douglas. 'Sin, conquest, servitude: English self-image in the chronicles of the early fourteenth century', Bc149, 146–68.
62. Moon, J.H. 'Thomas Rolff: an examination of the structure of society within the Kentish high weald based upon the study of a fifteenth-century parish clerk', *Archaeologia Cantiana* 112 (1994 for 1993), 193–212.
63. Moore, John S. 'Family-entries in English *libri vitae*, *c.*1050 to *c.*1530, part 1', *Nomina* 16 (1992/3), 99–128.
64. Neville, Cynthia J. 'Common knowledge of the common law in later medieval England', *Canadian J. of History* 29 (1994), 461–78.
65. O'Connor, Stephen. 'Adam Fraunceys and John Pyel: perceptions of status among merchants in fourteenth-century London', Bc16, 17–35.
66. O'Connor, Stephen. 'Joan Pyel (d. 1412)', Bc66, 71–75.
67. Palliser, D.M. 'Urban society' [in the fifteenth century], Bc68, 132–49.
68. Parsons, John Carmi. 'Introduction: family, sex, and power—the rhythms of medieval queenship', Bc50, 1–11.
69. Parsons, John Carmi. 'Mothers, daughters, marriage, power: some Plantagênet evidence, 1150-1500', Bc50, 63–78; 206–09.
70. Payling, Simon J. 'The waning of noble lordship in late fifteenth-century England' [review article], *Parliamentary History* 13 (1994), 322–32.
71. Prestwich, J.O. 'The career of Ranulf Flambard', Bc108, 299–310.

72. Purdon, Liam O. 'The rite of vassalage in *Havelok the Dane*', *Medievalia et Humanistica* ns 20 (1994), 25–39.
73. Rawcliffe, Carole. 'Margaret Stodeye, Lady Philipot (d. 1431)', Bc66, 85–98.
74. Renn, Derek F. 'Burhgeat and gonfanon: two sidelights from the Bayeux Tapestry', Bc69, 177–98.
75. Reynolds, Susan. *Fiefs and vassals*. Oxford; Oxford UP; 1994. Pp 544.
76. Roffe, David. 'Hereward "the Wake" and the barony of Bourne: a reassessment of a Fenland legend', *Lincolnshire History & Archaeology* 29 (1994), 7–10.
77. Rosser, Gervase. 'Going to the fraternity feast: commensality and social relations in late medieval England', *J. of British Studies* 33 (1994), 430-46.
78. Rubin, Miri. 'The poor' [in the fifteenth century], Bc68, 169–82.
79. Rutledge, Paul. 'The will of Oliver Wyth, 1291', Bc4, 9–30.
80. Schofield, Philip. 'Frankpledge lists as indices of migration and mortality: some evidence from Essex lists', *Local Population Studies* 52 (1994), 23–29.
81. Searle, Eleanor. 'Housed in abbeys: the Dacres of Cumberland and Sussex, part 1: the early Dacres and Lanercost priory', *Huntington Library Q.* 57 (1994), 153–65.
82. Smith, Mary Frances. 'Archbishop Stigand and the eye of the needle', Bc69, 199–219.
83. Stafford, Pauline. 'The portrayal of royal women in England, mid-tenth to mid-twelfth centuries', Bc50, 143–67; 217–20.
84. Stratford, Jenny. 'Joan Buckland (d. 1462)', Bc66, 113–28.
85. Sullivan, Matthew. 'Biographical notes on Robert Mannyng of Brunne and Peter Idley, the adaptor of Robert Mannyng's *Handlyng synne*', *Notes & Queries* 239 (1994), 302–04.
86. Sutton, Anne F. 'Alice Claver, silkwoman (d. 1489)', Bc66, 129–42.
87. Sutton, Anne F. 'Caxton was a mercer: his social milieu and friends', Bc97, 118–48.
88. Sutton, Anne F. 'Lady Joan Bradbury (d. 1530)', Bc66, 208–38.
89. Swanson, R.N. 'Clergy and manorial society in late medieval Staffordshire', *Staffordshire Studies* 5 (1993), 13–34.
90. Thomas, Hugh M. 'Portrait of a medieval anti-Semite: Richard Malebisse *vero agnomine mala bestia*', *Haskins Soc. J.* 3 (1993), 1–15.
91. Thomas, Hugh M. 'Subinfeudation and alienation of land, economic development, and the wealth of nobles on the honor of Richmond, 1066 to 1300', *Albion* 26 (1994), 397–417.
92. Turner, Ralph V. 'Toward a definition of the *curialis*: educated court cleric courtier, administrator, or "new man"?', *Medieval Prosopography* 15 (1994), 3–35.
93. Veale, Elspeth. 'Matilda Penne, skinner (d. 1392/3)', Bc66, 47–54.
94. Virgoe, Roger. 'The will of Hugh atte Fenne, 1476', Bc4, 31–57.
95. Ward, Jennifer C. 'Elizabeth de Burgh, lady of Clare (d. 1360)', Bc66, 28–45.
96. Williams, John Bryan. 'Judhael of Totnes: the life and times of a post-Conquest baron', Bc69, 271–89.

97. Wolf, Armin. 'Reigning queens in medieval Europe: when, where and why', Bc50, 169–88; 220.
98. Wood, Robert A. 'Poor widows, *c.*1393–1415', Bc66, 55–69.

(h) *Naval and Military*

1. Ayton, Andrew. *Knights and warhorses: military service and the English aristocracy under Edward III.* Woodbridge; Boydell; 1994. Pp xiv, 304.
2. Ayton, Andrew. 'English armies in the fourteenth century', Bc78, 21–38.
3. Ayton, Andrew. 'The English army and the Normandy campaign of 1346', Bc77, 253–68.
4. Bennett, Matthew. 'The development of battle tactics in the hundred years war', Bc78, 1–20.
5. Boardman, A.W. *The battle of Towton.* Stroud; Sutton; 1994. Pp 192.
6. Church, S.D. 'The earliest English muster roll, 18/19 December 1215', *Historical Research* 67 (1994), 1–17.
7. Coulson, Charles. 'Freedom to crenellate by licence: an historiographical revision', *Nottingham Medieval Studies* 38 (1994), 86–137.
8. Coulson, Charles. 'The castles of the Anarchy', Bc70, 67–92.
9. Curry, Anne E. 'English armies in the fifteenth century', Bc78, 39–68.
10. Davies, C.S.L. 'The alleged "sack of Bristol": international ramifications of Breton privateering, 1484–1485', *Historical Research* 67 (1994), 230-39.
11. de Renaucourt, J.-L. 'Operations navales en Manche en 1066', *Causeries Lyonsaises* 16 (1992), 1–30.
12. Fauchon, Jacques. 'La guerre en Avranchin, (1380-1381): le débarquement anglais de 1380', *Revue de l'Avranchin et du Pays de Granville* 69 (1992), 239–45.
13. Friel, Ian. 'Winds of change? Ships and the hundred years war', Bc78, 183–193.
14. Hardy, Robert. 'The longbow', Bc78, 161–81.
15. Hobbs, Doreen. 'Royal ships and their flags in the late fifteenth and early sixteenth centuries', *Mariner's Mirror* 80 (1994), 388–394.
16. Hughes, Michael. 'The fourteenth-century French raids on Hampshire and the Isle of Wight', Bc78, 121–43.
17. Hutchinson, Gillian. *Medieval ships and shipping.* Leicester; Leicester UP; 1994. Pp 256.
18. Jamieson, Neil. 'The recruitment of northerners for service in English armies in France, 1415–1450', Bc16, 102–15.
19. Jones, Michael C.E. 'War and fourteenth-century France', Bc78, 103–20.
20. Jones, Michael K. 'Richard III as a soldier', Bc52, 93–112.
21. Jones, Michael K. 'The relief of Avranches (1439): an English feat of arms at the end of the hundred years war', Bc97, 42–55.
22. Kenyon, John R. 'Coastal artillery fortification in England in the late fourteenth and early fifteenth centuries', Bc78, 145–49.
23. Massey, Robert. 'Lancastrian Rouen: military service and property holding, 1419–1449', Bc77, 269–86.
24. McGlynn, Sean. 'The myths of medieval warfare', *History Today* 44/1 (1994), 28–34.

25. Menard, Christian. 'La guerre de cent ans en Vexin: l'occupation anglaise, 1415–1441', *Les Cahiers de la Société Historique et Géographique du Bassin de l'Epte* 29 (1992), 1–11.
26. Meron, Theodor. *Henry's wars and Shakespeare's laws: perspectives on the law of war in the later middle ages*. Oxford; Oxford UP; 1993. Pp xi, 237.
27. Milner, John D. 'The English enterprise in France, 1412–1413', Bc16, 80-101.
28. Morillo, Stephen. *Warfare under the Anglo-Norman kings, 1066–1135*. Woodbridge; Boydell; 1994. Pp xii, 207.
29. Prestwich, J.O. 'Military intelligence under the Norman and Angevin kings', Bc22, 1–30.
30. Prestwich, Michael. 'The English medieval army to 1485', Bc109, 1–25.
31. Salamagne, Alain. 'L'attaque des places-fortes au XVè siècle à travers l'exemple des guerres anglo- et franco-bourguignonnes', *Revue Historique* 289 (1993), 65–113.
32. Schwetman, John W. 'The appearance of Saint George above the English troops at Agincourt: the source of a detail in the historical record', *Notes & Queries* 239 (1994), 304–07.
33. Smith, Robert D. 'Artillery and the hundred years war: myth and interpretation', Bc78, 151–160.
34. Steffy, J. Richard. *Wooden ship building and the interpretation of shipwrecks*. College Station (Texas); Texas A&M UP; 1994; Pp 314.
35. Suppe, Frederick C. *Military institutions on the Welsh Marches: Shropshire, AD 1066–1300*. Woodbridge; Boydell; 1994. Pp xii 203.
36. Thomas, Avril. 'Interconnections between the lands of Edward I: a Welsh-English mercenary force in Ireland, 1285–1304', *B. of the Board of Celtic Studies* 40 (1993), 135–47.
37. Vale, Malcolm. 'The war in Aquitaine', Bc78, 69–82.

(i) *Intellectual and Cultural*

1. Acker, Paul; Amino, Eriko. 'The book of palmistry', Bc9, 141–83.
2. Baldwin, David. 'The author of the "second continuation" of the Croyland chronicle: a fifteenth-century mystery solved?', *East Midland Historian* 4 (1994), 16–19.
3. Bennett, Michael J. 'Education and advancement' [in the fifteenth century], Bc68, 79–96.
4. Bowers, Roger. 'The musicians of the Lady chapel of Winchester cathedral priory, 1402–1539', *J. of Ecclesiastical History* 45 (1994), 210-37.
5. Brown, Michelle P. 'The role of the wax tablet in medieval literacy: a reconsideration in light of a recent find from York', *British Library J.* 20 (1994), 1–16.
6. Brown, Peter. 'The seven planets', Bc9, 3–21.
7. Brown, Shirley Ann; Herren, Michael W. 'The *Adelae comitissae* of Baudri of Bourgeuil and the Bayeux Tapestry', Bc69, 55–73.
8. Cochrane, Louise. *Adelard of Bath: the first English scientist*. London; British Museum; 1994. Pp 125.
9. Conti, Alessandro D. 'Logica intensionale e metafisica dell'essenza in John

Wyclif' [Intentional logic and the metaphysics of being in John Wycliffe], *B. dell'Istituto Storico Italiano per il medio Evo e Archivio Muratoriano* 99 (1993), 159–219.

10. Cylkowski, David G. 'A middle English treatise on horticulture: *Godfridus super Palladium*', Bc9, 301–29.
11. Dahood, Roger. 'Hugh de Morville, William of Canterbury and anecdotal evidence for English language history', *Speculum* 69 (1994), 40-56.
12. Delany, Sheila. *The naked text*. Berkeley (Ca); California UP; 1993. Pp 280.
13. Etzkorn, G.J. 'Ockham at a provincial Chapter, 1323: a prelude to Avignon', *Archivum Franciscanum Historicum* 83 (1990), 557–67.
14. Foster, Meryl. 'Custodians of St Cuthbert: the Durham monks' views of their predecessors, 1083–*c*.1200', Bc108, 53–65.
15. Fries, Maureen. 'How many roads to Camelot: the married knight in Malory', Bc100, 196–207.
16. George, David (ed.). *Lancashire: records of early English drama*. Toronto; Toronto UP; 1992. Pp cxxv, 471.
17. Greatrex, Joan. 'The English cathedral priories and the pursuit of learning in the later middle ages', *J. of Ecclesiastical History* 45 (1994), 396–411.
18. Gullick, Michael. 'The scribes of the Durham cantor's book (Durham Dean and Chapter library, MS B.IV.24) and the Durham martyrology scribe', Bc108, 93–109.
19. Gwara, Scott. 'The continuance of Aldhelm studies in post-Conquest England and glosses to the *Prosa de virginitate* in Hereford Cath. Lib. MS P.I.17', *Scriptorium* 48 (1994), 18–38.
20. Hanna, Ralph, III. 'Henry Daniel's *Liber uricrisiarum*, book 1, chapters 1–3', Bc9, 185–218.
21. Henry, Avril. 'Lichfield cathedral MS 16: its illuminated borders and original order', *Scriptorium* 48 (1994), 39–61.
22. Hiley, David. 'Changes in English chant repertories in the eleventh century as reflected in the Winchester sequences', Bc69, 137–54.
23. Howlett, David R. 'The shape and meaning of the Old English poem "Durham"', Bc108, 485–95.
24. Hunt, Tony. *Medieval surgery*. Woodbridge; Boydell; 1994. Pp xvi, 104.
25. Hunt, Tony (ed.). *Anglo-Norman medicine, vol. 1: Roger Frugard's 'Chirurgia' and the ' Practica brevis' of Platearius*. Woodbridge; Boydell; 1994. Pp viii, 336.
26. Jones, Peter Murray. 'Information and science' [in the fifteenth century], Bc68, 97–111.
27. Keiser, George R. 'Epilepsy: the falling evil', Bc9, 219–244.
28. Kennedy, Edward Donald. 'The stanziac *Morte Arthur*: the adaptation of a French romance for an English audience', Bc100, 91–112.
29. Lawrence, Anne. 'Anglo-Norman book production', Bc77, 79–93.
30. Lawrence, Anne. 'The artistic influence of Durham manuscripts', Bc108, 451–69.
31. Lester, Geoffrey A. 'The earliest English sailing directions', Bc9, 330-59.
32. Matheson, Lister M.; Shannon, Ann. 'A treatise on the elections of times', Bc9, 23–59.

33. McKendrick, Scot. 'The *Romuléon* and the manuscripts of Edward IV', Bc97, 149–69.
34. Means, Laurel. 'A translation of Martin of Spain's *De geomancia*', Bc9, 61–121.
35. Meehan, Bernard. 'Durham twelfth-century manuscripts in Cistercian houses', Bc108, 439–49.
36. Mooney, Linne R. 'Diet and bloodletting: a monthly regimen', Bc9, 245–61.
37. Moore, Bruce. 'The banns in medieval English drama', *Leeds Studies in English* 24 (1993), 91–122.
38. Noble, Peter S. 'Romance in England and Normandy in the twelfth century', Bc77, 69–78.
39. Offler, H.S. (ed.). *William Ockham: opera politica, iv.* Oxford; Oxford UP for British Academy; 1994. Pp 480.
40. Padel, O.J. 'The nature of Arthur', *Cambrian Medieval Celtic Studies* 27 (1994), 1–31.
41. Pearsall, Derek. *William Langland: Piers Plowman.* Exeter; Exeter UP; revised edn. 1994. Pp 416.
42. Pickett, Joseph P. 'A translation of the "Canutus" plague treatise', Bc9, 263–82.
43. Piper, Alan. 'The Durham cantor's book (Durham, Dean and Chapter library, MS B.IV.24)', Bc108, 79–92.
44. Pollard, Anthony J. 'Dominic Mancini's account of the events of 1483', *Nottingham Medieval Studies* 38 (1994), 152–63.
45. Roest, B. 'Johannes Quidort en Willem van Ockham over oorsprong en functie van wereldijk gezag' [John Quidort and William of Ockham on the origins and functions of secular authority], *Groniek* 110 (1990), 35–50.
46. Rusche, Philip G. 'A twelfth-century English fragment of Justinus' *Epitome*', *Scriptorium* 48 (1994), 140-46.
47. Saint, Andrew; Darley, Gillian (ed.). *The chronicles of London.* London; Weidenfeld & Nicolson; 1994. Pp 320.
48. Salisbury, Joyce E. *The beast within: animals in the middle ages.* London; Routledge; 1994. Pp 256.
49. Scanlon, Larry. *Narrative, authority and power: medieval exemplum and the Chaucerian tradition.* Cambridge; Cambridge UP (Cambridge Studies in Medieval Literature 20); 1994. Pp 378.
50. Shichtman, Martin B. 'Politicizing the ineffable: the *Quest del saint grall* and Malory's "Tale of the sankgreal"', Bc100, 163–79.
51. Stephens, Alan D. 'The Germanic heroic tradition and the cultural context of the Bayeux Tapestry', *Medieval History* 3 (1993), 51–58.
52. Stratford, Jenny. 'The royal library in England before the reign of Edward IV', Bc97, 187–97.
53. Sutton, Anne F. 'The court and its culture in the reign of Richard III', Bc52, 75–92.
54. Taavitsainen, Irma. 'A zodiacal lunary for medical professionals', Bc9, 283–300.
55. Thompson, M.W. 'Medieval technology in 1492', *Medieval History* 3 (1993), 64–70.

56. Voigts, Linda Ehrsam. 'The golden table of Pythagoras', Bc9, 123–39.
57. Williams, Steven J. 'Roger Bacon and his edition of the pseudo-Aristotelian *Secretum secretorum*', *Speculum* 69 (1994), 57–73.

(j) *Visual Arts*

1. Alexander, Jonathan. 'The pulpit with the four doctors at St James's, Castle Acre, Norfolk', Bc97, 198–206.
2. Blair, John. 'A Romanesque timber beakhead from Cameley, Somerset', *Antiquaries J.* 71 (1993 for 1991), 252–58.
3. Bliss, S. 'A question of identity?: the column figures on the west portal of Rochester cathedral', *Archaeologia Cantiana* 112 (1994 for 1993), 167–91.
4. Cambridge, Eric. 'Early Romanesque architecture in north-east England: a style and its patrons', Bc108, 141–60.
5. Cherry, John. 'The seal of Haltemprice priory', Bc40, 14–23.
6. Coldstream, Nicola. *The Decorated style: architecture and ornament, 1240-1360.* London; British Museum; 1994. Pp 208.
7. Crook, John. 'The architectural setting of the cult of St Cuthbert in Durham cathedral (1093–1200)', Bc108, 235–50.
8. Deighton, Alan. 'The literary context of the wall-painting at Idsworth, Hampshire', *Antiquaries J.* 73 (1994 for 1993), 69–75.
9. Dennison, Lynda. 'The Apocalypse, British Library, Royal MS. 19 B. XV: a reassessment of its artistic context in early fourteenth-century English manuscript illumination', *British Library J.* 20 (1994), 35–54.
10. Dumville, David N. 'Anglo-Saxon books: treasure in Norman hands?', Bc69, 83–99.
11. Fernie, E.C. 'Architecture and the effects of the Norman Conquest', Bc77, 105–16.
12. Fernie, E.C. 'The architectural influence of Durham cathedral', Bc108, 269–79.
13. French, T.W. 'St Anthony's Hall: an architectural and functional reconsideration', *York Historian* 11 (1994), 2–8.
14. Grant, Lindy. 'Architectural relationships between England and Normandy, 1100-1204', Bc77, 117–129.
15. Grape, Wolfgang. *The Bayeux Tapestry.* Munich; Prestel; 1994. Pp 175.
16. Grundy, Thirlie. 'The misericords of Carlisle cathedral: their carver and his psyche', *T. of the Cumberland & Westmorland Antiquarian & Arch. Soc.* 94 (1994), 91–103.
17. Harrison, S.A. 'Observations on the architecture of the Galilee chapel' [Durham], Bc108, 213–34.
18. Harvey, Anthony; Mortimer, Richard (ed.). *The funeral effigies of Westminster abbey.* Woodbridge; Boydell; 1994. Pp xiv, 195.
19. Haward, Birkin. *Suffolk medieval church arcades, 1150-1550: a measured drawing survey with notes and analysis.* Ipswich; Suffolk Inst. of Archaeology & History; 1993. Pp 480.
20. Heslop, T.A. 'Towards an iconology of croziers', Bc40, 36–45.
21. Kauffmann, C.M. 'Art and popular culture: new themes in the Holkham bible picture book', Bc40, 46–69.

22. Kemp, Brian R. 'English church monuments during the period of the hundred years war', Bc78, 195–211.
23. Kidson, Peter. 'Gervase, Becket, and William of Sens', *Speculum* 68 (1993), 969–91.
24. Kötzsche, Lieselotte. 'Die trauernden Frauen: zum Londoner Passionskästchen' [Grieving women: the London Passion casket], Bc40, 80-90.
25. Leah, Mark. *The late Saxon and medieval pottery industry of Grimston, Norfolk: excavations, 1962–1992*. Norwich; East Anglian Archaeology (report 64); 1993. Pp x, 131.
26. Marner, Dominic. 'The bible of Hugh of le Puiset (Durham, Dean and Chapter library, MS A.II.1)', Bc108, 471–84.
27. Martindale, Andrew. 'St Stephen's chapel, Westminster and the Italian experience', Bc40, 102–12.
28. McAleer, J. Philip. 'The west front of Durham cathedral: the beginning of a British tradition', Bc108, 185–212.
29. Park, David. 'Simony and sanctity: Herbert Losinga, St Wulfstan of Worcester and wall-paintings in Norwich cathedral', Bc40, 157–70.
30. Parker, Elizabeth C.; Little, Charles T. *The cloisters cross: its art and meaning*. London; Miller; 1994. Pp 316.
31. Rampton, Martha. 'The significance of the banquet scene in the Bayeux Tapestry', *Medievalia et Humanistica* ns 21 (1994), 33–53.
32. Reynolds, Catherine. 'English patrons and French artists in fifteenth-century Normandy', Bc77, 299–313.
33. Ritoók, Pál. 'The architecture of the Knights Templars in England', Bc126, 167–78.
34. Rogers, Nicholas. 'The artist of Trinity B.11.7 and his patrons', Bc97, 170-86.
35. Russo, Thomas E. 'The Romanesque rood screen of Durham cathedral: context and form', Bc108, 251–68.
36. Shakesby, Doreen. 'The crosses of Queen Eleanor', *Medieval History* 3 (1993), 26–29.
37. Thurlby, Malcolm. 'The roles of the patron and the master mason in the first design of the Romanesque cathedral of Durham', Bc108, 161–84.
38. Thurlby, Malcolm. 'The Romanesque apse vault at Peterborough cathedral', Bc40, 171–86.
39. Turner-Rugg, Alison. 'Medieval pottery in Hertfordshire: a gazetteer of the principal collections', *Hertfordshire Archaeology* 11 (1993), 30-53.
40. Vince, Alan G.; Wilmott, Tony. 'A lost tile pavement at Tewkesbury abbey and an early fourteenth-century tile factory', *Antiquaries J.* 71 (1993 for 1991), 138–73.
41. Wayment, Hilary. 'Ten Carmelite roundels at Queens' college, Cambridge', *Cambridge Antiquarian Soc. P.* 82 (1993), 139–56.
42. Williamson, Paul. 'Ivory carvings in English treasuries before the Reformation', Bc40, 187–202.
43. Wood, Rita. 'The Romanesque doorways of Yorkshire, with special reference to that at St Mary's church, Riccall', *Yorkshire Arch. J.* 66 (1994), 59–90.

44. Woodman, Francis. 'Hardley, Norfolk, and the rebuilding of its chancel', Bc40, 203–10.
45. Zarnecki, George. 'The priory church of Shobdon and its founder', Bc40, 211–20.

(k) *Topography*

1. Anon (ed.). 'The castle that Eudo built', *Colchester Archaeologist* 7 (1993–4), 1–7.
2. Ayers, Brian S. *Excavations at Fishergate, Norwich, 1985*. Norwich; East Anglian Archaeology (report 68); 1994. Pp x, 94.
3. Bacchus, D. 'Excavations at the south-east bastion, Rochester city wall', *Archaeologia Cantiana* 110 (1992), 129–38.
4. Blair, John. 'Hall and chamber: English domestic planning, 1000-1250', Bc107, 1–21.
5. Bond, C.J. 'Water management in the urban monastery', Bc64, 43–78.
6. Britnell, Richard H. 'Rochester bridge, 1381–1530', Bc12, 41–106.
7. Brooks, Howard; Andrews, David D. 'Great Bardfield church unmasked: the Norman church beneath the 14th-century embellishment', *Essex Archaeology & History* 3rd ser. 25 (1994), 265–70.
8. Butler, Lawrence A.S. *Pickering castle, north Yorkshire*. London; English Heritage; 1993. Pp 24.
9. Butler, Lawrence A.S. 'The archaeology of urban monasteries in Britain', Bc64, 79–86.
10. Champness, John. *Lancaster castle: a brief history*. Preston; Lancashire County Council; 1993. Pp 48.
11. Coplestone-Crow, Bruce. 'The castle of Ewyas Harold and its military arrangements in the Norman period', *Herefordshire Arch. News* 57 (1992), 7–11.
12. Coplestone-Crow, Bruce; Taylor, Elizabeth. 'Pembridge castle', *Herefordshire Arch. News* 62 (1994), 28–31.
13. Crook, John. 'The medieval roof of Marwell Hall, Hampshire', *Antiquaries J*. 73 (1994 for 1993), 37–68.
14. Cullum, Patricia H. 'St Leonard's hospital, York: the spatial and social anaylsis of an Augustinian hospital', Bc64, 11–18.
15. Currie, Christopher K. 'Earthworks at Compton Bassett, with a note on Wiltshire fishponds', *Wiltshire Arch. & Natural History Magazine* 87 (1994), 96–101.
16. Dixon, Philip. '*Mota, aula et turris*: the manor houses of the Anglo-Scottish border', Bc107, 22–48.
17. Dixon, Philip; Lott, B. 'The courtyard and the tower: contexts and symbols in the development of late medieval great houses', *J. of the British Arch. Association* 146 (1993), 93–101.
18. Dixon, Philip; Marshall, Pamela. 'The great tower in the 12th century: the case of Norham castle', *Arch. J*. 150 (1994 for 1993), 410-32.
19. du Boulay, F.R.H. *Medieval Bexley*. London; London Borough of Bexley, Libraries Dept.; 2nd edn. 1993. Pp x, 59.
20. Dyer, Christopher C. 'The great fire of Shipston-on-Stour', *Warwickshire History* 8 (1992–3), 179–94.

21. Evans, A.K.B. 'Cirencester abbey: from heyday to dissolution', *T. of the Bristol & Gloucestershire Arch. Soc.* 111 (1993), 115–42.
22. Fraser, R.; Maxwell, R.; Vaughan, Janet E. 'Excavation adjacent to Close Gate, Newcastle, 1988–1989', *Archaeologia Aeliana* 5th ser. 22 (1994), 85–151.
23. Fulford, Michael G. 'Early motte evidence at Pevensey', *Sussex Past & Present* 72 (1994), 7.
24. Gauthiez, Bernard. 'Paris, un Rouen Capétien?: développements comparés de Rouen et Paris sous les règnes de Henri II et Philippe-Auguste', Bc69, 117–36.
25. Gibson, Adrian V.B. 'The constructive geometry in the design of the thirteenth-century bars at Cressing Temple', *Essex Archaeology & History* 3rd ser. 25 (1994), 107–12.
26. Hall, Michael. 'Rockingham castle, Northants', *Country Life* 188/15 (1994), 62–65.
27. Hall, Michael. 'Stokesay castle, Shropshire', *Country Life* 188/13 (1994), 72–77.
28. Hall, Teresa. 'Witchampton: village origins', *Dorset Natural History & Arch. Soc. P.* 115 (1994 for 1993), 121–32.
29. Herring, P. 'St Michael's Mount: recent and future work', *Cornish Archaeology* 32 (1993), 153–59.
30. Higham, Mary. 'The boundary of Burton-in-Lonsdale chase', *Yorkshire Arch. J.* 66 (1994), 91–105.
31. Hislop, Malcolm J.B. 'Master John of Burcestre and the castles of Stafford and Maxstoke', *T. of the South Staffordshire Arch. & Historical Soc.* 33 (1991–2), 14–20.
32. Hooke, Della; Taylor, Ruth. 'The Augustinian priory of Studley, Warwickshire', *Birmingham & Warwickshire Arch. Soc. T.* 98 (1994), 73–90.
33. Hudson, Timothy P. 'Arundel's Blackfriars located', *Sussex Arch. Collections* 131 (1993), 114–18.
34. Isserlin, Raphael M.J. 'A medieval doorway at St Andrew's church, Sandon, Chelmsford', *Essex Archaeology & History* 3rd ser. 25 (1994), 279–85.
35. Isserlin, Raphael M.J. 'Investigations at St Michael's church, Great Sampford, 1992', *Essex Archaeology & History* 3rd ser. 25 (1994), 270-79.
36. Kay, Richard E. 'A motte and bailey and an ancient church site at Aberllyfni', *Herefordshire Arch. News* 61 (1994), 26–27.
37. Kay, Richard E.; Remfry, Paul M. 'Castell Tynboeth', *Herefordshire Arch. News* 60 (1993), 27–30.
38. Kenyon, John R. *Raglan castle*. Cardiff; Cadw; 2nd edn. 1994. Pp 56.
39. Leah, Mark. 'Excavations and watching brief at Castle Acre, 1985–1986', *Norfolk Archaeology* 41 (1993), 494–507.
40. Leyland, Martin. 'The origins and development of Durham castle', Bc108, 407–24.
41. Lilley, Keith D. 'A Warwickshire medieval borough: Brinklow and the contribution of town-plan analysis', *Birmingham & Warwickshire Arch. Soc. T.* 98 (1994), 51–60.

42. Lowther, Pamela; Ebbatson, Linda; Ellison, Margaret; Millett, Martin. 'The city of Durham: an archaeological survey', *Durham Arch. J.* 9 (1993), 27–119.
43. Marks, Richard. 'Archives and the visual arts: Potsgrove church, its fourteenth-century glazing and other fittings', Bc6, 39–53.
44. Matarasso, Francois. *The English castle*. London; Cassell; 1993. Pp 224.
45. Moorhouse, Stephen. 'Pottery and glass in the medieval monastery', Bc64, 127–48.
46. Munby, Julian. *Stokesay castle, Shropshire*. London; English Heritage; 1993. Pp 24.
47. Munby, Julian. 'Manorial building in timber in central and southern England, 1200-1550', Bc107, 49–64.
48. Mynard, Dennis C. *Excavations on medieval sites in Milton Keynes*. Aylesbury; Buckinghamshire Arch. Soc. (monograph 6); 1994. Pp 192.
49. Nolan, John; Vaughan, Janet E.; Graves, C. Pamela; Dobney, Keith; Jaques, Deborah. 'The town wall, Newcastle upon Tyne: excavations at Oxford Street and Croft Street, 1987–1989', *Archaeologia Aeliana* 5th ser. 21 (1993), 93–149.
50. O'Connor, T.P. 'Bone assemblages from monastic sites: many questions but few data', Bc64, 107–11.
51. Oosthuizen, Susan. 'Isleham: a medieval inland port', *Landscape History* 15 (1993), 29–35.
52. Palliser, D.M. 'The topography of monastic houses in Yorkshire towns', Bc64, 3–9.
53. Partridge, J. 'The excavation of a lift-shaft at Colchester castle', *Essex Archaeology & History* 3rd ser. 24 (1993), 234–36.
54. Remfry, Paul M. *Clun castle, 1066 to 1282*. Malvern; SCS; 1994. Pp 25.
55. Remfry, Paul M. *Hopton castle, 1066 to 1305*. Malvern; SCS; 1994. Pp 12.
56. Remfry, Paul M. *Radnor castle, 1066 to 1282*. Malvern; SCS; 1994. Pp 24.
57. Remfry, Paul M. 'Five castles in Clun lordship', *Herefordshire Arch. News* 61 (1994), 15–23.
58. Remfry, Paul M. 'Some additional notes on Aberllyfni castle', *Herefordshire Arch. News* 62 (1994), 12.
59. Renn, Derek F. *Goodrich castle, Herefordshire*. London; English Heritage; 1993. Pp 24.
60. Roberts, Edward. 'William of Wykeham's house at East Meon, Hants', *Arch. J.* 150 (1994 for 1993), 456–81.
61. Ross, M.S. 'Melbury Abbas: medieval pottery in perspective', *Dorset Natural History & Arch. Soc. P.* 115 (1994 for 1993), 111–19.
62. Ryder, Peter F. 'The two towers of Hexham', *Archaeologia Aeliana* 5th ser. 22 (1994), 185–217.
63. Salter, Mike. *Midland castles*. Birmingham; Quercus; 1993. Pp 96.
64. Saul, Nigel. 'Medieval Britain', Bc124, 115–206.
65. Saunders, Andrew. *Tynemouth priory, castle and twentieth-century fortifications, Tyne and Wear*. London; English Heritage; 1993. Pp 24.
66. Schofield, John. *Medieval London houses*. Yale; Yale UP; 1994. Pp 288.
67. Schofield, John. 'Medieval and later towns', Bc85, 195–214.
68. Schofield, John; Vince, Alan G. *Medieval towns*. London; Leicester UP; 1994. Pp xii, 243.

69. Stocker, David. 'Recording worked stone', Bc64, 19–27.
70. Stopford, J. 'The potential for studies of medieval tiles', Bc64, 87–105.
71. Summerson, Henry R.T. *Dunstanburgh castle, Northumberland.* London; English Heritage; 1993. Pp 24.
72. Swan, Vivien G.; Jones, Bridgett E.A.; Grady, Damian. 'Bolesford, north riding of Yorkshire: a lost wapentake centre and its landscape', *Landscape History* 15 (1993), 13–28.
73. Taylor, Christopher C. 'Medieval rural settlement: changing perceptions', *Landscape History* 14 (1992), 5–17.
74. Thompson, M.W. 'The place of Durham among Norman episcopal palaces and castles', Bc108, 425–36.
75. Trench, John Chenevix. 'Another thirteenth-century house for Buckinghamshire?', *Records of Buckinghamshire* 34 (1994 for 1992), 14–29.
76. Turnbull, Percival; Walsh, Deborah. 'Recent work at Egremont castle', *T. of the Cumberland & Westmorland Antiquarian & Arch. Soc.* 94 (1994), 77–89.
77. Vincent, Nicholas. 'The early years of Keynsham abbey', *T. of the Bristol & Gloucestershire Arch. Soc.* 111 (1993), 95–113.
78. Ward, Simon W. 'The monastic topography of Chester', Bc64, 113–26.
79. Watkin, T. Elfin; Hewett, Cecil A. 'The spire of All Saints' church, Maldon, Essex', *T. of the Ancient Monuments Soc.* 38 (1994), 105–46.
80. Whittick, Christopher. 'Dallingridge's Bay and Bodiam castle millpond: elements of a medieval landscape', *Sussex Arch. Collections* 131 (1993), 119–23.
81. Wrathmell, Stuart. 'Rural settlements in medieval England: perspectives and perceptions', Bc85, 178–94.

F. ENGLAND AND WALES 1500-1714

See also Aa2,21,57,81,89,101,145–147,b7,23,c1,32,48; Ba104,117,b35,63,109,123, 134,156,183,203,270,c10,17,32,39,49,51,56,82,92,104–105,115,118–119,138, 150,d1,9–10,16–19,24–29,37,41–42,44,46–48,50-55,58–59,61,66–67,75,79, 91,96–97,102,109,114; Ea16,e7,21,f24,g81,h15,i4,j19,k47,67; Ga13,c7,d21, f16

(a) *General*

1. Anon. 'Introduction' [Reformation to revolution], Bc103, 1–12.
2. Appel, Sabine. *Elisabeth I von England: die Biographie* [Elizabeth I of England: a biography]. Esslingen; Bechtle; 1994. Pp 314.
3. Bucholz, R.O. *The Augustan court: Queen Anne and the decline of court culture.* Stanford (Ca); Stanford UP; 1993. Pp xvi, 418.
4. Gibson, James M. 'Rochester bridge, 1530-1660', Bc12, 107–59.
5. Hornby, Eileen. 'Elizabeth Burnet 1661–1709: family, friends, religion and politics in the reign of Queen Anne', *Hatcher R.* 37 (1994), 31–39.
6. Hughes, Ann. 'Local history and the origins of the civil war', Bc103, 252–71.

7. Jones, J. Gwynfor. *Early modern Wales, c.1525–1640*. Basingstoke; Macmillan; 1994. Pp 270.
8. Kemeys, Brenda; Raggatt, Joyce. *The queen who survived: the life of Katherine Parr*. London; Adelphi; 1992. Pp 145.
9. Key, Jane. 'The letters and will of Lady Dorothy Bacon, 1597–1629', Bc4, 77–112.
10. Manning, Brian; Thompson, Willie. *A bourgeois revolution?: seventeenth-century England*. London; Pluto; 1993. Pp 64.
11. Pasmore, Stephen. *The life and times of Queen Elizabeth I at Richmond palace*. Richmond; Richmond Local History Soc.; 1992. Pp 72.
12. Riden, Dorothy M (ed.). 'The autobiography of Leonard Wheatcroft of Ashover, 1627–1706', Bc104, 73–120.
13. Russell, Conrad. 'England in 1637', Bc103, 116–41.
14. Scott, Jonathan. 'Restoration process: or, if this isn't a party, we're not having a good time', *Albion* 25 (1993), 619–37.
15. Sharpe, Kevin; Lake, Peter. 'Introduction' [culture and politics in early Stuart England], Bc17, 1–20; 321–25.
16. Skinner, Raymond J. 'The Willoughbys of Brook Hall, Westbury, and Wardour castle', *Wiltshire Arch. & Natural History Magazine* 87 (1994), 116–26.
17. Snowden, Keith. *Katharine Parr our northern queen: the life and northern associations of Queen Katharine Parr, the last wife of King Henry VIII*. Pickering; Castleden; 1994. Pp 64.
18. Turvey, R.K. 'A note on the date of birth of Sir John Perrot', *National Library of Wales J.* 28 (1994), 233–38.

(b) *Politics*

1. Alsop, James D. 'The Act for the Queen's regal power, 1554', *Parliamentary History* 13 (1994), 261–76.
2. Ashton, Robert. *Counter-revolution: the second civil war and its origins, 1646–1648*. London; Yale UP; 1994. Pp 512.
3. Basing, Patricia. 'Robert Beale and the queen of Scots', *British Library J.* 20 (1994), 65–82.
4. Brigden, Susan. 'Henry Howard, earl of Surrey, and the "conjured league"', *Historical J.* 37 (1994), 507–37.
5. Carlin, Norah. 'Liberty and fraternities in the English revolution: the politics of London artisans' protests, 1635–1659', *International R. of Social History* 39 (1994), 223–54.
6. Cust, Richard. 'News and politics in early seventeenth-century England', Bc103, 232–51.
7. de Krey, Gary S. 'Party lines: a reply', *Albion* 25 (1993), 639–43.
8. de Krey, Gary S. 'The first Restoration crisis: conscience and coercion in London, 1667–1673', *Albion* 25 (1993), 565–80.
9. Egloff, Carol S. 'John Hobart of Norwich and the politics of the Cromwellian protectorate', *Norfolk Archaeology* 42 (1994), 38–56.
10. Elton, Sir Geoffrey. 'Parliament in the reign of Elizabeth I', Bc103, 97–116.

11. Fox, Adam. 'Ballads, libels and popular ridicule in Jacobean England', *Past & Present* 145 (1994), 47–83.
12. Fox, Jean. 'Seal, Kemsing and Ightham—1560 to 1650', *Archaeologia Cantiana* 112 (1994 for 1993), 213–51.
13. Greaves, Richard L. 'Great Scott!: the Restoration in turmoil, or, Restoration crises and the emergence of party', *Albion* 25 (1993), 605–18.
14. Gurney, John. 'Gerrard Winstanley and the Digger movement in Walton and Cobham', *Historical J.* 37 (1994), 775–802.
15. Hackett, Helen. *Virgin mother, maiden queen: Elizabeth I and the cult of the Virgin Mary.* London; Macmillan; 1994. Pp 304.
16. Hammer, Paul E.J. 'The uses of scholarship: the secretariat of Robert Devereux, second earl of Essex, *c.*1585–1601', *English Historical R.* 109 (1994), 26–51.
17. Harris, Tim. 'Party turns?: or, Whigs and Tories get off Scott free', *Albion* 25 (1993), 581–90.
18. Harris, Tim. 'Sobering thoughts, but the party is not yet over: a reply', *Albion* 25 (1993), 645–47.
19. Haynes, Alan. *The Gunpowder plot.* Stroud; Sutton; 1994. Pp 192.
20. Hirst, Derek. 'The English republic and the meaning of Britain', *J. of Modern History* 66 (1994), 451–86.
21. Houlbrooke, Ralph A. 'Henry VIII's wills: a comment', *Historical J.* 37 (1994), 891–99.
22. Ives, E.W. 'Henry the Great?', *Historian* 43 (1994), 3–8.
23. Ives, E.W. 'Henry VIII's will: the protectorate provisions of 1546–1547', *Historical J.* 37 (1994), 901–14.
24. Jones, J. Gwynfor. 'John Penry: government, order and the "perishing souls" of Wales', *T. of the Honourable Soc. of Cymmrodorion* (1993), 47–81.
25. Keeler, Mary F. 'The committee for privileges of the House of Commons, 1604–1610 and 1614', *Parliamentary History* 13 (1994), 147–69.
26. Knights, Mark. *Politics and opinion in crisis, 1678–1681.* Cambridge; Cambridge UP; 1994. Pp 300.
27. Lee, Maurice, Jr. 'The Buccleuch marriage contract: an unknown episode in Scottish politics', *Albion* 25 (1993), 395–418.
28. Liedl, Janice. 'The penitent pilgrim: William Calverley and the Pilgrimage of Grace', *Sixteenth Century J.* 25 (1994), 585–94.
29. Loades, David M. *Essays in the reign of Edward VI.* Bangor; Headstart; 1994. Pp 242.
30. Loades, David M. *The politics of marriage: Henry VIII and his queens.* Stroud; Sutton; 1994. Pp viii, 179.
31. Lockyer, Roger; O'Sullivan, Dan. *Tudor England, 1485–1603.* London; Longman; 1993. Pp xii, 177.
32. Lynch, Michael. *The interregnum, 1649–1660.* London; Hodder & Stoughton; 1994. Pp 120.
33. Marshall, Alan. *Intelligence and espionage in the reign of Charles II, 1660-1685.* Cambridge; Cambridge UP; 1994. Pp 300.
34. Marshall, Rosalind K. *Mary I.* London; HMSO; 1993. Pp ix, 150.

35. Maslen, Kate. *The life of Francis Freeman of Marlborough (c.1600-1671).* Woking; Cleaver; 1994. Pp 79.
36. Metz, Karl Heinz. *Oliver Cromwell: zur Geschichte eines schliesslichen Helden* [the history of a complete hero]. Göttingen; Muster-Schmidt; 1993. Pp 132.
37. Miller, John. *An English absolutism?: the later Stuart monarchy, 1660-1688.* London; Historical Association; 1992. Pp 36.
38. Morrill, John. 'Reconstructing the history of early Stuart parliaments' [review article], *Archives* 21 (1994), 67–72.
39. Morrill, John. 'The coming of war', Bc103, 142–56.
40. Mullett, Michael. *James II and English politics, 1678–1688.* London; Routledge; 1993. Pp xv, 85.
41. Newman, William R. 'George Starkey and the selling of secrets', Bc118, 193–211.
42. Ollard, Richard. *Cromwell's earl.* London; HarperCollins; 1994. Pp 384.
43. Oster, Malcolm. 'Virtue, providence and political neutralism: Boyle and interregnum politics', Bc10, 19–36.
44. Redworth, Glyn. 'Of pimps and princes: three unpublished letters from James I and the prince of Wales relating to the Spanish match', *Historical J.* 37 (1994), 401–09.
45. Robertson, Mary L. 'A reply to Helen Speight', *Historical J.* 37 (1994), 639–41.
46. Robinson, W.R.B. 'Some Welsh members of Henry VIII's household in the 1520s', *B. of the Board of Celtic Studies* 40 (1993), 157–70.
47. Rowse, A.L. *Regicides.* London; Duckworth; 1994. Pp 184.
48. Russell, Conrad. 'The Anglo-Scottish union, 1603–1643: a success?', Bc89, 238–56.
49. Smith, David L. *Constitutional royalism and the search for settlement, c.1640-1649.* Cambridge; Cambridge UP; 1994. Pp 230.
50. Speight, Helen M. '"The politics of good governance": Thomas Cromwell and the government of the southwest of England', *Historical J.* 37 (1994), 623–38.
51. Spurr, John. 'Perjury, profanity and politics', *Seventeenth Century* 8 (1993), 29–50.
52. Stater, Victor L. *Noble government: Stuart lord lieutenancy and the transformation of English politics.* Athens (Ga); Georgia UP.; 1994. Pp 256.
53. Stone, Brian. *Derbyshire in the civil war.* Cromford; Scarthin; 1992. Pp 155.
54. Stoyle, Mark. *Loyalty and locality: popular allegiance in Devon during the English civil war.* Exeter; Exeter UP; 1994. Pp 346.
55. Taylor, Stephen. 'Plus ça change ...?: new perspectives on the revolution of 1688' [review article], *Historical J.* 37 (1994), 457–70.
56. Troost, W. 'Willem III en de *exclusion crisis*, 1679–1681', *Bijdragen en mededelingen betreffende de Geschiedenis der Nederlanden* 107 (1992), 28–46.
57. Underdown, David. 'Popular politics before the civil war', Bc103, 208–31.
58. von Greyerz, Kaspar. *England im Jahrhundert der Revolutionen, 1603–1714* [England in the century of revolution] Stuttgart; Ulmer; 1994. Pp 287.
59. Whitehead, Bertrand T. *Brags and boasts: propaganda in the year of the Armada.* Stroud; Sutton; 1994. Pp 240.

(c) *Constitution, Administration and Law*

1. Andrews, Sue. *The people of Hadleigh pay their taxes: the lay subsidy returns for 1524*. Ipswich; Andrews; 1994. Pp 15.
2. Arkell, Tom. 'Printed instructions for administering the hearth tax', Bc39, 38–64.
3. Ashley, Roger. 'War in the Ordnance Office: the Essex connection and Sir John Davis', *Historical Research* 67 (1994), 337–45.
4. Bernard, G.W.; Hoyle, Richard W. 'The instructions for the levying of the Amicable Grant, March 1525', *Historical Research* 67 (1994), 190-202.
5. Braddick, Michael J. *Parliamentary taxation in seventeenth-century England: local administration and response*. London; Royal Historical Soc. (studies in history 70); 1994. Pp x, 342.
6. Carter, Patrick. 'Clerical taxation during the civil war and interregnum', *Historical Research* 67 (1994), 119–33.
7. Carter, Patrick. 'The records of the court of first fruits and tenths, 1540-1554', *Archives* 21 (1994), 57–66.
8. Champion, W.A. 'Litigation in the boroughs: the Shrewsbury *curia parva*, 1580-1730', *J. of Legal History* 15 (1994), 201–22.
9. Dwelly, Edward; Webb, Adrian James (ed.). *Exchequer depositions by commission, 1565–1610, for Somerset*. Weston-super-Mare; Galloway; 1992. Pp 40.
10. Gowing, Laura. 'Language, power and the law: women's slander litigation in early-modern London', Bc105, 26–47.
11. Hollis, Daniel W., III. 'The crown lands and the financial dilemma in Stuart England', *Albion* 26 (1994), 419–42.
12. Hoyle, Richard W. 'Crown, parliament and taxation in sixteenth-century England', *English Historical R.* 109 (1994), 1174–96.
13. Jack, Sybil M. 'In search of the custom of the Exchequer', *Parergon* 11 (1993), 89–105.
14. Jay, Stewart. 'Servants of monarchs and lords: the advisory role of early English judges', *American J. of Legal History* 38 (1994), 118–96.
15. Johnson, D.A. 'A Staffordshire quarter sessions fine book, 1572', Bc113, 53–65; 233–37.
16. Jones, N.G. '*Tyrrel's case* (1557) and the use upon a use', *J. of Legal History* 14 (1993), 75–93.
17. Lang, R.G. (ed.). *Two Tudor subsidy assessment rolls for the City of London, 1541 and 1542*. London; London Record Soc. vol. 29; 1993. Pp lxviii, 425.
18. Nicholson, Bradley J. 'Legal borrowing and the origins of slave law in the British colonies', *American J. of Legal History* 38 (1994), 38–54.
19. Prichard, M.J.; Yale, D.E.C. *Hale and Fleetwood on Admiralty jurisdiction*. London; Selden Soc.; 1993. Pp ccl, 420.
20. Rosenheim, James. 'A reply to Richard Greaves and Jonathan Scott', *Albion* 25 (1993), 649–51.
21. Rosenheim, James. 'Documenting authority: text and magistracy in Restoration society', *Albion* 25 (1993), 591–604.
22. Salt, S.P. 'Sir Simonds d'Ewes and the levying of ship money, 1635–1640', *Historical J.* 37 (1994), 253–87.

23. Stretton, Tim. 'Women, custom and equity in the Court of Requests', Bc105, 170-90.
24. Warneford, Francis E. (ed.). *Star chamber suits of John and Thomas Warneford*. Devizes; Wiltshire Record Soc. vol. 48; 1993. Pp xix, 108.
25. Webb, Adrian James (ed.). *1649 assessment of the hundred of North Petherton and the borough of Bridgwater*. Weston-super-Mare; Galloway; 1993. Pp 53.
26. Webb, Adrian James (ed.). *The 1601 and 1628 subsidy returns for south Somerset*. Weston-super-Mare; Galloway; 1993. Pp vi, 47.
27. Weikel, Ann (ed.). *The court rolls of the manor of Wakefield, 1537–1539*. Leeds; Yorkshire Arch. Soc.; 1993. Pp xxvii, 279.

(d) *External Affairs*

1. Appleby, J.C. 'Thomas Mun's West Indies venture, 1602–1603', *Historical Research* 67 (1994), 101–10.
2. Bosher, J.F. 'The Franco-Catholic danger, 1660-1715', *History* 79 (1994), 5–30.
3. Deloisin, Charles Giry. 'Money and early Tudor diplomacy: the English pensioners of the French king (1475–1547)', *Medieval History* 3 (1993), 128–46.
4. Eysturlid, Lee W. '"Where everything is weighed in the scales of material interest": Anglo-Turkish trade, piracy, and diplomacy in the Mediterranean during the Jacobean period', *J. of European Economic History* 22 (1993), 613–25.
5. Hair, P.E.H.; Alsop, J.D. *English seamen and traders in Guinea, 1553–1565: the new evidence of their wills*. Lampeter; Mellen; 1992. Pp xi, 394.
6. Jansson, Maija; Rogozhin, Nikolai (ed.). *England and the north: the Russian embassy of 1613–1614*. Philadelphia; Memoirs of the American Philosophical Soc. 210; 1994. Pp xxxvi, 236.
7. Johnson, Donald S. *Charting the sea of darkness: the four voyages of Henry Hudson*. New York; McGraw Hill; 1993. Pp xiii, 242.
8. Kekewich, Margaret Lucille (ed.). *Princes and peoples, France and the British Isles, 1620-1714: an anthology of primary sources*. Manchester; Manchester UP; 1994. Pp 272.
9. Laurie, Bruce. *The life of Richard Kane: Britain's first lieutenant-governor of Minorca*. Rutherford (NJ); Fairleigh Dickinson UP; 1994. Pp 291.
10. Niederkorn, Jan Paul. 'Ein englischer Vermittlungsversuch zwischen Kaiser und Sultan während des "langen Turkenkriegs"' [An English attempt at negotiation between the Emperor and the Sultan during the long Turkish war], *Mitteilungen des Österreichischen Staatsarchives* 43 (1992), 159–80.
11. Pflederer, Richard. 'Dutch maps and English ships in the eastern seas', *History Today* 44/1 (1994), 35–41.
12. Rehberg, R. 'Aufbruch zur Weltmacht: zur aussenpolitischen Dimension der englischen Revolution im 17. Jahrhundert' [Becoming a world power: towards a foreign policy perspective on the English revolution of the 17th century], *Zeitschrift für Geschichtswissenschaft* 40 (1992), 1031–39.
13. Richardson, Glenn. 'Good friends and brothers?: Francis I and Henry VIII', *History Today* 44/9 (1994), 20-26.

14. Richardson, Joan A. Carew. 'The piratical Carews of Dorset and Hampshire', *Catholic Ancestor* 5 (1994), 31–37.
15. Thompson, Roger. '*The uprooted* or "worlds in motion": East Anglian founders of New England, 1629–1640', *Parergon* 11 (1993), 1–15.
16. Wernham, Richard Bruce. *The return of the armadas: the last years of the Elizabethan war against Spain, 1595–1603*. Oxford; Oxford UP; 1994. Pp xiv, 452.
17. Wheeler, James S. 'English financial operations during the first Dutch war, 1652–1654', *J. of European Economic History* 23 (1994), 329–43.

(e) *Religion*

1. Albin, Hugh O. (ed.). *Thomas More and Canterbury*. Bath; Downside; 1994. Pp viii, 136.
2. Anon. 'Fr. Henry Palgrave and the Willoughbys of Norfolk, I', *Catholic Ancestor* 5 (1994), 94–103.
3. Anon. 'The Kent/Sussex border at the change of religion', *Catholic Ancestor* 5 (1994), 109–18.
4. Ash, Patricia Mary. *Blessed Edward Oldcorne, SJ: a saint and martyr of Worcestershire*. Cambridge; Castano; 1992. Pp 45.
5. Aston, Margaret. *Faith and fire: popular and unpopular religion, 1350-1600*. London; Hambledon; 1994. Pp 360.
6. Birrell, T.A. 'English Counter-Reformation book culture', *Recusant History* 22 (1994), 113–22.
7. Bowker, Margaret. 'Historical survey, 1450-1750' [Lincoln cathedral], Bc5, 164–209.
8. Bray, Gerald (ed.). *Documents of the English Reformation*. London; Lutterworth; 1994. Pp 672.
9. Carrafiello, Michael L. 'English Catholicism and the Jesuit mission of 1580-1581', *Historical J*. 37 (1994), 761–74.
10. Carrafiello, Michael L. '*Rebus sic stantibus* and English Catholicism, 1606–1610', *Recusant History* 22 (1994), 29–40.
11. Carrafiello, Michael L. 'Robert Parsons and equivocation, 1606–1610', *Catholic Historical R*. 79 (1993), 671–80.
12. Chibi, Andrew A. 'Henry VIII and his marriage to his brother's wife: the sermon of Bishop John Stokesley of 11 July 1535', *Historical Research* 67 (1994), 40-56.
13. Chibi, Andrew A. '"Turpitudinem uxoris fratris tui non revelavit": John Stokesley and the divorce question', *Sixteenth Century J*. 25 (1994), 387–97.
14. Coggins, James Robert. *John Smyth's congregation: English separatism, Mennonite influence and the elect nation*. Scottdale (Pa); Herald; 1991. Pp 239.
15. Cooper, T.N. 'Oligarchy and conflict: Lichfield cathedral clergy in the early sixteenth century', *Midland History* 19 (1994), 40-57.
16. Copson, Stephen. 'The identification of a parliamentary army chaplain: John Moore of Leicestershire', *Leicestershire Arch. & Historical Soc. T*. 68 (1994), 94–97.
17. Craig, John. 'The Brownists in Bury St Edmunds', Bc38, 142–43.

18. Craig, John; Litzenberger, Caroline. 'Wills as religious propaganda: the testament of William Tracy', *J. of Ecclesiastical History* 44 (1993), 415–31.
19. Cunnane, James. 'Ceredigion and the old faith', *Ceredigion* 12/2 (1994), 3–34.
20. Daniell, David. *William Tyndale*. London; Yale UP; 1994. Pp 368.
21. Daniell, David (ed.). *Tyndale's Old Testament*. London; Yale UP; 1992. Pp 688.
22. Davis, Edward B. 'The anonymous works of Robert Boyle and the *Reasons why a Protestant should not turn Papist* (1687)', *J. of the History of Ideas* 55 (1994), 611–29.
23. Davis, J.C. 'Against formality: one aspect of the English revolution', *T. of the Royal Historical Soc.* 6th ser. 3 (1993), 265–88.
24. Dickens, A.G. *Late monasticism and the Reformation*. London; Hambledon; 1994. Pp xiii, 222.
25. Dickens, A.G. 'The early expansion of Protestantism in England, 1520-1558', Bc103, 157–78.
26. Donagan, Barbara. 'Did ministers matter?: war and religion in England, 1642–1649', *J. of British Studies* 33 (1994), 119–56.
27. Dorman, Marianne (ed.). *The liturgical sermons of Lancelot Andrewes, vol. 2: the paschal and pentecostal sermons*. Edinburgh; Pentland; 1993. Pp xii, 353.
28. Draper, Gillian. 'The first hundred years of Quakerism in Kent, part 1', *Archaeologia Cantiana* 112 (1994 for 1993), 317–40.
29. English, Barbara; Hoyle, Richard W. 'What was in St Mary's Tower: an inventory of 1610', *Yorkshire Arch. J.* 65 (1993), 91–94.
30. Evans, E.D. 'A Llanuwchllyn ecclesiastical dispute', *Merioneth Historical & Record Soc. J.* 11 (1993), 414–32.
31. Fincham, Kenneth (ed.). *Visitation articles and injunctions of the early Stuart church*, vol. 1. Woodbridge; Boydell; 1994. Pp xxvi, 227.
32. Foster, Andrew. *The Church of England, 1570-1640*. Harlow; Longman; 1994. Pp 144.
33. Freeman, Thomas. 'Notes on a source for John Foxe's account of the Marian persecution in Kent and Sussex', *Historical Research* 67 (1994), 203–11.
34. Goldie, Mark; Spurr, John. 'Politics and the Restoration parish: Edward Fowler and the struggle for St Giles Cripplegate', *English Historical R.* 109 (1994), 572–96.
35. Gordon, Peter. 'William Castell of Courteenhall: a seventeenth-century pioneer of missionary work', *Northamptonshire Past & Present* 8 (1993/4), 354–62.
36. Gosling, M. 'Berkshire and Oxfordshire Catholics and the Lenten assizes of 1588', *Oxoniensia* 58 (1993), 253–62.
37. Gray, Madeleine. 'The diocese of Llandaff in 1563', *J. of Welsh Religious History* 2 (1994), 31–95.
38. Gray, Madeleine. 'The diocese of St Asaph in 1563', *J. of Welsh Religious History* 1 (1993), 1–40.
39. Gregory, Brad S. 'The "true and zealouse service of God": Robert Parsons,

Edmund Bunny, and *The first booke of the Christian exercise*', *J. of Ecclesiastical History* 45 (1994), 238–68.
40. Greig, Martin. 'Heresy hunt: Gilbert Burnet and the convocation controversy of 1701', *Historical J.* 37 (1994), 569–92.
41. Hall, Basil. 'Martin Bucer in England', Bc51, 144–60.
42. Hamilton, Alistair. 'The English interest in the Arabic-speaking Christians', Bc92, 30-53.
43. Hansford-Miller, Frank. *A history and geography of English religion. vol. 13: religion and the civil war.* Canterbury; Abcado; 1994. 113 leaves.
44. Hansford-Miller, Frank. *The history and geography of English religion, vol. 16: the rise of nonconformity.* Canterbury; Abcado; 1993. Pp 100.
45. Hardman-Moore, Susan. 'Popery, purity and providence: deciphering the New England experiment', Bc89, 257–89.
46. Harris, Ian. 'The politics of Christianity', Bc138, 197–216.
47. Hay, George. 'An English bishop in the Volscians', Bc114, 23–30.
48. Hill, L.M. 'The Marian "experience of defeat": the case of Sir John Bourne', *Sixteenth Century J.* 25 (1994), 531–49.
49. Hobbs, Gerald. 'Martin Bucer and the Englishing of the psalms: pseudonymity in the service of early English Protestant piety', Bc51, 161–75.
50. Holland, Susan. 'Archbishop Abbot and the problem of "Puritanism"', *Historical J.* 37 (1994), 23–43.
51. Houlbrooke, Ralph A. 'The letters and will of Thomas Grene (d. 1545), rector of Poringland', Bc4, 59–76.
52. Houliston, Victor. 'The fabrication of the myth of Father Parsons', *Recusant History* 22 (1994), 141–51.
53. Howard-Drake, Jack. *Oxford church courts: depositions, 1542–1550.* Oxford; Oxfordshire County Council; 1991. Pp ix, 44.
54. Hoyle Richard W.; Summerson, Henry R.T. 'The earl of Derby and the deposition of the abbot of Furness in 1514', *Northern History* 30 (1994), 184–92.
55. Hughes, Ann. *Godly reformation and its opponents in Warwickshire, 1640-1662.* Stratford-upon-Avon; Dugdale Soc. (occasional papers 35); 1993. Pp 29.
56. Hughes, Ann. 'Religion and society in Stratford upon Avon, 1619–1638', *Midland History* 19 (1994), 58–84.
57. Ingle, H. Larry. *First among friends: George Fox and the creation of Quakerism.* Oxford; Oxford UP; 1994. Pp ix, 407.
58. Ives, E.W. 'Ann Boleyn and the early Reformation in England: the contemporary evidence', *Historical J.* 37 (1994), 389–400.
59. Jackson, Peter. 'Nonconformity and the Compton census in late seventeenth-century Devon', Bc39, 117–29.
60. James, Christine. '*Ban wedy i dynny* [a lesson drawn]: medieval Welsh law and early Protestant propaganda', *Cambrian Medieval Celtic Studies* 27 (1994), 61–86.
61. Jelsma, A.J.; Boersma, O. (ed.). *Acta van het consistorie van de Nederlandse gemeente te Londen, 1559–1585* [Acts of the Consistory of the Dutch Congregation in London 1559–1585]. The Hague; Instituut voor Nederlandse Geschiedenis; 1993. Pp xxxvi, 869.

62. Jones, D.R.L. 'Fame and obscurity: Samuel Jones of Brynllywarch', *J. of Welsh Religious History* 1 (1993), 41–65.
63. Knott, John Ray. *Discourses of martyrdom in English literature, 1563–1694.* Cambridge; Cambridge UP; 1993. Pp xv, 278.
64. Lake, Peter. 'Calvinism and the English Church', Bc103, 179–207.
65. Lamont, William. 'The two "national churches" of 1691 and 1829', Bc89, 335–52.
66. Little, Bryan. 'The vicars apostolic (1688): twenty years of anticipation', Bc114, 5–8.
67. Littleton, Charles. 'The French Church of London in European Protestantism: the role of Robert le Maçon, dit de la Fontaine, minister, 1574–1611', *Huguenot Soc. P.* 26 (1994), 45–57.
68. Litzenberger, Caroline. 'Richard Cheyney, bishop of Gloucester: an infidel in religion?', *Sixteenth Century J.* 25 (1994), 567–84.
69. Loach, Jennifer. 'Mary Tudor and the re-Catholicisation of England', *History Today* 44/11 (1994), 16–22.
70. Loades, David M.; Duffy, Eamon. 'The stripping of the altars: the academic debate', *Medieval History* 3 (1993), 185–91.
71. MacCulloch, Diarmaid. 'Two dons in politics: Thomas Cranmer and Stephen Gardiner, 1503–1533', *Historical J.* 37 (1994), 1–22.
72. Marsh, Christopher W. *The Family of Love in English society, 1550-1630.* Cambridge; Cambridge UP; 1994. Pp xvii, 305.
73. Marshall, Peter. *The Catholic priesthood and the English Reformation.* Oxford; Oxford UP; 1994. Pp xi, 271.
74. McClendon, Muriel C. '"Against God's word": government, religion and the crisis of authority in early Reformation Norwich', *Sixteenth Century J.* 25 (1994), 353–69.
75. McCoog, Thomas M. *English and Welsh Jesuits, 1555–1650, part 1: A-F.* London; Catholic Record Soc. vol. 74; 1994. Pp 184.
76. McCoog, Thomas M. '"The flower of Oxford": the role of Edmund Campion in early recusant polemics', *Sixteenth Century J.* 24 (1993), 899–913.
77. Milton, Anthony. *Catholic and reformed: Roman and Protestant churches in English Protestant thought, 1600-1640.* Cambridge; Cambridge UP; 1994. Pp 300.
78. Milton, Anthony. '"The unchanged peacemaker?": John Dury and the politics of irencism in England, 1628–1643', Bc118, 95–118.
79. Morgan, Nicholas. *Lancashire Quakers and the establishment, 1660-1730.* Halifax; Ryburn; 1993. Pp 320.
80. Morrill, John. 'A British patriarchy?: ecclesiastical imperialism under the early Stuarts', Bc89, 209–37.
81. Mount, Harry. 'Egbert van Heemskerck's *Quaker meetings* revisited', *J. of the Warburg & Courtauld Inst.* 56 (1993), 209–28.
82. Mullins, Daniel J. 'Homily for the tercentenary of the western district', Bc114, 1–4.
83. Parrey, Yvonne. '"Devoted disciples of Christ": early sixteenth-century religious life in the nunnery at Amesbury', *Historical Research* 67 (1994), 240-48.

84. Piepho, Lee. 'Mantuan's Eclogues in the English Reformation', *Sixteenth Century J.* 25 (1994), 623–32.
85. Popkin, Richard H. 'Hartlib, Dury and the Jews', Bc118, 118–37.
86. Rhodes, J.T. 'English books of martyrs and saints of the late sixteenth and early seventeenth centuries', *Recusant History* 22 (1994), 7–25.
87. Rhodes, J.T. 'Syon abbey and its religious publications in the sixteenth century', *J. of Ecclesiastical History* 44 (1993), 11–25.
88. Rose, Craig. 'Providence, Protestant union and godly reformation in the 1690s', *T. of the Royal Historical Soc.* 6th ser. 3 (1993), 151–69.
89. Rummel, Erika. 'New perspectives on the controversy between Erasmus and Lee', *Nederlands Archief voor Kerkgeschiedenis* 74 (1994), 226–32.
90. Sharpe, Kevin. 'Archbishop Laud', Bc103, 71–77.
91. Sheils, W.J. 'Provincial preaching on the eve of the civil war: some West Riding fast sermons', Bc89, 290-312.
92. Sheridan, Ann. 'Zachary Mayne, 1631–1694', *Devonshire Association Report & T.* 126 (1994), 181–97.
93. Smallbone, Ken. *James Potter, Quaker: a record of dissent in the 17th and 18th centuries*. York; Sessions; 1992. Pp v, 254.
94. Sprunger, Keith L. *Trumpets from the tower: English Puritan printing in the Netherlands, 1600-1640*. Leiden; Brill; 1994. Pp xvi, 240.
95. Terrar, Edward. 'A seventeenth-century theology of liberation: antinomianism and labour theory of value in the beliefs of English Catholic labouring people, 1639–60', *J. of Religious History* 17 (1993), 297–320.
96. Trowell, Stephen. 'George Keith: post-Restoration Quaker theology and the experience of defeat', *B. of the John Rylands University Library of Manchester* 76/1 (1994), 119–37.
97. Trueman, Carl R. *Luther's legacy: salvation and English reformers, 1525–1556*. Oxford; Oxford UP; 1994. Pp xxii, 307.
98. Tyacke, Nicholas. 'Puritanism, Arminianism and counter-revolution', Bc103, 53–70.
99. Underwood, Andrew. 'The case against Hugh Reeve, a seventeenth-century recusant', Bc6, 72–84.
100. Wabuda, Susan. 'Bishops and the provision of homilies, 1520 to 1547', *Sixteenth Century J.* 25 (1994), 551–66.
101. Wabuda, Susan. 'Equivocation and recantation during the English Reformation: the "subtle shadows" of Dr Edward Crome', *J. of Ecclesiastical History* 44 (1993), 224–42.
102. Walsham, Alexandra. '"The fatall vesper": providentialism and anti-Popery in late Jacobean London', *Past & Present* 144 (1994), 36–87.
103. White, Peter. 'The *via media* in the early Stuart church', Bc103, 78–96.
104. Whiteman, Anne. 'The Compton census of 1676', Bc39, 78–96.
105. Williams, Glanmor. 'The Edwardian Reformation in Wales', *J. of Welsh Religious History* 2 (1994), 14–30.
106. Williams, Michael E. 'William Allen: the sixteenth-century Spanish connection', *Recusant History* 22 (1994), 123–40.
107. Wojcik, Jan W. 'The theological context of Boyle's *Things above reason*', Bc10, 139–55.

108. Woods, William. 'Publications connected with the Dutch church in Norwich', Bc38, 29–36.
109. Woodward, Donald. '"Here comes a chopper to chop off his head": the execution of three priests at Newcastle and Gateshead, 1592–1594', *Recusant History* 22 (1994), 1–6.
110. Wykes, David L. 'Friends, parliament and the Toleration Act', *J. of Ecclesiastical History* 45 (1994), 42–63.
111. Yule, George. 'James VI and I: furnishing the churches in his two kingdoms', Bc89, 182–208.

(f) *Economic Affairs*

1. Appleby, J.C. 'A Guinea venture, *c.*1657: a note on the early English slave trade', *Mariner's Mirror* 79 (1993), 84–87.
2. Armitage, David. '"The projecting age": William Paterson and the Bank of England', *History Today* 44/6 (1994), 5–10.
3. Bettey, J.H. 'Manorial stewards and the conduct of manorial affairs', *Dorset Natural History & Arch. Soc. P.* 115 (1994 for 1993), 15–19.
4. Budd, Brian D. 'The bricks of the Marie Rose', *Medieval History* 3 (1993), 100-17.
5. Carlos Ann M. 'Bonding and the agency problem: evidence from the Royal African Company, 1672–1691', *Explorations in Economic History* 31 (1994), 313–35.
6. Coney, Audrey. 'Fish, fowl and fen: landscape and economy in seventeenth-century Martin Mere', *Landscape History* 14 (1992), 51–64.
7. Craton, Michael. 'Property and propriety: land tenure and slave property in the creation of a British West Indian plantocracy, 1612–1740', Bc150, 497–529.
8. Freear, J. 'Robert Loder, Jacobean management accountant', Bc35, 473–89.
9. Garnier, Richard. 'Thomas Tompion: the Huguenot connection', *Huguenot Soc. P.* 26 (1994), 1–13.
10. Grassby, Richard. *The English gentleman in trade: the life and works of Sir Dudley North, 1641–1691*. Oxford; Oxford UP; 1994. Pp xvi, 390.
11. Hoechstetter-Muller, A. 'Die "Company of mines royal" und die Kupferbergwerke in Keswick, Cumberland, zur Zeit Joachim und Daniel Hoechstetters, (1526–1580)' [The "Company of Mines Royal" and the copper mines in Keswick, Cumberland in the time of Joachim and Daniel Hoechstetters, (1526–1580)], Bc146, 75–91.
12. Hogarth, Sylvia D.; Webb, Christopher C. 'The account book of the York Company of Silkweavers, 1611–1700', *Yorkshire Arch. J.* 66 (1994), 191–213.
13. Hughes, Paul. 'Thomas Surbey's 1699 survey of the rivers Ouse and Humber', *Yorkshire Arch. J.* 66 (1994), 149–90.
14. Kiernan, David T. (ed.). 'Lawrence Oxley's accounts, 1672–1681', Bc104, 121–82.
15. Manterfield, John B. 'Edward Pawlett of Grantham: a provincial bookseller, 1660-1687', *Lincolnshire History & Archaeology* 29 (1994), 11–16.
16. McKenzie, D.F. 'The London book trade in 1644', Bc25, 131–52.

17. McRae, Andrew. 'To know one's own: estate surveying and the representation of the land in early modern England', *Huntington Library Q.* 56 (1993), 332–57.
18. Shammas, Carole. 'The decline of textile prices in England and British America prior to industrialization', *Economic History R.* 2nd ser. 47 (1994), 483–507.
19. Spence, Richard T. 'Mining and smelting in Yorkshire by the Cliffords, earls of Cumberland, in the Tudor and early Stuart period', *Yorkshire Arch. J.* 64 (1992), 157–83.
20. Spicer, A. 'Valenciennes et Southampton: un nouvel éclairage sur leurs liens au XVIème siècle', *Valentiana* 10 (1992), 33–44.
21. Tittler, Robert. 'Money-lending in the west midlands: the activities of Joyce Jefferies, 1638–1649', *Historical Research* 67 (1994), 249–63.
22. Zahedieh, Nuala. 'London and the colonial consumer in the late seventeenth century', *Economic History R.* 2nd ser. 47 (1994), 239–61.
23. Zell, Michael. *Industry in the countryside: Wealden society in the sixteenth century*. Cambridge; Cambridge UP; 1994. Pp xv, 257.

(g) *Social History (General)*

1. Ahern, Sheila. 'The apprehension of Nicholas Jennings: a study of an Elizabethan conman', *Parergon* 11 (1993), 17–25.
2. Amussen, Susan Dwyer. '"Being stirred to much unquietness": violence and domestic violence in early modern England', *J. of Women's History* 6 (1994), 70-89.
3. Arnold, Janet. 'The kirtle, or surcoat, and mantle of the most noble Order of the Garter worn by Christian IV, king of Denmark and Norway', *Antiquaries J.* 72 (1994 for 1992), 141–67.
4. Atkinson, J.A. (ed.). *Darlington wills and inventories, 1600-1625*. Durham; Surtees Soc. vol. 201; 1993. Pp x, 234.
5. Barnard, Toby C. 'The Hartlib circle and the cult and culture of improvement in Ireland', Bc118, 281–98.
6. Baskerville, Steven W. 'The family in Puritan political theology', *J. of Family History* 18 (1993), 157–77.
7. Beare, Thomas. *The bailiff of Blackmoor, 1586: an examination of the history, laws and customs of medieval and sixteenth-century tinners*. Camborne; Penhellick; 1994. Pp xx, 138.
8. Beaufort, D.A. 'The medical practitioners of western Sussex in the early modern period: a preliminary survey', *Sussex Arch. Collections* 131 (1993), 139–51.
9. Beckett, J.V.; Polak, J.P. (ed.). 'The Scarsdale surveys of 1652–1662', Bc104, 3–72.
10. Ben-Amos, Ilana Krausman. *Adolescence and youth in early modern England*. London; Yale UP; 1994. Pp x, 335.
11. Bettey, J.H. 'Lord Shaftesbury's estate and gardening notebook, 1675', *Somerset & Dorset Notes & Queries* 33 (1994), 295–98.
12. Binns, Jack. 'Sir Hugh Cholmley: Whitby's benefactor or beneficary', *Northern History* 30 (1994), 86–104.

13. Carlson, Eric Josef. *Marriage and the English Reformation.* Oxford; Blackwell; 1994. Pp xi, 276.
14. Churches, Christine. '"Equity against a purchaser shall not be": a seventeenth-century case study in landholding and indebtedness', *Parergon* 11 (1993), 69–87.
15. Cooper, Janet; Goose, Nigel R.; Thornton, C.C. 'Tudor and Stuart Colchester', Bc86, 67–132.
16. Coster, W. *Kinship and inheritance in early modern England: three Yorkshire parishes.* York; Borthwick Inst. (Borthwick paper 83); 1993. Pp vi, 27.
17. Dingwall, Helen M. *Physicians, surgeons and apothecaries: medical practice in seventeenth-century Edinburgh.* Edinburgh; Canongate; 1994. Pp 256.
18. Dolan, Frances E. *Dangerous familiars: representations of domestic crime in England, 1550-1700.* Ithaca (NY); Cornell UP; 1994. Pp xiii, 253.
19. Emmison, Frederick G. (ed.). *Essex wills, vol. 9: the bishop of London's commissary court, 1569–1578.* Chelmsford; Essex Record Office (publications 127); 1994. Pp xvii, 274.
20. Erickson, Amy Louise. *Women and property in early modern England.* London; Routledge; 1993. Pp xiii, 306.
21. Esser, Raingard. 'The "politijcke mannen" [political men]: a particular institution of the Norwich strangers' communities', Bc38, 21–28.
22. Feingold, Mordechai. 'Patrons and professors: the origins and motives for the endowment of university chairs—in particular the Laudian professorship of Arabic', Bc92, 109–28.
23. Fletcher, Anthony. 'The Protestant idea of marriage in early modern England', Bc89, 161–81.
24. Gaskill, Malcolm. 'Witchcraft and power in early modern England: the case of Margaret Moore', Bc105, 125–45.
25. Gatley, David Alan; Midgley, Margaret; Bateman, Sheena. 'The Stoke-upon-Trent parish listing, 1701', Bc113, 171–225; 251–53.
26. Gibbs, Graham C. 'Abel Boyer: the making of a British subject, 1699–1706', *Huguenot Soc. P.* 26 (1994), 14–44.
27. Graf, Holger Th. 'Leicestershire small towns and pre-industrial urbanization', *Leicestershire Arch. & Historical Soc. T.* 68 (1994), 98–120.
28. Groves, Jill. *The impact of civil war on a community: Northenden and Etchells in Cheshire, 1642–1660.* Sale; Northern Writers Advisory Services; 1992. Pp 33.
29. Hardie, Caroline. 'Survey of resource management: a single class assessment', Bc144, 39–48.
30. Harley, David. '"Bred up in the study of that faculty": licensed physicians in north-west England, 1660-1760', *Medical History* 38 (1994), 398–420.
31. Harley, David. 'The beginnings of the tobacco controversy: Puritanism, James I, and the royal physicians', *B. of the History of Medicine* 67 (1993), 28–50.
32. Heal, Felicity; Holmes, Clive. *The gentry in England and Wales, 1500-1700.* London; Macmillan; 1994. Pp 400.
33. Hindle, Steve. 'The shaming of Margaret Knowsley: gossip, gender and the experience of authority in early modern England', *Continuity & Change* 9 (1994), 391–419.

34. Houston, Rab A. *Social change in the age of the Enlightenment: Edinburgh, 1660-1760*. Oxford; Oxford UP; 1994. Pp x, 443.
35. Hudson, Geoffrey L. 'Negotiating for blood money: war widows and the courts in seventeenth-century England', Bc105, 146–69.
36. Hunt, John Dixon. 'Hortulan affairs', Bc118, 321–43.
37. Ingram, Martin. '"Scolding women cucked or washed": a crisis in gender relations in early modern England?', Bc105, 48–80.
38. Jenner, Mark. '"Another *epocha*?" Hartlib, John Lanyon and the improvement of London in the 1650s', Bc118, 343–56.
39. Johnson, Marion. *Lords of the manor of Ockbrook: archives, 1583–1605*. Borrowash; Johnson; 1994. Pp 40.
40. Johnson, Marion. *Yeomen of Elizabethan Ockbrook: archives of the sixteenth century*. Borrowash; Johnson; 1994. Pp 40.
41. Jones, J. Gwynfor. 'Sir John Wynn, junior, of Gwydir and Llanfrothen and the "grand tour", 1613–1614', *Merioneth Historical & Record Soc. J.* 11 (1993), 379–413.
42. Jones, J. Gwynfor. 'Sir John Wynn, junior, of Gwydir and Llanfrothen: letters relating to his "Grand tour", 1613–1614, part 1', *Merioneth Historical & Record Soc. J.* 12 (1994), 17–28.
43. Kettle, Ann J. 'Matthew Cradock's book of remembrance, 1614–1615', Bc113, 67–169; 239–49.
44. Key, Newton E. 'The political culture and political rhetoric of county feasts and feast sermons, 1654–1714', *J. of British Studies* 33 (1994), 223–56.
45. Lee, Maurice, Jr. *Heiresses of Buccleuch: marriage, money and politics in seventeenth-century Britain*. Edinburgh; Canongate; 1994. Pp 228.
46. Levin, Carole. *'The heart and stomach of a king': Elizabeth I and the politics of sex and power*. Philadelphia; Pennsylvania State UP; 1994. Pp x, 243.
47. Lowe, Roger. *The diary of Roger Lowe of Ashton-in-Makerfield, Lancashire, 1663–1678, including a record of burials at Winwick church, 1666–1671*. Wigan; Picks; 1994. Pp 110.
48. Manning, Roger Burrow. *Hunters and poachers: a cultural and social history of unlawful hunting in England, 1485–1640*. Oxford; Oxford UP; 1993. Pp xi, 255.
49. Marcombe, David. *English small town life: Retford, 1520-1642*. Nottingham; Department of Adult Education, University of Nottingham; 1993. Pp xii, 319.
50. Murphy, Martin (ed.). 'William Atkins, a relation of the journey from St Omers to Seville, 1662', Bc93, 191–288.
51. Oestmann, Cord. *Lordship and community: the Lestrange family and the village of Hunstanton, Norfolk, in the first half of the sixteenth century*. Woodbridge; Boydell; 1994. Pp xii, 289.
52. Osborne, Simon. 'The war, the people and the absence of the Clubmen in the midlands, 1642–1646', *Midland History* 19 (1994), 85–104.
53. Ovenden, Richard. 'Jaspar Gryffyth and his books', *British Library J.* 20 (1994), 107–39.
54. Patton, Brian. 'The women are revolting? Women activism and popular satire in the English revolution', *J. of Medieval & Renaissance Studies* 23 (1993), 69–86.

55. Pelling, Margaret. 'Apprenticeship, health and social cohesion in early modern London', *History Workshop* 37 (1994), 33–56.
56. Prest, Wilfred. 'Hale as husband', *J. of Legal History* 14 (1993), 142–44.
57. Razzell, Peter (ed.). *Journals of two travellers in Elizabethan and early Stuart England: Thomas Platter and Horatio Busino*. Llanybydder; Caliban; 1994. Pp 200.
58. Reed, Michael. 'Osterley Park in 1668: the probate inventory of Sir William Waller', *London & Middlesex Arch. Soc. T.* 42 (1994 for 1991), 115–20.
59. Sharpe, Jim. 'Women, witchcraft and the legal process', Bc105, 106–24.
60. Shields, David S. 'Anglo-American clubs: their wit, their heterodoxy, their sedition', *William & Mary Q.* 3rd ser. 51 (1994), 293–304.
61. Shields, Elizabeth L. 'The members of parliament for the city of York, 1485–1515', *York Historian* 11 (1994), 9–22.
62. Sinclair, Iona (ed.). *The pyramid and the urn: life in letters of a Restoration squire—William Lawrence of Shurdington, 1636–1697*. Stroud; Sutton; 1994. Pp 176.
63. Smith, Hassell. 'The gardens of Sir Nicholas and Sir Francis Bacon: an enigma resolved and a mind explored', Bc89, 125–60.
64. Spence, Richard T. 'A noble funeral in the great civil war', *Yorkshire Arch. J.* 65 (1993), 115–23.
65. Staves, Susan. 'Chattel property rules and the construction of Englishness, 1660-1800', *Law & History R.* 12 (1994), 123–53.
66. Stevens, Douglas J. *Lawyer Lashbrooke: a seventeenth-century west-country rogue*. Winsford; Hulse; 1994. Pp 43.
67. Storey, Matthew (ed.). *Two East Anglian diaries, 1641–1679: Isaac Archer and William Coe*. Woodbridge; Boydell; 1994. Pp x, 296.
68. Swain, J.T. 'The Lancashire witch trials of 1612 and 1634 and the economics of witchcraft', *Northern History* 30 (1994), 64–85.
69. Tate, William Edward; Edmonds, Edward L. 'A commentary on Richard Brathwaite's *Barnabae itinerarium*, 1638', *Yorkshire Arch. J.* 65 (1993), 125–42.
70. Thomas, Keith. 'Cleanliness and godliness in early modern England', Bc89, 56–83.
71. Thomas, M. Wynn. *Morgan Llwyd: ei gyfeillion a'i gyfnod* [Morgan Llwyd: his friends and his time]. Cardiff; Wales UP; 1991. Pp x, 198.
72. Thompson, Janet A. *'Her good name and credit': women and witchcraft in seventeenth-century Devon*. Frankfurt-am-Main; Lang; 1992. Pp 260.
73. Todd, Barbara J. 'Demographic determinism and female agency: the remarrying widow reconsidered ... again', *Continuity & Change* 9 (1994), 421–50.
74. Tyson, Blake. 'The Elizabethan farmstead at St Lawrence's vicarage, Appleby', *T. of the Cumberland & Westmorland Antiquarian & Arch. Soc.* 94 (1994), 121–34.
75. van Strien, C.D. 'John Talman en andere britse toeristen in Leiden en omstreken rond 1700' [John Talman and other British travellers in Leiden and its environs around 1700], *Jaarboekje voor Geschiedenis en Oudheidkunde van Leiden en Omstreken* 82 (1990), 31–60.
76. Vickerstaff, J.J. *A great revolutionary deluge?: education and the Reformation in county Durham*. Middlesbrough; Teesside UP; 1992. Pp 56.

77. von Friedeburg, Robert. *Sundenzucht und sozialer Wandel: Earls Colne (England), Ipswich und Springfield (Neuengland) c.1524–1690 im Vergleich* [The control of sin and social change: Earls Colne (England), Ipswich and Springfield (New England) c.1524–1690 compared]. Stuttgart; Steiner; 1993. Pp 311.
78. Walker, Garthine. 'Women, theft and the world of stolen goods', Bc105, 81–105.
79. Walker, Garthine; Kermode, Jennifer I. 'Introduction' [women, crime, courts], Bc105, 1–25.
80. Warmington, A.R. '"Madd, Bedlam madd": an incident in Gloucester's 17th-century municipal history reconsidered', *T. of the Bristol & Gloucestershire Arch. Soc.* 111 (1993), 165–73.
81. Weinstein, Rosemary. *Tudor London*. London; HMSO; 1994. Pp x, 110.
82. Westhauser, Karl E. 'Friendship and family in early modern England: the sociability of Adam Eyre and Samuel Pepys', *J. of Social History* 27 (1994), 517–36.
83. Woodward, Donald. 'The determination of wage rates in the early modern north of England', *Economic History R.* 2nd ser. 47 (1994), 22–43.

(h) *Social Structure and Population*

1. Alexander, James. 'The City revealed: an analysis of the 1692 poll tax and the 1693 4s aid in London', Bc39, 181–200.
2. Andrews, I.K.D. *Poole census, 1574*. Poole; Poole Borough Council; 1992. Pp 25.
3. Arkell, Tom. 'A method for estimating population totals from the Compton census returns', Bc39, 97–116.
4. Arkell, Tom. 'An examination of the poll taxes of the later seventeenth century, the Marriage Duty Act and Gregory King', Bc39, 142–80.
5. Arkell, Tom; Schurer, Kevin. 'Hearth tax and Compton census: introducing the documents', Bc39, 31–37.
6. Arkell, Tom; Schurer, Kevin. 'Introduction: surveying the people', Bc39, 1–5.
7. Arkell, Tom; Schurer, Kevin. 'Poll taxes and the Marriage Duty Act: introducing the documents', Bc39, 130-41.
8. Armstrong, David M. 'Birth, marriage and death in Elizabethan Cumbria', *Local Population Studies* 53 (1994), 29–41.
9. Bedells, John (ed.). *The visitation of the county of Huntingdon 1684 by Sir Henry St George*. London; Harleian Soc. ns 13; 1994. Pp xi, 128.
10. Boulton, Jeremy. 'The Marriage Duty Act and parochial registration in London, 1695–1706', Bc39, 222–52.
11. Cooper, Sheila. 'Household form and composition in King's Lynn: a reconstruction based on the poll taxes of 1689–1702', Bc39, 201–21.
12. Cotton, William Alfred. *Bromsgrove and the poll tax of 1690: a record of poll tax payers in the parish of Bromsgrove, Worcestershire: transcript of the notes of W.A. Cotton with additional information extracted from the writings of John Lacey in 1778*. Bromsgrove; Bromsgrove Branch of the Birmingham & Midland Soc. for Genealogy and Heraldry; [1993?]. Pp 51.
13. Dingwall, Helen M. *Late seventeenth-century Edinburgh: a demographic study*. Aldershot; Scolar; 1994. Pp xi, 324.

14. Galley, Chris. 'A never-ending succession of epidemics?: mortality in early-modern York', *Social History of Medicine* 7 (1994), 29–57.
15. Hebden, J.R. (ed.). *The hearth tax list for Barkston Ash and Osgoldcross wapentakes, West Riding of Yorkshire: Lady Day, 1672*. Ripon; Ripon Historical Soc.; 1994. Pp 99.
16. Hebden, J.R. (ed.). *The hearth tax list for Staincliffe and Ewcross wapentakes, West Riding of Yorkshire: Lady Day, 1672*. Ripon; Ripon Historical Soc.; 1992. Pp 104.
17. Hebden, J.R. (ed.). *The hearth tax list for York city parishes and Ainsty wapentake, Lady Day, 1672*. Ripon; Ripon Historical Soc.; 1992; Pp 72.
18. Husbands, Chris. 'Hearths, wealth and occupations: an exploration of the hearth tax in the later seventeenth century', Bc39, 65–77.
19. Jewers, Arthur J.; Phipps, H.R.; Webb, Adrian James (ed.). *Marriage allegation bonds of the bishops of Bath & Wells from their commencement to the year 1775, with amendments or variant readings taken from the Wells book of marriage allegations, 1672–1676, and an index of occupations*. Weston-super-Mare; Galloway; 1993. Microfiche edn.
20. Laslett, Peter. 'Natural and political observations on the population of late seventeenth-century England: reflections on the work of Gregory King and John Graunt', Bc39, 6–30.
21. Moore, John S. 'Canterbury visitations and the demography of mid-Tudor Kent', *Southern History* 15 (1993), 36–85.
22. Moore, John S. 'Jack Fisher's 'flu: a virus still virulent', *Economic History R*. 2nd ser. 47 (1994), 359–61.
23. Nurse, Bernard. 'Occupations in parish registers: the evidence from Newport, Essex', *Local Population Studies* 52 (1994), 39–42.
24. Phillips, C.B. 'The plague in Kendal in 1598: some new evidence', *T. of the Cumberland & Westmorland Antiquarian & Arch. Soc.* 94 (1994), 135–42.
25. Phipps, H.R.; Webb, Adrian James (ed.). *Abstracts of Bath & Wells diocesan records: marriage licences and allegations, 1583–1681, with a surname index to parties*. Weston-super-Mare; Galloway; 1994. Microfiche edn.
26. Porter, Stephen. 'From death to burial in seventeenth-century England', *Local Historian* 23 (1993), 199–204.
27. Schurer, Kevin. 'Variations in household structure in the late seventeenth century: toward a regional analysis', Bc39, 253–78.
28. Squibb, George Drewry (ed.). *The visitation of Oxfordshire, 1669 and 1675*. London; Harleian Soc. ns 12; 1993. Pp xv, 137.
29. Tarver, Anne. 'Barrow upon Soar: some aspects of the social and economic structure of a late sixteenth-century village', *East Midland Historian* 4 (1994), 20-27.
30. Wrightson, Keith; Schulze, W. (ed.). 'Zwei Wege zur Erfassung der englischen Sozialstruktur des 16. und 17. Jahrhunderts' [Two approaches to the understanding of English social structure in the sixteenth and seventeenth centuries], *Ständische Gesellschaft & soziale Mobilität*, ed. W. Schulze (Munich; Oldenbourg; 1988), 187–203.
31. Zell, Michael. 'Fisher's 'flu and Moore's probates: quantifying the mortality crisis of 1556–1560', *Economic History R*. 2nd ser. 47 (1994), 354–58.

(i) *Naval and Military*

1. Alsop, James D.; Dick, K.R. 'The origin of public tendering for Royal Navy provisions, 1689–1720', *Mariner's Mirror* 80 (1994), 395–402.
2. Atkin, Malcolm. 'David Papillon and the civil war defences of Gloucester', *T. of the Bristol & Gloucestershire Arch. Soc.* 111 (1993), 147–64.
3. Atkin, Malcolm; Howes, Russell. 'The use of archaeology and documentary sources in identifying the civil war defences of Gloucester', *Post-Medieval Archaeology* 27 (1993), 15–41.
4. Ball, R.M. 'After Edgehill fight', *Historical Research* 67 (1994), 111–16.
5. Brammer, Betty. *Winceby and the battle.* Boston; Kay; 1994. Pp 102.
6. Buchet, Christian. 'The Royal Navy and the Caribbean, 1689–1763', *Mariner's Mirror* 80 (1994), 30-44.
7. Chandler, David. 'The great captain-general, 1702–1714', Bc109, 69–92.
8. Childs, John. 'The Restoration army, 1660-1702', Bc109, 48–69.
9. Donagan, Barbara. 'Atrocity, war crime, and treason in the English civil war', *American Historical R.* 99 (1994), 1137–66.
10. Fissel, Mark Charles. *The bishops' wars: Charles I's campaigns against Scotland, 1638–1640.* Cambridge; Cambridge UP; 1994. Pp xv, 336.
11. Foard, Glenn. *Col. John Pickering's regiment of foot, 1644–1645.* Whitstable; Pryor; 1994. Pp 160.
12. Foard, Glenn. 'The civil war defences of Northampton', *Northamptonshire Past & Present* 9 (1994), 4–44.
13. Gentles, Ian. 'The choosing of officers for the New Model Army', *Historical Research* 67 (1994), 264–85.
14. Gruffydd, R. Geraint. 'Edward Turberville: bardd Catholig' [Edward Turberville, Catholic poet], Bc102, 103–22.
15. Hebb, David Delison. *Piracy and the English government, 1616–1642.* Aldershot; Scolar; 1994. Pp xv, 303.
16. Heumann, J. Mark. 'Fairfax and the Lifeguard's colors', *Albion* 26 (1994), 443–60.
17. Hitchcock, R.F. 'Cavendish's last voyage: the charges against Davis', *Mariner's Mirror* 80 (1994), 259–69.
18. Hoyle, Richard W. (ed.). *The military survey of Gloucestershire, 1522.* Bristol; Bristol & Gloucestershire Arch. Soc.; 1993. Pp liii, 265.
19. Isemonger, Paul Lewis. *The English civil war: a living history.* Stroud; Sutton; 1994. Pp xv, 144.
20. Kitson, Frank. *Prince Rupert: portrait of a soldier.* London; Constable; 1994. Pp 356.
21. le Fevre, Peter. '"Meer laziness" or incompetence: the earl of Torrington and the battle of Beachy Head', *Mariner's Mirror* 80 (1994), 290-97.
22. Lee, C.D. 'The battle of Beachy Head: Lord Torrington's conduct', *Mariner's Mirror* 80 (1994), 270-89.
23. Morrison-Low, A.D. 'Early navigational instruments in Scotland: icons and survivals', Bc36, 218–31.
24. Peachey, Stuart. *The mechanics of infantry combat in the first English civil war.* Bristol; Stuart; 1992. Pp 28.

25. Pilkington, Colin. *To play the man: the story of Lady Derby and the siege of Lathom House, 1643–1645*. Preston; Carnegie; 1991. Pp 128.
26. Porter, Stephen. *Destruction in the English civil wars*. Stroud; Sutton; 1994. Pp xii, 180.
27. Powell, W.R. 'Silas Taylor of Harwich (1624–1678): naval affairs, espionage, and local history', *Essex Archaeology & History* 3rd ser. 25 (1994), 174–84.
28. Rogers, Hugh Cuthbert Basset. *Generals-at-sea: naval operations during the English civil war and the three Anglo-Dutch wars*. Bromley; Galago; 1992. Pp 161.
29. Roy, Ian. 'Towards the standing army, 1485–1660', Bc109, 25–48.
30. Smith, Victor T.C. 'Cockham wood fort', *Archaeologia Cantiana* 112 (1994 for 1993), 55–75.
31. Williams, Noel St John. *Redcoats and courtesans: the birth of the British army, 1660-1690*. London; Brassey's; 1994. Pp xiv, 273.
32. Willmoth, Frances. 'Mathematical sciences and military technology: the Ordnance office in the reign of Charles II', Bc49, 117–31.

(j) *Political Thought and History of Ideas*

1. Aarsleff, Hans. 'Locke's influence', Bc115, 252–89.
2. Arneil, Barbara. 'Trade, plantations, and property: John Locke and the economic defense of colonialism', *J. of the History of Ideas* 55 (1994), 591–609.
3. Ashcraft, Richard. 'Lockean ideas, poverty, and the development of liberal political theory', Bc150, 43–61.
4. Ashcraft, Richard. 'Locke's political philosophy', Bc115, 226–52.
5. Ayers, M.R. 'The foundations of knowledge and the logic of substance: the structure of Locke's general philosophy', Bc138, 49–74.
6. Beckles, Hilary. 'The concept of "white slavery" in the English Caribbean during the early seventeenth century', Bc150, 572–84.
7. Bennett, Jonathan. 'Locke's philosophy of mind', Bc115, 89–115.
8. Bolton, Martha Brandt. 'The real Molyneux question and the basis of Locke's answer', Bc138, 75–100.
9. Brown, Keith M. 'The vanishing emperor: British kingship and its decline, 1603–1707', Bc119, 58–91.
10. Bryden, D.J. 'A 1701 dictionary of mathematical instruments', Bc36, 365–82.
11. Burgess, Glenn. 'Protestant polemic: the Leveller pamphlets', *Parergon* 11 (1993), 45–67.
12. Capková, Dagmar. 'Comenius and his ideals: escape from the labyrinth', Bc118, 75–93.
13. Chappell, Vere. 'Introduction' [Locke], Bc115, 1–5.
14. Chappell, Vere. 'Locke on the freedom of the will', Bc138, 101–122.
15. Chappell, Vere. 'Locke's theory of ideas', Bc115, 26–56.
16. Chibi, Andrew A. 'The interpretation and use of divine and natural law in the first marriage crisis of Henry VIII', *Archiv für Reformationsgeschichte* 85 (1994), 265–86.
17. Christianson, Paul. 'Arguments on billeting and martial law in the parliament of 1628', *Historical J.* 37 (1994), 539–67.

18. Clericuzio, Antonio. 'New light on Benjamin Worsley's natural philosophy', Bc118, 236–47.
19. Clucas, Stephen. 'In search of "the true logick": methodological eclecticism among the "Baconian reformers"', Bc118, 51–75.
20. Cohen, I. Bernard. 'Harrington and Harvey: a theory of the state based on the new physiology', *J. of the History of Ideas* 55 (1994), 187–210.
21. Condren, Conal. *The language of politics in seventeenth-century England.* Basingstoke; Macmillan; 1994. Pp xi, 215.
22. Condren, Conal. 'The paradoxes of recontextualization in early modern intellectual history' [review article], *Historical J.* 37 (1994), 225–31.
23. Cowan, Edward J. 'The political ideas of a covenanting leader: Archibald Campbell, marquis of Argyll, 1607–1661', Bc119, 241–62.
24. Davis, Edward B. '"Parcere nominibus": Boyle, Hooke and the rhetorical interpretation of Descartes', Bc10, 157–75.
25. Faulkner, Robert K. *Francis Bacon and the politics of progress.* Savage (Md); Rowman & Littlefield; 1993. Pp 315.
26. Ford, John D. '*Lex, rex iusto posita*: Samuel Rutherford on the origins of government', Bc119, 262–90.
27. Gorham, Geoffrey. 'Mind-body dualism and the Harvey-Descartes controversy', *J. of the History of Ideas* 55 (1994), 211–34.
28. Greengrass, Mark; Leslie, Michael; Raylor, Timothy. 'Introduction' [Samuel Hartlib], Bc118, 1.
29. Guyer, Paul. 'Locke's philosophy of language', Bc115, 115–46.
30. Harris, Ian. *The mind of John Locke: a study of political theory in its intellectual setting.* Cambridge; Cambridge UP; 1994. Pp xv, 429.
31. Harvey, John H. 'Coronary flowers and their "Arabick" background', Bc92, 297–303.
32. Hotson, Howard. 'Philosophical pedagogy in reformed central Europe between Ramus and Comenius: a survey of the continental background of the "three foreigners"', Bc118, 27–50.
33. Houston, Alan Craig. '"A way of settlement": the Levellers, monopolies and the public interest', *History of Political Thought* 14 (1993), 381–420.
34. Hunter, Michael. 'The conscience of Robert Boyle: functionalism, "dysfunctionalism" and the task of historical understanding', Bc49, 147–59.
35. Jones, R. Brinley. *William Salesbury.* Cardiff; Wales UP; 1994. Pp 100.
36. Kitromilides, Paschalis M. 'John Locke and the Greek intellectual tradition: an episode in Locke's reception in south-east Europe', Bc138, 217–236.
37. Krook, Dorothea. *John Sergeant and his circle: a study of three seventeenth-century English Aristotelians*, ed. Beverley C. Southgate. Leiden; Brill; 1993. Pp xxiv, 177.
38. Lamont, William (ed.). *A holy commonwealth.* Cambridge; Cambridge UP; 1994. Pp xxxi, 254.
39. Losonsky, Michael. 'Locke on meaning and signification', Bc138, 123–142.
40. Macintosh, J.J. 'Locke and Boyle on miracles and God's existence', Bc10, 193–214.
41. Mason, Roger A. 'The Scottish Reformation and the origins of Anglo-British imperialism', Bc119, 161–87.
42. McCann, Edwin. 'Locke's philosophy of body', Bc115, 56–89.

43. McPhail, Bridget. 'Scotland's sovereignty asserted: the debate over the Anglo-Scottish Union of 1707', *Parergon* 11 (1993), 27–44.
44. Milton, J.R. 'Locke at Oxford', Bc138, 29–48.
45. Milton, J.R. 'Locke's life and times', Bc115, 5–26.
46. Milton, Philip. 'Hobbes, heresy and Lord Arlington', *History of Political Thought* 14 (1993), 501–46.
47. Moore, Tod. 'Recycling Aristotle: the sovereignty theory of Richard Hooker', *History of Political Thought* 14 (1993), 345–59.
48. Mulligan, Lotte. 'Robert Boyle, "right reason", and the meaning of metaphor', *J. of the History of Ideas* 55 (1994), 235–57.
49. Nelson, Scott A. *The discourses of Algernon Sidney*. Rutherford (NJ); Fairleigh Dickinson UP; 1993. Pp 176.
50. Newman, William. 'Arabo-Latin forgeries: the case of the *Summa perfectionis*', Bc92, 278–97.
51. Norbrook, David. 'Lucan, Thomas May and the creation of a republican literary culture', Bc17, 45–66; 331–36.
52. Nyland, Chris. 'John Locke and the social position of women', *History of Political Economy* 25 (1993), 39–63.
53. Perez-Ramos, Antonio. 'Francis Bacon and man's two-faced kingdom', Bc32, 140-66.
54. Peters, J.S. 'The Bank, the press and the "return of nature": on currency, credit, and literary property in the 1690s', Bc150, 365–88.
55. Pocock, J.G.A. 'England's Cato: the virtues and fortunes of Algernon Sidney' [review article], *Historical J.* 37 (1994), 915–35.
56. Prívratská, Jana; Prívratský, Vladimir. 'Language as the product and mediator of knowledge: the concept of J.A. Comenius', Bc118, 162–75.
57. Quinn, David Beers. *Thomas Harriot and the problem of America*. Oxford; Oriel College; 1992. Pp 16.
58. Rex, Richard. 'The new learning', *J. of Ecclesiastical History* 44 (1993), 26–44.
59. Reynolds, Noel B.; Hilton, John L. 'Thomas Hobbes and authorship of the *Horae subsecivae*', *History of Political Thought* 14 (1993), 361–80.
60. Rogers, G.A.J. 'Introduction' [Locke's philosophy], Bc138, 1–28.
61. Rosicka, Janina. 'John Locke and the Polish Enlightenment', Bc138, 237–252.
62. Russell, G.A. 'The impact of *Philosophus autodidactus*: Pocockes, John Locke and the Society of Friends', Bc92, 224–66.
63. Schneewind, J.B. 'Locke's moral philosophy', Bc115, 199–226.
64. Shanahan, Timothy. 'Teleological reasoning in Boyle's *Disquisition about final causes*', Bc10, 177–92.
65. Shapiro, Barbara. 'The concept "fact": legal origins and cultural diffusion', *Albion* 26 (1994), 1–25; 227–52.
66. Sharpe, Kevin. 'The king's writ: royal authors and royal authority in early modern England', Bc17, 117–38; 343–47.
67. Shepard, Amanda. *Gender and authority in sixteenth-century England*. Keele; Ryburn; 1994. Pp 224.
68. Simmons, A. John. *On the edge of anarchy: Locke, consent, and the limits of society*. Princeton (NJ); Princeton UP; 1993. Pp ix, 293.

69. Smuts, Malcolm. 'Court-centered politics and the uses of Roman historians, *c*.1590-1630', Bc17, 21–43; 325–31.
70. Sommerville, J.P. *King James VI and I: political writings*. Cambridge; Cambridge UP; 1994. Pp 384.
71. Sorell, Tom. 'Seventeenth-century materialism: Gassendi and Hobbes', Bc32, 235–72.
72. Springborg, Patricia. 'Hobbes, heresy, and the *Historia ecclesiastica*', *J. of the History of Ideas* 55 (1994), 553–71.
73. Spurr, John. 'Revolution and languages of liberty' [review article], *Historical J.* 37 (1994), 471–76.
74. Sugarman, David; Warrington, Ronnie. 'Land law, citizenship, and the invention of "Englishness": the strange world of the equity of redemption', Bc150, 111–43.
75. Sullivan, Vickie. 'The civic humanist portrait of Machiavelli's English successors', *History of Political Thought* 15 (1994), 73–96.
76. Tully, James. 'Rediscovering America: the *Two treatises* and aboriginal rights', Bc138, 165–196.
77. Walker, William. *Locke, literary criticism and philosophy*. Cambridge; Cambridge UP; 1994. Pp 300.
78. Weyer, Stefan. *Die Cambridge Platonists: Religion und Freiheit in England im 17. Jahrhundert* [The Cambridge Platonists: religion and freedom in seventeenth-century England]. Frankfurt-am-Main; Lang; 1993. Pp 181.
79. Winship, Michael P. 'Prodigies, Puritanism and the perils of natural philosophy: the example of Cotton Mather', *William & Mary Q.* 3rd ser. 51 (1994), 92–105.
80. Wolterstorff, Nicholas. 'Locke's philosophy of religion', Bc115, 172–99.
81. Wood, Ellen Merksins. 'Radicalism, capitalism and historical contexts: not only a reply to Richard Ashcraft on John Locke', *History of Political Thought* 15 (1994), 324–72.
82. Woolhouse, Roger. 'Locke's theory of knowledge', Bc115, 146–72.
83. Wormald, Brian Harvey Goodwin. *Francis Bacon: history, politics and science, 1561–1626*. Cambridge; Cambridge UP; 1993. Pp xii, 409.
84. Zagorin, Perez. 'The political beliefs of John Pym to 1629', *English Historical R.* 109 (1994), 867–90.

(k) *Arts and Cultural History*

1. Adamson, J.S.A. 'Chivalry and political culture in Caroline England', Bc17, 161–97; 349–57.
2. Anglo, Sydney. 'Tudor dynastic symbols', Bc82, 193–210.
3. Anon. *Renaissance commonplace books from the Huntington library: a listing and guide to the microfilm collection*. Marlborough; Matthew; 1994. Pp 20.
4. Archer, John Michael. *Sovereignty and intelligence: spying and court culture in the English Renaissance*. Cambridge; Stanford UP; 1993. Pp viii, 203.
5. Aston, Margaret. *The king's bedpost: Reformation and iconography in a Tudor group portrait*. Cambridge; Cambridge UP; 1993. Pp xii, 266.
6. Bawcutt, N.W. (ed.). *The Malone society collections*, vol. 15. Oxford; Oxford UP; 1994. Pp 203.

7. Bawcutt, N.W. (ed.). 'Ralph Crane's transcript of "A game at chess", Bodleian Manuscript Malone 25', Bc81, 1–109.
8. Beecher, Donald. 'Mannerist explorations in the English theatre: a case study of *The Changeling*', Bc82, 177–92.
9. Bellany, Alastair. '"Raylinge rymes and vaunting verse": libellous politics in early Stuart England, 1603–1628', Bc17, 285–310; 367–71.
10. Bonomi, Patricia U. 'Lord Cornbury redressed: the governor and the problem portrait', *William & Mary Q.* 3rd ser. 51 (1994), 106–18.
11. Butler, Martin. 'Ben Jonson and the limits of courtly panegyric', Bc17, 91–115.
12. Capp, Bernard. *The world of John Taylor, the water-poet, 1578–1653.* Oxford; Oxford UP; 1994. Pp 216.
13. Carlson, David A. 'Arthur before and after the revolution: the Blome-Stanby edition of Malory (1634) and *Brittain's Glory* (1684)', Bc100, 234–53.
14. Carnegie, David (ed.). 'The part of "Poore"', Bc81, 111–69.
15. Charney, Maurice. 'Imagery in Shakespeare's *Titus Andronicus*: blood, the wood and the sea', Bc82, 135–46.
16. Collinson, Patrick. 'Protestant culture and the cultural revolution', Bc103, 33–52.
17. Cullum, David; Wardley, Peter. 'The diffusion of the Hindu-Arabic numerical system: numeracy, literacy and historical analysis of writing skills in seventeenth-century west Cornwall', Bc135, 3–31.
18. Duffy, Maureen. *Henry Purcell.* London; Fourth Estate; 1994. Pp 288.
19. Dunn, Kevin. 'Milton among the monopolists: *Areopagitica*, intellectual property and the Hartlib circle', Bc118, 177–93.
20. Elliot, John (ed.). 'Musical and dramatic documents from the Middle Temple', Bc81, 171–94.
21. Finke, Laurie A. 'Spenser for hire: Arthurian history as cultural capital in *The fairie queen*', Bc100, 211–33.
22. Foakes, R.A. 'The image of the Swan theatre', Bc82, 337–59.
23. Friedman, Jerome. *Miracles and the pulp press during the English revolution: the battle of the frogs and Fairford's flies.* London; UCL; 1993. Pp xv, 304.
24. Gibson-Wood, Carol. 'The political background to Thornhill's paintings in St Paul's cathedral', *J. of the Warburg & Courtauld Inst.* 56 (1993), 229–37.
25. Glanville, Philippa. 'The Howard grace cup', *History Today* 44/10 (1994), 41–45.
26. Gruffydd, R. Geraint. 'Dr John Davies, "old man of Brecknock"', *Archaeologia Cambrensis* 141 (1992), 1–13.
27. Hall, Marie Boas. 'Arabick learning in the correspondence of the Royal Society, 1660-1677', Bc92, 147–58.
28. Hart, Vaughan. *Art and magic in the court of the Stuarts.* London; Routledge; 1994. Pp 280.
29. Harth, Phillip. *Pen for a party: Dryden's Tory propaganda in its contexts.* Princeton (NJ); Princeton UP; 1994. Pp xi, 341.
30. Hattaway, Michael. 'Blood is their argument: men of war and soldiers in Shakespeare and others', Bc89, 84–101.

31. Hepburn, Frederick. 'The portraiture of Lady Margaret Beaufort', *Antiquaries J.* 72 (1994 for 1992), 118–40.
32. Holt, P.M. 'Background to Arabic studies in seventeenth-century England', Bc92, 20-30.
33. Hopton, F.C. 'The buildings of Shaftesbury abbey in the mid-sixteenth century', *Dorset Natural History & Arch. Soc. P.* 115 (1994 for 1993), 1–13.
34. Howarth, David. 'William Trumbull and art collecting in Jacobean England', *British Library J.* 20 (1994), 140-62.
35. Jardine, Lisa; Sherman, William. 'Pragmatic readers: knowledge transactions and scholarly services in late Elizabethan England', Bc89, 102–24.
36. Johnston, Alexandra F. 'Performance practice informed by image: the iconography of the *Chester* pageant', Bc82, 245–63.
37. Keefer, Michael H. 'Self-image and spectacle in *Tamburlaine* and *Doctor Faustus*', Bc82, 109–34.
38. Koran, Joshua M. 'Robert Gomersall: a forgotten life', *Somerset & Dorset Notes & Queries* 33 (1994), 330-36.
39. Lake, A. Cynfael. 'Dau gywydd i'r Doctor Wiliam Glyn' [Two odes in honour of Dr Wiliam Glyn], *Caernarvonshire Historical Soc. T.* 54 (1993), 91–105.
40. Lake, Peter. 'Deeds against nature: cheap print, Protestantism and murder in early seventeenth-century England', Bc17, 257–83; 361–67.
41. Lake, Peter. 'Popular form, Puritan content?: two Puritan appropriations of the murder pamphlet from mid seventeenth-century London', Bc89, 313–34.
42. Laroque, Franois. '*Macbeth*: theatre of image to shadow theatre', Bc82, 147–76.
43. Lascombes, Andre. 'Image et culture commune à la Renaissance: de quelques exemples anglais et français' [Image and popular culture in the Renaissance: some English and French examples], Bc82, 9–31.
44. Lewalski, Barbara Kiefer. *Writing women in Jacobean England.* London; Harvard UP; 1993. Pp xii, 431.
45. Lindsay, Alexander. 'Some drafts by Richard Steele for *The Tatler*, *The Spectator*, and *The Guardian*', *British Library J.* 20 (1994), 163–73.
46. Loach, Jennifer. 'The function of ceremonial in the reign of Henry VIII', *Past & Present* 142 (1994), 43–68.
47. Loskoutoff, Yvan. 'Astrée à la licorne: l'éloge d'Elisabeth I concluant le troisième livre des *Tragiques* d'Agrippa d'Aubigne', *Bibliothèque d'Humanisme et Renaissance* 54 (1992), 373–84.
48. Marcus, Leah S. 'Politics and pastoral: writing the court on the countryside', Bc17, 139–59; 347–49.
49. McCann, Alison. 'The building of the sixteenth-century rood loft in Westbourne church', *Sussex Arch. Collections* 131 (1993), 124–25.
50. McGee, C.E. (ed.). 'Fireworks for Queen Elizabeth', Bc81, 195–205.
51. Mortensen, Preben. 'Shaftesbury and the morality of art appreciation', *J. of the History of Ideas* 55 (1994), 631–50.
52. Newman, John. 'Inigo Jones and the politics of architecture', Bc17, 229–55; 360-61.
53. Norland, Howard. 'The dramatic spectacle of *Wit's progress*', Bc82, 80-93.

54. Norris, H.T. 'Edmund Castell and his *Lexicon heptaglotton* (1669)', Bc92, 70-88.
55. Peacock, John. 'The politics of portraiture', Bc17, 199–228; 357–60.
56. Platt, Colin. *Great rebuildings of Tudor and Stuart England: revolutions in architectural taste*. London; UCL; 1994. Pp 208.
57. Potter, Loïs. 'Perspectives ey'd awry: artificiality as truth in English Renaissance drama', Bc82, 264–80.
58. Pugh, Thomas B. 'A portrait of Queen Anne of Denmark at Parham Park, Sussex', *Seventeenth Century* 8 (1993), 167–80.
59. Rivington, Charles A. *Pepys and the booksellers*. York; Sessions; 1993. Pp 80.
60. Roberts, Peter. 'Elizabethan players and minstrels and the legislation of 1572 against retainers and vagabonds', Bc89, 29–55.
61. Russell, G.A. 'Introduction' [the age of "Arabick"], Bc92, 1–20.
62. Salmon, Vivian. 'Arabists, and linguists in seventeenth-century England', Bc92, 54–70.
63. Schofield, John. 'Building in religious precincts in London at the Dissolution and after', Bc64, 29–41.
64. Scolnicov, Hanna. 'The woman in the window: a theatrical icon', Bc82, 281–305.
65. Sharpe, Allison E. 'The Paget memorial, St Mary's, Skirpenbeck: some problems considered', *Yorkshire Arch. J.* 65 (1993), 95–114.
66. Smith, Nigel. *Literature and revolution in England, 1640-1660*. London; Yale UP; 1994. Pp xiv, 425.
67. Strasser, Gerhard F. 'Closed and open languages: Samuel Hartlib's involvement with cryptology and universal languages', Bc118, 151–62.
68. Streitberger, W.R. *Court revels, 1485–1559*. Toronto; Toronto UP; 1994. Pp xvi, 454.
69. Streitberger, W.R. 'Records of royal banqueting houses and Henry VIII's timber lodging, 1543–1559', *J. of the Soc. of Archivists* 15 (1994), 187–202.
70. Tribble, Evelyn B. *Margins and marginality: the printed page in early modern England*. Charlottesville (Va); Virginia UP; 1994. Pp 194.
71. Wakefield, Colin. 'Arabic manuscripts in the Bodleian Library: the seventeenth-century collections', Bc92, 128–47.
72. White, Paul Whitefield. *Theatre and reformation: Protestantism, patronage, and playing in Tudor England*. Cambridge; Cambridge UP; 1993. Pp xvii, 286.
73. Wilson, Alison. 'An Elizabethan miniature in the Parker library, Corpus Christi college', *T. of the Cambridge Bibliographical Soc.* 4 (1994), 461–85.
74. Worden, Blair. 'Ben Jonson among the historians', Bc17, 67–89; 336–43.
75. Worrall, Andrew. 'Richard Barnfield, 1574–1627', *Shropshire History & Archaeology* 69 (1994), 120-25.

(l) *Science and Technology*

1. Alexander, Peter. 'Solidity and elasticity in the seventeenth century', Bc138, 143–164.

2. Baldwin, Martha R. 'Toads and plague: the amulet controversy in seventeenth-century medicine', *B. of the History of Medicine* 67 (1993), 227–47.
3. Barker, Richard. 'A manuscript of shipbuilding, *c.*1600, copied by Newton', *Mariner's Mirror* 80 (1994), 16–29.
4. Bird, Nicholas du Q. 'A sixteenth-century Glastonbury elixir', *Somerset & Dorset Notes & Queries* 33 (1994), 283–86.
5. Bracegirdle, Brian. 'Seventeenth-century simple microscopes', Bc36, 295–305.
6. Clericuzio, Antonio. 'Carneades and the chemists: a study of *The sceptical chymist* and its impact on seventeenth-century chemistry', Bc10, 79–90.
7. Clifton, Gloria C. 'The spectacle-makers' company and the origins of the optical instrument-making trade in London', Bc36, 341–64.
8. Cook, Harold J. 'Good advice and little medicine: the professional authority of early modern English physicians', *J. of British Studies* 33 (1994), 1–31.
9. Coughlan, Patricia. 'Natural history and historical nature: the project for a natural history of Ireland', Bc118, 298–319.
10. de Graeve, Jan. 'Francis Hall's sundial for the blind', Bc36, 216–17.
11. Figala, Karin; Petzold, Ulrich. 'Alchemy in the Newtonian circle: personal acquaintances and the problem of the late phase of Isaac Newton's alchemy', Bc49, 173–91.
12. Gatti, Hilary. *The natural philosophy of Thomas Harriot.* Oxford; Oriel College (Thomas Harriot lecture); 1993. Pp 28.
13. Gibson, James M. 'The 1566 survey of the Kent coast', *Archaeologia Cantiana* 112 (1994 for 1993), 341–53.
14. Gingerich, Owen. 'Astronomical paper instruments with moving parts', Bc36, 63–74.
15. Harley, David. 'Political post-mortems and morbid anatomy in seventeenth-century England', *Social History of Medicine* 7 (1994), 1–28.
16. Harvey, P.D.A. 'Estate surveyors and the spread of the scale-map in England, 1550-1580', *Landscape History* 15 (1993), 37–49.
17. Harwood, John T. 'Science writing and writing science: Boyle and rhetorical theory', Bc10, 37–56.
18. Henry, John. 'Boyle and cosmical qualities', Bc10, 119–38.
19. Home, R.W. 'Newton's subtle matter: the *Opticks* queries and the mechanical philosophy', Bc49, 193–202.
20. Hughes, John Trevor. *Thomas Willis, 1621–1675: his life and work.* London; Royal Soc. of Medicine Services; 1991. Pp xvi, 151.
21. Hunter, Michael. 'Robert Boyle reconsidered', Bc10, 1–18.
22. Kaplan, Barbara Beigun. *'Divulging of useful truths in physick': the medical agenda of Robert Boyle.* Baltimore (Md); Johns Hopkins UP; 1994. Pp xii, 216.
23. Keil, Inge. 'Technology transfer and scientific specialization: Johann Wiesel, optician of Augsburg, and the Hartlib circle', Bc118, 268–79.
24. Mercier, Raymond. 'English orientalists and mathematical astronomy', Bc92, 158–215.
25. Molland, George. 'The limited lure of Arabic mathematics', Bc92, 215–24.
26. Nance, Brian K. 'Determining the patient's temperament: an excursion in seventeenth-century medical semeiology', *B. of the History of Medicine* 67 (1993), 417–38.

27. Newman, William. 'Boyle's debt to corpuscular alchemy', Bc10, 107–18.
28. Nichols, Richard S. (ed.). *The diaries of Robert Hooke: Leonardo of London, 1635–1703*. Lewes; Book Guild; 1994. Pp 185.
29. Oster, Malcolm. 'Millenarianism and the new science: the case of Robert Boyle', Bc118, 137–49.
30. Principe, Lawrence M. 'Boyle's alchemical pursuits', Bc10, 91–105.
31. Pumfrey, Stephen. '"These 2 hundred years not the like published as Gellibrand has done de Magnete": the Hartlib circle and magnetic philosophy', Bc118, 247–68.
32. Sargent, Rose-Mary. 'Learning from experience: Boyle's construction of an experimental philosophy', Bc10, 57–78.
33. Shapin, Steven. *A social history of truth: civility and science in seventeenth-century England*. Chicago; Chicago UP; 1994. Pp 498.
34. Shapiro, Alan E. *Fits, passions and paroxysms: physics, method and chemistry and Newton's theories of coloured bodies and fits of easy reflection*. Cambridge; Cambridge UP; 1993. Pp xvii, 400.
35. Simcock, A.V. 'An equinoctial ring dial by Ralph Greatorex', Bc36, 201–15.
36. Vaughan, Denys. '"A very artificial workman": the altitude sundials of Humphrey Cole', Bc36, 191–200.
37. Wear, Andrew. 'English medical writers and their interest in classical Arabic medicine in the seventeenth century', Bc92, 266–78.
38. Webster, Charles. 'Benjamin Worsley: engineering for universal reform from the Invisible College to the Navigation Act', Bc118, 213–36.

G. BRITAIN 1714–1815

See also Aa6,28–29,42,48,144,b32,c59; Ba2,38,80,92,102,117,127,139,154,170, 183, 192,204,b26–27,35,40,54,63,71,76,109,123,142,148,156,188,192,203,234,258–259, 270,285,c19,44,57–62,90,110,134,137,d4–8,11–15,19–23,30–38,40–45,49–51,53–58,60–62,64–65,67,73,76–77,103,105–106,109,115; Fc8,e7,44,g34,65,h19,i6; He31, h21,k21

(a) *General*

1. Barry, Jonathan. 'The making of the middle class?' [review article], *Past & Present* 145 (1994), 194–208.
2. Birdwood, Vere. *So dearly loved, so much admired: letters to Hester Pitt, Lady Chatham, from her relations and friends, 1744–1801*. London; HMSO; 1994. Pp xxxii, 362.
3. Board, Beryl; Durgan, Shirley. 'Georgian Colchester', Bc86, 133–75.
4. Brent, Colin E. *Georgian Lewes, 1714–1830: the heyday of a county town*. Lewes; Brent; 1993. Pp iv 246.
5. Cannon, John. *Samuel Johnson and the politics of Hanoverian England*. Oxford; Oxford UP; 1994. Pp vii, 326.
6. Claeys, Gregory. 'The origins of the rights of labor: republicanism, com-

merce, and the construction of modern social theory in Britain, 1796–1805', *J. of Modern History* 66 (1994), 249–90.

7. Clark, J.C.D. *Samuel Johnson: literature, religion and English cultural politics from the Restoration to romanticism*. Cambridge; Cambridge UP; 1994. Pp xii, 270.
8. Corbin, Alain; Phelps, Jocelyn. *The lure of the sea: the discovery of the seaside in the western world, 1750-1840*. Cambridge; Polity; 1993. Pp 392.
9. Gerrard, Christine. *The patriot approach to Walpole: politics, poetry and national myth, 1725–1742*. Oxford; Oxford UP; 1994. Pp xiv, 273.
10. Hancox, Joy. *The queen's chameleon: the life of John Byrom, a study in conflicting loyalties*. London; Cape; 1994. Pp 275.
11. Hartley, B.R. 'Thomas Griffith of York, "once governor of the castle and now a debtor from the same"', *York Historian* 11 (1994), 40-55.
12. Harvey, A.D. *Sex in Georgian England: attitudes and prejudices from the 1720s to the 1820s*. London; Duckworth; 1994. Pp 205.
13. Hinchliffe, G. 'The Robinsons of Newby Park and Newby Hall, part 2: Tancred Robinson, baronet, Rear Admiral of the White, 1686–1754', *Yorkshire Arch. J.* 64 (1992), 185–202.
14. Hinchliffe, G. 'The Robinsons of Newby Park and Newby Hall, part 3: Thomas Robinson, Baron Grantham, (1695–1770) and his descendants', *Yorkshire Arch. J.* 65 (1993), 143–52.
15. Houston, Rab A. 'Eighteenth-century Scottish studies: out of the laager?', *Scottish Historical R.* 73 (1994), 64–81.
16. Martin, Joanna (ed.). *The Penrice letters, 1768–1795*. Swansea; South Wales Record Soc. no. 9; 1993. Pp 191.
17. McGregor, Margaret. 'Astry and Orlebar correspondence from two Ashton court papers, 1709–1721', Bc6, 116–28.
18. Meakin, David. 'Introduction' [the secular city], Bc134, 3–14.
19. Porter, Roy. 'Enlightenment London and urbanity', Bc134, 27–41.
20. Tillyard, S.K. *Aristocrats: Caroline, Emily, Louisa and Sarah Lennox, 1750-1832*. London; Chatto; 1994. Pp 462.
21. Whatley, Christopher A. 'Flat laager?' [eighteenth-century Scottish studies], *Scottish Historical R.* 73 (1994), 82–88.
22. Woolley, Dorothy. 'Thomas Hill of Tern, 1693–1782', *Archives* 21 (1994), 155–71.

(b) *Politics*

1. Black, Jeremy. 'Eighteenth-century English politics: recent work and current problems', *Albion* 25 (1993), 419–41.
2. Bourne, Kenneth; Taylor, W.B. (ed.). *The Horner papers: selections from the correspondence and miscellaneous writings of Francis Horner, MP, 1795–1817*. Edinburgh; Edinburgh UP; 1994. Pp xxix, 980.
3. Brooke, Alan; Kipling, Lesley. *Liberty or death: radicals, republicans and Luddites, 1793–1823*. Honley; Workers History; 1993. Pp 112.
4. Cruickshanks, Eveline. 'Lord Cowper, Lord Orrery, the duke of Warton and Jacobitism', *Albion* 26 (1994), 27–40.

5. Ditchfield, G.M. 'The House of Lords and the impeachment of Warren Hastings', *Parliamentary History* 13 (1994), 277–96.
6. Downie, J.A. 'Stating facts right about Defoe's *Review*' [reprint], Bc15, 8–22.
7. Drescher, Seymour. 'Whose abolition?: popular pressure and the ending of the British slave trade', *Past & Present* 143 (1994), 136–66.
8. Eastwood, David. *Governing rural England: tradition and transformation in local government, 1780-1840.* Oxford; Oxford UP; 1994. Pp xiii, 311.
9. Eastwood, David. 'Men, morals and the machinery of social legislation, 1790-1840', *Parliamentary History* 13 (1994), 190-205.
10. Farrell, Stephen. 'Division lists and the nature of the Rockingham Whig party in the House of Lords', *Parliamentary History* 13 (1994), 170-89.
11. Furbank, P.N.; Owens, W.R. 'Defoe and "Sir Andrew Politick"', *British J. for Eighteenth-Century Studies* 17 (1994), 27–39.
12. Ginter, Donald. 'Unofficial division lists of the British House of Commons, 1761–1820: a new data base and its implications', *Parliamentary History* 13 (1994), 11–31.
13. Graham, Jenny. 'The publication of part one of the *Rights of man*', *Enlightenment & Dissent* 12 (1993), 70-77.
14. Houston, Rab A. 'Popular politics in the reign of George II: the Edinburgh cordiners', *Scottish Historical R.* 72 (1993), 167–89.
15. Hume, David. *Political essays*, ed. Knud Haakonssen. Cambridge; Cambridge UP; 1994. Pp lxiv, 346.
16. Jones, Clyve. '1720-1723 and all that: a reply to Eveline Cruickshanks', *Albion* 26 (1994), 41–53.
17. Jones, Clyve. 'Whigs, Jacobites and Charles Spencer, third earl of Sunderland', *English Historical R.* 109 (1994), 52–73.
18. Lockwood, Thomas. 'The life and death of *Common sense*' [reprint], Bc15, 78–93.
19. Miller, Peter N. *Defining the common good: empire, religion and philosophy in eighteenth-century Britain.* Cambridge; Cambridge UP; 1994. Pp viii, 472.
20. Peters, Marie. 'The myth of William Pitt, earl of Chatham, great imperialist, part II: Chatham and imperial reorganization, 1763–1778', *J. of Imperial & Commonwealth History* 22 (1994), 393–431.
21. Pettit, Alexander. 'Propaganda, public relations, and the *Remarks on the Craftsman's vindication of his two Honble patrons, in his paper of May 22, 1731*', *Huntington Library Q.* 57 (1994), 45–59.
22. Priestley, Joseph. *Political writings*, ed. Peter N. Miller. Cambridge; Cambridge UP; 1993. Pp xxxix, 147.
23. Smyth, Jim. 'Republicans and conservatives: British politics and ideology, *c.*1750-1832', *Historical J.* 37 (1994), 697–704.
24. Speck, W.A. '*The Examiner* re-examined' [reprint], Bc15, 34–43.
25. Targett, Simon. 'Government and ideology during the age of Whig supremacy: the political argument of Sir Robert Walpole's newspaper propagandists', *Historical J.* 37 (1994), 289–317.
26. Thomas, M. Rhiannon. 'William Vaughan (1707–1775), Corssygedol', *Merioneth Historical & Record Soc. J.* 11 (1992), 255–71.

27. Thompson, E.P. 'Hunting the Jacobin fox' [John Thelwall], *Past & Present* 142 (1994), 94–140.
28. Varey, Simon. '*The Craftsman*' [reprint], Bc15, 58–77.
29. Wellenreuther, Hermann. 'Activities of an estate agent in mid eighteenth-century England: Robert Butcher and the town of Bedford', Bc6, 156–76.

(c) *Constitution, Administration and Law*

1. Barrell, John. 'Imagining the king's death: the arrest of Richard Brothers', *History Workshop* 37 (1994), 1–32.
2. Cairns, John W. 'Adam Smith and the role of the courts in securing justice and liberty', Bc110, 31–61.
3. Cockburn, J.S. 'Punishment and brutalization in the English Enlightenment', *Law & History R.* 12 (1994), 155–79.
4. Cotter, William R. 'The Somerset case and the abolition of slavery in England', *History* 79 (1994), 31–56.
5. Coy, Betty; Hill, Sheila; Sharpe, Vera. *Transcription of minutes of the corporation of Boston, vol. 5: 1717–1763*. Boston; History of Boston Project; 1993. Pp xix, 831.
6. Davey, B.J. *Rural crime in the eighteenth century: north Lincolnshire, 1740-1780*. Hull; Hull UP; 1994. Pp xv, 167.
7. Hoppit, Julian; Innes, Joanna; Styles, John. 'Towards a history of parliamentary legislation, 1660-1800', *Parliamentary History* 13 (1994), 312–21.
8. Johnson, Wilfred Harold. *Crime and disorder in late Georgian Alfriston*. Eastbourne; Downsway; 1994. Pp 95.
9. King, Peter; Noel, Joan. 'The origins of "the problem of juvenile delinquency": the growth of juvenile prosecutions in London in the late eighteenth and early nineteenth centuries', *Criminal Justice History* 14 (1993), 17–41.
10. Lindgren, J. Ralph. 'Adam Smith's treatment of criminal law', Bc110, 63–81.
11. MacKinnon, Kenneth A.B. 'Adam Smith on delictual liability', Bc110, 83–112.
12. Macnair, Michael. 'Sir Jeffrey Gilbert and his Treatises', *J. of Legal History* 15 (1994), 252–68.
13. McCann, Timothy J. 'On the alleged murder of his chaplain by Henry Browne, 5th viscount Montague of Cowdray', *Sussex Arch. Collections* 131 (1993), 126–28.
14. Meldrum, Tim. 'A women's court in London: defamation at the bishop of London's Consistory Court, 1700-1745', *London J.* 19 (1994), 1–20.
15. Oldham, James. 'Truth-telling in the eighteenth-century English courtroom', *Law & History R.* 12 (1994), 95–121.
16. Smith, Lionel D. 'The stockbroker and the solicitor-general: the story behind *Taylor v. Plumer*', *J. of Legal History* 15 (1994), 1–22.
17. Snyder, Henry L. 'Charges to grand juries: the evidence of the Eighteenth-Century Short-Title Catalogue', *Historical Research* 67 (1994), 286–300.

18. Sokol, Mary. 'Bentham and Blackstone on incorporeal hereditaments', *J. of Legal History* 15 (1994), 287–305.
19. Stevenson, John. 'An unbroken wave?' [review article], *Historical J.* 37 (1994), 683–95.

(d) *External Affairs*

1. Atkinson, Alan. 'The free-born Englishman transported: convict rights as a measure of eighteenth-century empire', *Past & Present* 144 (1994), 88–115.
2. Bickers, Robert A. 'Ritual and diplomacy', Bc44, 7–10.
3. Black, Jeremy. *British foreign policy in an age of revolutions, 1783–1793.* Cambridge; Cambridge UP; 1994. Pp xii, 559.
4. Black, Jeremy. 'From Pillnitz to Valmy: British foreign policy and revolutionary France, 1791–1792', *Francia* 21/2 (1994), 129–46.
5. Black, Jeremy. 'Omitted from the record: the duke of Newcastle's reassurance over British foreign policy in 1727', *Archives* 21 (1994), 80-82.
6. Black, Jeremy. 'The coming of war between Britain and France, 1792–1793', *Francia* 20/2 (1993), 69–108.
7. Bullion, John L. 'George III on empire, 1783', *William & Mary Q.* 3rd ser. 51 (1994), 305–10.
8. Cashin, Edward J. *Governor Henry Ellis and the transformation of British north America.* Athens (Ga); Georgia UP; 1994. Pp 320.
9. Clark, Jennifer. 'The war of 1812: American nationalism and rhetorical images of Britain', *War & Soc.* 12 (1994), 1–26.
10. Fischer, David Hackett. *Paul Revere's ride.* Oxford; Oxford UP; 1994. Pp xviii, 445.
11. Greene, Jack P. 'The Jamaica privilege controversy, 1764–1766: an episode in the process of constitutional definition in the early modern British empire', *J. of Imperial & Commonwealth History* 22 (1994), 16–53.
12. Gurtler, Gernot O. '"Under the shadow of high diplomacy": some aspects on the bishop of Namur's political and religious career, including the "Mémoire instructif" of his secret mission to Vienna in 1734', *T. of the Cumberland & Westmorland Antiquarian & Arch. Soc.* 94 (1994), 143–69.
13. Heilingsetzer, Georg. 'Koalition gegen Revolution: England, Österreich und das Problem einer europäischen Friedensordnung, 1789–1814' [Coalition against revolution: England, Austria and the problem of a European peace settlement], *Mitteilungen des Instituts für Österreichische Geschichtsforschung* 101 (1993), 360-82.
14. Hevia, James L. 'The Macartney embassy in the history of Sino-western relations', Bc44, 57–79.
15. Ingersoll, Thomas N. '"Releese us out of the Cruell Bondegg": an appeal from Virginia in 1723', *William & Mary Q.* 3rd ser. 51 (1994), 777–82.
16. MacDonogh, Katherine. 'A sympathetic ear: Napoleon, Elba and the British', *History Today* 44/2 (1994), 29–35.
17. Marshall, P.J. 'Britain and China in the late eighteenth century', Bc44, 11–29.
18. Obeyesekere, Gananath. *The apotheosis of Captain Cook: European mythmaking in the Pacific.* Princeton (NJ); Princeton UP; 1993. Pp xvii, 251.

19. O'Shaughnessy, Andrew J. 'The Stamp Act crisis in the British Caribbean', *William & Mary Q.* 3rd ser. 51 (1994), 203–26.
20. Purdue, A.W. 'John and Harriet Carr: a brother and sister from the north-east on the grand tour', *Northern History* 30 (1994), 122–38.
21. Richardson, David. 'Cape Verde, Madeira and Britain's trade to Africa, 1698–1740', *J. of Imperial & Commonwealth History* 22 (1994), 1–15.
22. Shunhong, Zhang. 'Historical anachronism: the Qing court's perception of and reaction to the Macartney embassy', Bc44, 31–42.
23. Thompson, Martyn P. 'Ideas of Europe during the French revolution and Napoleonic wars', *J. of the History of Ideas* 55 (1994), 37–58.
24. Thorp, Jennifer D. (ed.). *The Acland journal: Lady Harriet Acland and the American war*. Winchester; Hampshire County Council (Hampshire record ser.); 1993. Pp xxxviii, 67.
25. Wang, Tseng-Tsai. 'The Macartney mission: a bicentennial review', Bc44, 43–56.
26. Welsh, Jennifer M. *Edmund Burke and international relations: the commonwealth of Europe and the crusade against the French revolution*. London; Macmillan; 1994. Pp 200.
27. Woodfine, Philip (ed.). 'Horace Walpole and British relations with Spain, 1738', Bc93, 289–328.

(e) *Religion*

1. Almond, Philip C. *Heaven and hell in Enlightenment England*. Cambridge; Cambridge UP; 1994. Pp xiii, 218.
2. Berman, David. *George Berkeley: idealism and the man*. Oxford; Oxford UP; 1994. Pp xi, 230.
3. Blackwood, B. Gordon. 'Lancashire Catholics, Protestants and Jacobites during the 1715 rebellion', *Recusant History* 22 (1994), 41–59.
4. Boyce, Douglas. 'The diary of the Reverend John Robinson', *Lincolnshire History & Archaeology* 29 (1994), 28–38.
5. Brain, Timothy J. 'Richard Watson: Westmorland "statesman"?', *T. of the Cumberland & Westmorland Antiquarian & Arch. Soc.* 94 (1994), 171–200.
6. Brown-Lawson, A. *John Wesley and the Anglican evangelicals of the eighteenth century*. Edinburgh; Pentland; 1994. Pp xiv, 409.
7. Bushby, David. 'Samuel Bennett of Tempsford', Bc6, 61–71.
8. Cornwall, Robert D. 'Advocacy of the independence of the Church from the state in eighteenth-century England: a comparison of a nonjuror and a nonconformist view', *Enlightenment & Dissent* 12 (1993), 12–27.
9. Currie, David A. 'Cotton Mather's *Bonifacius* in Britain and America', Bc91, 73–89.
10. Davies, Glanville J. 'The rural parish church in Dorset in the eighteenth century', *Dorset Natural History & Arch. Soc. P.* 115 (1994 for 1993), 21–28.
11. Ditchfield, G.M. 'Public and parliamentary support for the unitarian petition of 1792', *Enlightenment & Dissent* 12 (1993), 28–48.
12. Edwards, Pamela J. 'Coleridge reconsidered: religion, metaphysics and political ideas', *Enlightenment & Dissent* 12 (1993), 63–9.
13. Fielden, Keith (ed.). *Caleb's letters: fifty newly discovered letters written by*

a Todmorden Methodist to his son and his family, describing events in Todmorden during the period 1799–1814. London; Fielden; 1993. Pp 84.
14. Fishburn, Janet F. 'Gilbert Tennent, established "dissenter"', *Church History* 63 (1994), 31–49.
15. Foley, Brian C. 'A priest describes to the Cardinal Prefect of Propaganda his mission in the Isle of Man', *Recusant History* 22 (1994), 60-62.
16. Goldie, Mark. 'Bishop Hay, Bishop Geddes and the Scottish Catholic Enlightenment', *Innes R.* 45 (1994), 82–86.
17. Gooch, Leo. '"What shall we do with the wanton student?": tutoring the Catholic gentry in the eighteenth century', *Recusant History* 22 (1994), 63–74.
18. Haydon, Colin. *Anti-Catholicism in eighteenth-century England, a political and social study*. Manchester; Manchester UP; 1993. Pp xi, 276.
19. Hempton, David. 'Evangelicalism in English and Irish society, 1780-1840', Bc91, 156–76.
20. Holtby, R.T. 'Thomas Herring as archbishop of York, 1743–1747', *Northern History* 30 (1994), 105–21.
21. Houston, Rab A. '"Bustling artisans": church patronage at South Leith in the 1740s and 1750s', *Albion* 26 (1994), 55–77.
22. Jacob, William. 'Evidence for dissent in Norfolk, 1711–1800, from the records of the diocese of Norwich', Bc38, 37–45.
23. Lambert, Frank. *'Pedlar in divinity': George Whitefield and the transatlantic revival, 1739–70*. Princeton (NJ); Princeton UP; 1994. Pp. xiii, 238.
24. Lilly, Derek B.; Lilly, Jane. *St Andrew's Clevendon churchwardens' accounts, 1730-1801*. Weston-super-Mare; Galloway; 1994. Pp 54.
25. Logan, Beryl. *A religion without talking: religious belief and natural belief in Hume's philosophy of religion*. Frankfurt-am-Main; Lang; 1993. Pp xii, 184.
26. Mills, Frederick V. 'The Society in Scotland for Propagating Christian Knowledge in British north America, 1730-1775', *Church History* 63 (1994), 15–30.
27. Noll, Mark A. 'Revolution and the rise of evangelical social influence in north Atlantic societies', Bc91, 113–36.
28. O'Brien, Susan. 'Eighteenth-century publishing networks in the first years of transatlantic evangelicalism', Bc91, 38–57.
29. Pettit, Alexander. 'The Francis Hare controversy of 1732', *British J. for Eighteenth-Century Studies* 17 (1994), 41–53.
30. Schmidt, Leigh E. 'Time, celebration and the Christian year in eighteenth-century evangelicalism', Bc91, 90-109.
31. Stell, Christopher. 'Wesley's chapel, City Road, Islington: anniversary address, 1993', *T. of the Ancient Monuments Soc.* 38 (1994), 15–29.
32. Stephens, John 'Richard Price, *A sermon on Revelation 21.3*, *Enlightenment & Dissent* 12 (1993), 78–91.
33. Stewart, James A. 'The Clan Ranald and Catholic missionary successes, 1715–1745', *Innes R.* 45 (1994), 29–46.
34. Stout, Harry S. 'George Whitefield in three countries', Bc91, 58–72.
35. Walsh, John. '"Methodism" and the origins of English-speaking evangelicalism', Bc91, 19–37.
36. Ward, W.R. (ed.). *Parson and parish in eighteenth-century Surrey: replies to*

bishops' visitations. Guildford; Surrey Record Soc. vol. 34; 1994. Pp xxvi, 198.

37. Ward, W.R.; Heitzenrater, Richard P. (ed.). *The works of John Wesley, vol. 22: journals and diaries, 1765–1775*. London; SPCK; 1994. Pp xii, 480.
38. Williams, James. 'A 1784–1786 examination roll for the parish of Ruthwell', *Dumfriesshire & Galloway Natural History & Antiquarian Soc. T.* 3rd ser. 67 (1992), 89–92.
39. Wynne, Thomas. 'The conversion of Alexander Cameron', *Innes R.* 45 (1994), 178–87.
40. Young, B.W. '"The soul-sleeping system": politics and heresy in eighteenth-century England', *J. of Ecclesiastical History* 45 (1994), 64–81.
41. Young, B.W. 'William Law and the Christian economy of salvation', *English Historical R.* 109 (1994), 308–22.

(f) *Economic Affairs*

1. Bebb, Prudence. *Shopping in Regency York, 1811 to 1820: butcher, baker, candlestick-maker*. York; Sessions; 1994. Pp vi, 85.
2. Beckett, J.V. 'Aristocratic financial troubles and the operation of the land market: the sale of Astwell and Falcutt in 1774–1778', *Northamptonshire Past & Present* 8 (1993/4), 378–82.
3. Beckett, J.V. 'Gosfield Hall: a country estate and its owners, 1715–1825', *Essex Archaeology & History* 3rd ser. 25 (1994), 185–92.
4. Berg, Maxine; Hudson, Pat. 'Growth and change: a comment on the Crafts-Harley view of the industrial revolution', *Economic History R.* 2nd ser. 47 (1994), 147–49.
5. Brown, Vivienne. *Adam Smith's discourse: canonicity, commerce and conscience*. London; Routledge; 1994. Pp 224.
6. Buck, Anne. 'Mantuamakers and milliners: women making and selling clothes in eighteenth-century Bedfordshire', Bc6, 142–55.
7. Church, Roy A. 'Introduction' [coal and iron industries], Bc61, ix-xxxvii.
8. Clay, C.G.A. 'Henry Hoare, banker, his family and the Stourhead estate', Bc55, 113–38.
9. Clemens, Picht. *Die Zukunft des Handels in einer kommerziellen Gesellschaft: zur wirtschaftspolitischen Publizistik Englands im 18. Jahrhundert* [The future of trade in a commercial society: journalism on economic policy in eighteenth-century England]. Göttingen; Vandenhoeck & Ruprecht (German Historical Inst. London vol. 33); 1993.
10. Cookson, Gillian. 'Innovation, diffusion, and mechanical engineers in Britain, 1780-1850', *Economic History R.* 2nd ser. 47 (1994), 749–53.
11. Crafts, N.F.R.; Mills, Terence C. 'Trends in real wages in Britain, 1750-1913', *Explorations in Economic History* 31 (1994), 176–94.
12. Crompton, Gerald W. 'Canals and the industrial revolution', *J. of Transport History* 3rd ser. 14 (1993), 93–110.
13. Cuenca Esteban, Javier. 'British textile prices, 1770-1831: are British growth rates worth revising once again?', *Economic History R.* 2nd ser. 47 (1994), 66–105.

14. Cullen, L.M. *Smuggling and the Ayrshire economic boom of the 1760s and 1770s*. Ayr; Ayrshire Arch. & Natural History Soc. (monograph 14); 1994. Pp 55.
15. de Vries, Jan. 'The industrial revolution and the industrious revolution', *J. of Economic History* 54 (1994), 249–70.
16. Donnelly, Tom. 'Merchant accounts and Scottish trade, 1680-1730: the Charteris and Clephane papers', *Business Archives* 68 (1994), 5–14.
17. Durie, Alastair J. '"Gentlemen pretty much strangers to the Baltic trade": the Edinburgh Roperie and Sailcloth Company, 1750-1802', *Scottish Industrial History* 14–15 (1992), 27–34.
18. Elder, Melinda. *The slave trade and the economic development of eighteenth-century Lancaster*. Keele; Ryburn; 1994. Pp 232.
19. Evans, Chris. 'Iron puddling: the quest for a new technology in eighteenth-century industry', *Llafur* 6/3 (1994), 44–57.
20. Fabian, Bernhard. 'The first English bookshop on the continent', Bc25, 49–64.
21. Forrester, D.A.R. 'Early canal company accounts: financial and accounting aspects of the Forth and Clyde navigation, 1768–1816', Bc35, 297–326.
22. Freeman, Michael. 'Popular attitude to turnpikes in early eighteenth-century England', *J. Historical Geography* 19 (1993), 33–47.
23. Gherity, James A. 'An early publication by Adam Smith', *History of Political Economy* 25 (1993), 241–82.
24. Hancock, David. '"Domestic bubbling": eighteenth-century London merchants and individual investment in the funds', *Economic History R.* 2nd ser. 47 (1994), 679–702.
25. Harris, Ron. 'The Bubble Act: its passage and its effect on business organization', *J. of Economic History* 54 (1994), 610-27.
26. Hoppit, Julian; Wrigley, E.A. 'Introduction' [industrial revolution in Britain], Bc57, ix-xl.
27. Houston, Rab A. 'Fraud in the Scottish linen industry: Edinburgh's charity workhouse, 1745–1758', *Archives* 21 (1994), 43–56.
28. Hughes, J.R.T. 'A world elsewhere: the importance of starting English', Bc55, 63–87.
29. Jenkins, D.T. 'Introduction' [textile industries], Bc59, ix-xliii.
30. Keirn, Tim. 'Monopoly, economic thought, and the Royal African Company', Bc150, 427–66.
31. Klein, Lawrence E. 'Property and politeness in the early eighteenth-century Whig moralists: the case of *The Spectator*', Bc150, 221–33.
32. Lieberman, David. 'Property, commerce, and the common law: attitudes to legal change in the eighteenth century', Bc150, 144–60.
33. Linsley, Stafford M. 'A north-eastern agent and bottle salesman in East Anglia, 1781', *Norfolk Archaeology* 42 (1994), 89–97.
34. Linsley, Stafford M. 'Tyne crossings at Hexham up to 1795', *Archaeologia Aeliana* 5th ser. 22 (1994), 235–53.
35. McClure, David. *Tolls and tacksmen: eighteenth-century roads in the county of John Loudon McAdam*. Ayr; Ayrshire Arch. & Natural History Soc. (Ayrshire monographs 13); 1994. Pp 56.
36. McConnell, Anita. 'From craft workshop to big business: the London

scientific instrument trade's response to increasing demand, 1750-1820', *London J.* 19 (1994), 36–53.

37. Mepham, M.J. 'The Scottish Enlightenment and the development of accounting', Bc35, 268–93.
38. Morgan, Kenneth. 'Bristol West India merchants in the eighteenth century', *T. of the Royal Historical Soc.* 6th ser. 3 (1993), 185–208.
39. Munckton, Thelma. *Chancellor's farm accounts, 1766–1767.* Weston-super-Mare; Galloway; 1994. Pp xiv, 42.
40. Nyland, Chris. 'Adam Smith, stage theory, and the status of women', *History of Political Economy* 25 (1993), 617–40.
41. O'Brien, P.K. 'Introduction' [industrial revolution in Europe], Bc58, ix-lviii.
42. O'Brien, P.K.; Heath, D. 'English and French landowners, 1688–1789', Bc55, 1–22.
43. Pearson, Mowbray (ed.). *More frost and snow: the diary of Janet Burnet, 1758–1795.* Edinburgh; Canongate; 1994. Pp viii, 127.
44. Pollard, Sidney. 'Introduction' [metal fabrication and engineering industries], Bc60, ix-xxxvii.
45. Sharpe, Pamela. 'De-industrialization and re-industrialization: women's employment and the changing character of Colchester, 1700-1850', *Urban History* 21 (1994), 77–96.
46. Solkin, David. 'ReWrighting Shaftesbury: the *Air Pump* and the limits of commercial humanism', Bc150, 234–53.
47. Taylor, W.B. 'The workshops and manufactories of York in the second half of the eighteenth century', *York Historian* 10 (1992), 18–30.
48. Tibbles, Anthony (ed.). *Transatlantic slavery: against human dignity.* London; HMSO; 1994. Pp 168.
49. Truckell, A.E. 'Arbigland accounts, 1751–1759', *Dumfriesshire & Galloway Natural History & Antiquarian Soc. T.* 3rd ser. 67 (1992), 81–86.
50. Turner, Michael. 'Common property and property in common' [review article], *Agricultural History R.* 42 (1994), 158–62.
51. Ville, Simon P. 'Transport and the industrial revolution' [review article], *J. of Transport History* 3rd ser. 13 (1992), 180-85.
52. Wade-Martins, Susanna; Williamson, Tom. 'Floated water-meadows in Norfolk: a misplaced innovation', *Agricultural History R.* 42 (1994), 20-37.
53. Watney, Bernard M. 'Excavations at the Longton Hall porcelain manufactory, part 3: the porcelain and other ceramic finds', *Post-Medieval Archaeology* 27 (1993), 57–109.
54. Werner, Alex. 'The port of London, 1750-1806', Bc121, 39–62.
55. Whatley, Christopher A. 'Women and the economic transformation of Scotland, *c.*1740-1830', *Scottish Economic & Social History* 14 (1994), 19–40.
56. Whyld, Martin. 'John Badder of Cossall', *East Midland Historian* 4 (1994), 28–31.
57. Wicklen, S.I. 'Thomas Roberts (1760-1811): Caernarfon's first printer, 1796–1811', *Caernarvonshire Historical Soc. T.* 55 (1994), 25–45.
58. Wilkins, Frances. *George Moore and friends: the letters from a Manx merchant, 1750-1760.* Kidderminster; Wyre Forest; 1994. Pp 320.
59. Wilson, G. 'Public hire chairs in Weymouth', *Somerset & Dorset Notes & Queries* 33 (1994), 289–94.

60. Withers, Charles W.J. 'On Georgics and geology: James Hutton's *Elements of agriculture* and agricultural science in eighteenth-century Scotland', *Agricultural History R.* 42 (1994), 38–48.

(g) *Social Structure and Population*

1. Beckett, J.V. 'Love and marriage among the English aristocracy: the Buckingham-Chandos alliance of 1796', *Records of Buckinghamshire* 34 (1994 for 1992), 101–06.
2. Davies, Glanville J. 'Mary Ozard and the "affair of the bastard"', *Dorset Natural History & Arch. Soc. P.* 115 (1994 for 1993), 168–69.
3. d'Cruze, Shani. 'Care, diligence and "usfull pride" [sic]: gender, industrialization and the domestic economy, *c.*1770 to *c.*1840', *Women's History R.* 3 (1994), 315–45.
4. Hales, Jane. *On the parish: recorded lives of the poor of Holt and district, 1780-1835*, ed. Susan Yaxley. Dereham; Larks; 1994. Pp 70.
5. Kelch, Ray A.; Wootten, William W. 'A Northamptonshire courtship in the eighteenth century', *Northamptonshire Past & Present* 9 (1994), 47–54.
6. May, Keith Le. 'London records of poor relief for French Protestants, 1750-1850', *Huguenot Soc. P.* 26 (1994), 71–82.
7. Pooley, Colin G.; d'Cruze, Shani. 'Migration and urbanization in north-west England, *c.*1760-1830', *Social History* 19 (1994), 339–58.
8. Razzell, Peter. 'Introduction' [population history], Bc88, 1–5.
9. Razzell, Peter. 'Recent research on eighteenth-century population history', Bc88, 207–29.
10. Smail, John. 'Local disputes and the making of Halifax's middle class, 1748–1776', *T. of the Halifax Antiquarian Soc.* ns 2 (1994), 29–41.
11. Sokoll, Thomas. *Household and family among the poor: the case of two Essex communities in the late eighteenth and early nineteenth centuries.* Bochum; Brockmeyer; 1993. Pp xxiv, 383.
12. Szreter, Simon. 'Mortality in England in the eighteenth and the nineteenth centuries: a reply to Sumit Guha', *Social History of Medicine* 7 (1994), 269–82.
13. Tinley, Ruth; Mills, Dennis. 'Population turnover in an eighteenth-century Lincolnshire parish in comparative context', *Local Population Studies* 52 (1994), 30-38.
14. van Mesdag, Martin. 'Tweehonderd Schotten in een Hollands schip naar Canadese wildernis' [Two hundred Scots in a Dutch ship bound for the Canadian wilderness], *Spiegel der Zeilvaart* 17/9 (1993), 8–12.
15. Wells, Roger. 'Migration, the law and parochial policy in eighteenth and early nineteenth-century southern England', *Southern History* 15 (1993), 87–139.

(h) *Naval and Military*

1. Adam, Robin. 'The *Mercer* affair', *Mariner's Mirror* 80 (1994), 147–58.
2. Annand, A.Mck. 'Lord Kilmarnock's Horse Grenadiers (later Foot Guards), in the army of Prince Charles Edward, 1745–1746', *J. of the Soc. for Army Historical Research* 72 (1994), 70-75.

3. Barr, William; Williams, Glyndwr (ed.). *Voyages to Hudson Bay in search of a north-west passage, 1741–1747, vol. 1: voyage of Christopher Middleton, 1741–1742.* London; Hakluyt Soc. 2nd ser. vol. 177; 1994. Pp xii, 333.
4. Bartlett, C.J. 'Gentlemen versus democrats: cultural prejudice and military strategy in Britain in the war of 1812', *War in History* 1 (1994), 140-59.
5. Cavaliero, Roderick. *Admiral Satan: Suffren's naval campaigns in the Indies.* London; Tauris; 1993. Pp 224.
6. Cole, Richard C. 'A British military family in New York, 1755–1791', *Genealogists' Magazine* 24 (1994), 381–88.
7. Cormack, Andrew. 'Captain John Stewart, 25th foot the Edinburgh regiment, circa 1790', *J. of the Soc. for Army Historical Research* 72 (1994), 1–7.
8. Fisher, Robin. *Vancouver's voyage: charting the north-west coast, 1791–1795.* Seattle (Wa); Washington UP; 1992. Pp xii, 131.
9. Flayhart, William Henry. *Counterpoint to Trafalgar: the Anglo-Russian invasion of Naples, 1805–1806.* Columbia (SC); South Carolina UP; 1992. Pp xi, 198.
10. Gardiner, Robert. *The heavy frigate: 18–pounder frigates, vol. 1, 1778–1800.* London; Conway; 1994. Pp 127.
11. Gates, David. 'The transformation of the army, 1783–1815', Bc109, 133–60.
12. Godbold, Steve. 'A Napoleonic coastal gun battery: excavations at Bathside Bay, Harwich, 1990-1991', *Essex Archaeology & History* 3rd ser. 25 (1994), 193–218.
13. Goddard, J.C. 'An insight into the life of Royal Navy surgeons during the Napoleonic war, part 2', *J. of the Royal Navy Medical Service* 78/1 (1992), 27–36.
14. Gutridge, Tony. 'Aspects of naval prize agency, 1793–1815', *Mariner's Mirror* 80 (1994), 45–53.
15. Guy, Alan J. 'The army of the Georges, 1714–1783', Bc109, 92–111.
16. Harfield, Alan. 'The siege of Meester Cornelis, Java, August 1811', *J. of the Soc. for Army Historical Research* 72 (1994), 133–41.
17. Hayter, Tony. 'The army and the first British empire, 1714–1783', Bc109, 111–33.
18. Haythornthwaite, Philip J. *Armies of Wellington.* London; Arms & Armour; 1994. Pp 318.
19. Haythornthwaite, Philip J. 'Standing regulations of the London and Westminster Light Horse Volunteers, 1804', *J. of the Soc. for Army Historical Research* 72 (1994), 219–31.
20. Janzen, Olaf Uwe. 'Showing the flag: Hugh Palliser in western Newfoundland, 1764', *Northern Mariner* 3 (1993), 3–14.
21. Kleine, E.H. 'Under Napoleon's standard', *Hawick Arch. Soc. T.* (1993), 10-16.
22. Lanning, George. 'Thomas Hardy and *The alarm*', *Dorset Natural History & Arch. Soc. P.* 115 (1994 for 1993), 170-72.
23. Mackesy, Piers. *British victory in Egypt, 1801: the end of Napoleon's conquest.* London; Routledge; 1994. Pp 288.
24. McCorry, Helen (ed.). '"Besides, he was very drunk at the time ...": desertion and discipline, north Britain, 1751–1753', *J. of the Soc. for Army Historical Research* 72 (1994), 142–59.

25. Naish, John. 'The achievements of Captain George Vancouver: a reassessment after 200 years', *Mariner's Mirror* 80 (1994), 418–30.
26. Packer, Maurice. 'A scandalous naval history: *The naval Atlantis*', *Factotum* 38 (1994), 12–17.
27. Raban, Peter. 'The profits of privateering: a Guernsey fleet, 1756–1762', *Mariner's Mirror* 80 (1994), 298–311.
28. Ranlet, Philip. 'The fate of the *Washington*, 1775–1776: a precedent for future British conduct', *American Neptune* 54 (1994), 194–98.
29. Raynor, David; Skinner, Andrew. 'Sir James Steuart: nine letters on the American conflict, 1775–1778', *William & Mary Q.* 3rd ser. 51 (1994), 755–76.
30. Strawson, John. *Duke and the emperor: Wellington and Napoleon*. London; Constable; 1994. Pp 356.

(i) *Intellectual and Cultural*

1. Adam, Keith. 'Introduction: living with the legend', *Architectural History* 4 (1993), 1–9.
2. Anon (ed.). 'Sir Hans Sloane's catalogues' [in the British Museum], Bc90, 291–94.
3. Archibald, Marion. 'Coins and medals' [in the British Museum], Bc90, 150-166.
4. Bentman, R. 'Thomas Gray and the poetry of "hopeless love"', *J. of the History of Sexuality* 3 (1993), 203–22.
5. Bermingham, Ann. 'The picturesque and ready-to-wear femininity', Bc41, 81–119.
6. Black, Jeremy. 'The language of licentiousness', *The Scriblerian* 26/2 (1994), 153–9.
7. Bloom, Edward A.; Bloom, Lillian D. (ed.). *The Piozzi letters: correspondence of Hester Lynch Piozzi, 1784–1821 (formerly Mrs Thrale), volume 1: 1784–1791*. Newark (NJ); Delaware UP; 1989. Pp 417.
8. Bloom, Edward A.; Bloom, Lillian D. (ed.). *The Piozzi letters: correspondence of Hester Lynch Piozzi, 1784–1821 (formerly Mrs Thrale), volume 2: 1792–1798*. Newark (NJ); Delaware UP; 1991. Pp 592.
9. Bloom, Edward A.; Bloom, Lillian D. (ed.). *The Piozzi letters: correspondence of Hester Lynch Piozzi, 1784–1821 (formerly Mrs Thrale), volume 3: 1799–1804*. Newark (NJ); Delaware UP; 1993. Pp 529.
10. Bohls, Elizabeth A. 'Aesthetics and orientalism in Lady Mary Wortley Montagu's letters', *Studies in Eighteenth-Century Culture* 23 (1994), 179–205.
11. Borsay, Peter. 'Image and counter-image in Georgian Bath', *British J. for Eighteenth-Century Studies* 17 (1994), 165–79.
12. Borsay, Peter. 'The London connection: cultural diffusion and the eighteenth-century provincial town', *London J.* 19 (1994), 21–35.
13. Briggs, Peter M. '"News from the little world": a critical glance at eighteenth-century British advertising', *Studies in Eighteenth-Century Culture* 23 (1994), 29–45.
14. Brown, Iain Gordon. 'A bibliophile's bagnio: Sir James Clerk's pantheon for Penicuik', Bc139, 135–50.

15. Brown, Iain Gordon. 'Architects or gentlemen? Adam heraldry and its implications', *Architectural History* 4 (1993), 82–92.
16. Brown, Yu-Ying. 'Japanese books and manuscripts' [in the British Museum], Bc90, 278–90.
17. Cameron, Neil. 'Adam and "Gothick" at Yester chapel', *Architectural History* 4 (1993), 39–44.
18. Carré, Jacques. *Lord Burlington, (1694–1753): le connaisseur, le mécène, l'architecte.* Clermont-Ferrand; Adosa; 1993. 2 vol.
19. Caygill, Marjorie. 'Sloane's will and the establishment of the British Museum', Bc90, 45–68.
20. Chapman, John. 'Enclosure commissioners as landscape planners', *Landscape History* 15 (1993), 51–55.
21. Charlesworth, Michael. 'The ruined abbey: picturesque and Gothic values', Bc41, 62–80.
22. Cherry, John. 'Medieval and later antiquities' [in the British Museum], Bc90, 198–221.
23. Claeys, Gregory. *Utopias of the British Enlightenment.* Cambridge; Cambridge UP; 1994. Pp 288.
24. Clemit, Pamela. 'Godwin's educational theory: *The Enquirer*', *Enlightenment & Dissent* 12 (1993), 3–11.
25. Collett-White, James. '*My choice*: a poem written in 1751 by Mary Orlebar', Bc6, 129–41.
26. Copley, Stephen. 'William Gilpin and the black-lead mine', Bc41, 42–61.
27. Copley, Stephen; Garside, Peter. 'Politics of the picturesque', Bc41, 1–12.
28. Craik, Roger. *James Boswell, 1740-1795: the Scottish perspective.* Edinburgh; HMSO; 1994. Pp ix, 195.
29. Crimmins, James E. 'Bentham's political radicalism re-examined', *J. of the History of Ideas* 55 (1994), 259–81.
30. Daniels, Stephen; Watkins, Charles. 'Picturesque landscaping and estate management: Uvedale Price and Nathaniel Kent at Foxley', Bc41, 13–41.
31. Davies, Caryl. '"Romantic Jones": the picturesque and politics on the South Wales circuit, 1775–1781', *National Library of Wales J.* 28 (1994), 255–78.
32. Donoghue, E. 'Imagined more than women: lesbians as hermaphrodites, 1671–1766', *Women's History R.* 2 (1993), 199–216.
33. Downie, J.A.; Corns, Thomas N. 'Early eighteenth-century periodicals' [reprint], Bc15, 1–7.
34. Eimer, Christopher. *Medallic portraits of the duke of Wellington.* London; Spink; 1994. Pp 152.
35. Emerson, Richard. 'Robert Adam and John Clerk of Eldin: from primitive hut to temple of religion', Bc139, 165–80.
36. Evensky, Jerry. 'Setting the scene: Adam Smith's moral philosophy', Bc110, 7–29.
37. Evensky, Jerry. 'The role of law in Adam Smith's moral philosophy: natural jurisprudence and utility', Bc110, 199–219.
38. Everett, Nigel. *The Tory view of landscape.* London; Yale UP; 1994. Pp 224.
39. Franklin, Colin. *Lord Chesterfield: his character and 'Characters'.* Aldershot; Scolar; 1993. Pp xvi, 150.

40. Garside, Peter. 'Picturesque figure and landscape: Meg Merrilies and the gypsies', Bc41, 145–74.
41. Gascoigne, John. *Joseph Banks and the English Enlightenment: useful knowledge and polite culture*. Cambridge; Cambridge UP; 1994. Pp 344.
42. Goulstone, John. '"Long Robin" and the cricketing Colchins', *Genealogists' Magazine* 24 (1994), 389–93.
43. Gow, Ian; Simpson, James. '8 Queen Street Edinburgh: restoring an Adam house', *Architectural History* 4 (1993), 58–65.
44. Griffiths, Beryl H. 'Rhys Jones (1713–1801) o'r Blaenau' [from Blaenau Ffestiniog], *Merioneth Historical & Record Soc. J.* 11 (1993), 433–45.
45. Harris, John. *The Palladian revival: Lord Burlington, his villa and garden at Chiswick*. London; Yale UP; 1994. Pp 224.
46. Harris, John. 'A Roger Morris-John Adam episode in astylar Palladianism', Bc139, 128–34.
47. Harris, Michael. 'Literature and commerce in eighteenth-century London: the making of *The Champion*' [reprint], Bc15, 94–115.
48. Harrison, Brett. 'Thorp Arch Hall, 1719–1756: "dabling a little in mortar"', Bc112, 1–39.
49. Herbert, Nicholas. '"The only resource for honest poverty": charity schoolmasters of the Stroud region, 1760-1830', *T. of the Bristol & Gloucestershire Arch. Soc.* 111 (1993), 175–89.
50. Higgins, Ian. *Swift's politics: a study in disaffection*. Cambridge; Cambridge UP; 1994. Pp xi, 232.
51. Horden, John. *John Freeth (1731–1808): political ballad-writer and innkeeper*. Oxford; Leopard's Head; 1993. Pp xvi, 205.
52. Imprey, Oliver. 'Oriental antiquities' [in the British Museum], Bc90, 222–227.
53. Irving, Howard. 'Classic and Gothic: Charles Burney on "ancient music"', *Studies in Eighteenth-Century Culture* 23 (1994), 243–63.
54. James, F.E. 'Cirencester grammar school, the Rev. Dr John Washbourn and some medical pupils', *T. of the Bristol & Gloucestershire Arch. Soc.* 111 (1993), 191–99.
55. James, Frank A.J.L.; Field, J.V. 'Frankenstein and the spark of being', *History Today* 44/9 (1994), 47–53.
56. Jenkins, Ian. 'Classical antiquities: Sloane's "repository of time"' [in the British Museum], Bc90, 167–73.
57. Johnsen, Berit Eide. 'North Sea cultural influence in southern Norway through Norwegian prisoners of war in Great Britain, 1807–1814', Bc121, 237–63.
58. Jones, Vivien. '"The coquetry of nature": politics and the picturesque in women's fiction', Bc41, 120-44.
59. Kay, William. 'Robert Adam: some responses to a Scottish background', *Architectural History* 4 (1993), 23–38.
60. Kelly, Gary. *Women, writing, and revolution, 1790-1827*. Oxford; Oxford UP; 1993. Pp 328.
61. King, J.H.C. 'Ethnographic collections: collecting in the context of Sloane's catalogue of "miscellanies"' [in the British Museum], Bc90, 228–44.
62. Klein, Lawrence E. *Shaftesbury and the culture of politeness: moral*

discourse and cultural politics in early eighteenth-century England. Cambridge; Cambridge UP; 1994. Pp xiii, 217.

63. Knight, Charles A. '*The Spectator*'s generalizing discourse' [reprint], Bc15, 44–57.
64. Knox, Tim. 'The great pagoda at Kew', *History Today* 44/7 (1994), 22–26.
65. Kowaleski-Wallace, Beth. 'Tea, gender, and domesticity in eighteenth-century England', *Studies in Eighteenth-Century Culture* 23 (1994), 131–45.
66. MacGregor, Arthur. 'Egyptian antiquities' [in the British Museum], Bc90, 174–179.
67. MacGregor, Arthur. 'Prehistoric and Romano-British antiquities' [in the British Museum], Bc90, 180-197.
68. MacGregor, Arthur. 'The life and character of Sir Hans Sloane', Bc90, 11–44.
69. MacInnes, Ranald. 'Robert Adam's public buildings', *Architectural History* 4 (1993), 10-22.
70. Mallory, Robin Paul. 'Introduction' [Adam Smith], Bc110, 1–5; 151.
71. Mannings, David. 'Jonathan Richardson, Thomas Gray, and the genealogy of art', *J. of the History of Ideas* 55 (1994), 405–20.
72. Mason, Adair Stuart. *George Edwards: the bedell and his birds.* London; Royal College of Physicians; 1992. Pp viii, 71.
73. Maurer, Michael. *'O Britannien, von deiner Freiheit einen Hut voll': Deutsche Reiseberichte des 18. Jahrhunderts* [German travel writings in the eighteenth-century]. Munich; Beck; 1992.
74. Mautner, Thomas (ed.). *Francis Hutcheson: two texts on human nature.* Cambridge; Cambridge UP; 1993. Pp 240.
75. Mendus, Susan. 'When the kissing had to stop: passion in the thought of Mary Wollstonecraft', *Historian* 42 (1994), 13–16.
76. Miles, Robert. *Ann Radcliffe: the great enchantress.* Manchester; Manchester UP; 1994. Pp 208.
77. Misenheimer, James B., Jr. 'Dr Johnson, Warren Cordell and the love of books', Bc25, 87–103.
78. Modiano, Raimonda. 'The legacy of the picturesque: landscape, property and the ruin', Bc41, 196–219.
79. Müller, Karis. 'Physics and the deity: the ideas of R. Boscovich and J. Priestley', *Enlightenment & Dissent* 12 (1993), 49–62.
80. Nenadic, Stana. 'Middle-rank consumers and domestic culture in Edinburgh and Glasgow, 1720-1840', *Past & Present* 145 (1994), 122–56.
81. Nicholson, Colin. *Writing and the rise of finance: capital satires of the early eighteenth century.* Cambridge; Cambridge UP; 1994. Pp xv, 219.
82. Nickson, M.A.E. 'Books and manuscripts' [in the British Library], Bc90, 263–77.
83. Noblett, William. 'Dru Drury's letters (1770-1775) to the Cambridge bookseller, John Woodyer', *T. of the Cambridge Bibliographical Soc.* 4 (1994), 539–47.
84. Philp, Mark (ed.). *The political and philosophical writings of William Godwin.* London; Pickering & Chatto; 1993. 7 vols.
85. Pittock, Murray. *Poetry and Jacobite politics in eighteenth-century Britain and Ireland.* Cambridge; Cambridge UP; 1994. Pp xiii, 254.

86. Punter, David. 'The picturesque and the sublime: two worldscapes', Bc41, 220-39.
87. Redford, Bruce (ed.). *The letters of Samuel Johnson, vol. 4: 1782–1784; vol. 5: appendices and comprehensive index.* Oxford; Oxford UP; 1994. 2 vols. Pp xix, 462; xv, 174.
88. Ritvo, Harriet. 'Possessing Mother Nature: genetic capital in eighteenth-century Britain', Bc150, 413–26.
89. Rogers, Pat. '*The Rambler* and the eighteenth-century periodical essay: a dissenting view' [reprint], Bc15, 116–29.
90. Rose, Mark. *Authors and owners: the invention of copyright.* London; Harvard UP; 1993. Pp x, 176.
91. Rowlands, John. 'Prints and drawings' [in the British Museum], Bc90, 245–62.
92. Ruchinskaya, Tatiana. 'Record linkage: the Scottish architect William Hastie and the reconstruction of Moscow after the fire of 1812', *The Historian* 44 (1994), 16–18.
93. Sanderson, Margaret H.B. 'Trivial pursuit? Portrait of the artist in letters and diaries', *Architectural History* 4 (1993), 66–81.
94. Sell, Alan P.F. 'Philosophy in the eighteenth-century dissenting academies of England and Wales', *History of Universities* 11 (1992), 75–122.
95. Simpson, James. 'An account of the reconstruction of William Adam's design for Gladney House, with some observations on other early projects', Bc139, 91–98.
96. Slade, Harry Gordon. 'John Elphinstone and the castle of Glamis', Bc139, 119–27.
97. Stevenson, Janet H. 'The Nesbitts of Norwood House: a footnote to Junius', *Surrey Arch. Collections* 82 (1994), 169–79.
98. Stoker, David. 'Benjamin Mackerell: antiquary, librarian, and plagiarist', *Norfolk Archaeology* 42 (1994), 1–12.
99. Thaddeus, Janice Farrar. 'Elizabeth Hamilton's domestic politics', *Studies in Eighteenth-Century Culture* 23 (1994), 265–84.
100. Thompson, E.P. *Witness against the beast: William Blake and the moral law.* Cambridge; Cambridge UP; 1993. Pp xxi, 234.
101. Tiggesbaumker, Gunter. *Lady Elisabeth Craven, Ansbachs letzte Markgrafin [last countess of Ansbach], (1750-1828).* Weidenbach; Schmidt; 1994. Pp 23.
102. Todd, William B. 'Editing Adam Smith and Edmund Burke: some preliminary considerations', Bc25, 1–16.
103. Townsend, John Rowe (ed.). *Trade and plumb-cake for ever, huzza!: the life and work of John Newbery, 1713–1767: author, bookseller, entrepreneur and pioneer of publishing for children.* Cambridge; Colt; 1994. Pp xv, 173.
104. Walker, Diane A. 'The Bute mausoleum at St Margaret's church, Roath', *Arch. J.* 150 (1994 for 1993), 482–97.
105. Walker, Frank Arneil. 'Robert Adam at Walkinshaw', Bc139, 151–64.
106. Wawn, Andrew. 'Shrieks at the stones: the vikings, the Orkneys and the Scottish Enlightenment', Bc23, 408–22.
107. Whale, John. 'Romantics, explorers and picturesque travellers', Bc41, 175–95.

108. White, Stephen K. *Edmund Burke: modernity, politics and aesthetics.* London; Sage; 1994. Pp 97.
109. Whittingham, Selby. 'Turner's Lincolnshire connections: prospects, progeny and politics', *Lincolnshire History & Archaeology* 29 (1994), 39–44.
110. Winton, Calhoun. '*The Tatler*: from half-sheet to book' [reprint], Bc15, 23–33.
111. Worrall, David. 'Agrarians against the picturesque: ultra-radicalism and the revolutionary politics of land', Bc41, 240-60.
112. Zebrowski, Martha K. 'Richard Price: British Platonist of the eighteenth century', *J. of the History of Ideas* 55 (1994), 17–35.

(j) *Science*

1. Allen, Elizabeth; Turk, J.L.; Murley, Sir Reginald (ed.). *The case books of John Hunter, FRS.* London; Royal Soc. of Medicine Services; 1993. Pp xxv, 699.
2. Anderson, R.G.W. 'Joseph Black and his chemical furnace', Bc36, 118–26.
3. Bates, David L. 'All manner of natural knowledge: the Northampton Philosophical Society', *Northamptonshire Past & Present* 8 (1993/4), 363–77.
4. Bates, David L. 'When earth her entrails shook: the earthquake of 1750', *Northamptonshire Past & Present* 9 (1994), 55–68.
5. Bennett, J.A. 'Equipping the Radcliffe observatory: Thomas Hornsby and his instrument-makers', Bc36, 232–41.
6. Borsay, Anne. 'Using the records of an eighteenth-century infirmary', *Archives* 21 (1994), 172–82.
7. Cannon, John. 'Botanical collections' [in the British Museum], Bc90, 136–149.
8. Chapman, Allan. 'Scientific instruments and industrial innovation: the achievement of Jesse Ramsden', Bc36, 418–30.
9. Clutton-Brock, Juliet. 'Vertebrate collections' [in the British Museum], Bc90, 77–92.
10. Day, Michael. 'Humana: anatomical, pathological and curious human specimens in Sloane's museum' [in the British Museum], Bc90, 69–76.
11. Fitton, Mike; Gilbert, Pamela. 'Insect collections' [in the British Museum], Bc90, 112–122.
12. Frost, Alan. *Sir Joseph Banks and the transfer of plants to and from the south Pacific, 1786–1798.* Melbourne; Colony; 1993. Pp viii, 61.
13. Leopold, J.H. 'Some notes on Benjamin Ayres', Bc36, 395–402.
14. Morton, Alan Q.; Wess, Jane A. *Public and private science: the King George III collection.* Oxford; Oxford UP; 1993. Pp 720.
15. Porter, Roy. 'Health care in Enlightenment England: knowledge, power, and the market', Bc131, 95–111.
16. Thackray, John. 'Mineral and fossil collections' [in the British Museum], Bc90, 123–35.
17. Way, Kathie. 'Invertebrate collections' [in the British Museum], Bc90, 93–111.

18. Winston, Mark. 'The Bethel at Norwich: an eighteenth-century hospital for lunatics', *Medical History* 38 (1994), 27–51.
19. Wood, Paul B. *Science in the Scottish Enlightenment: Thomas Reid's papers on the life sciences*. Edinburgh; Edinburgh UP; 1994. Pp 272.

H. BRITAIN 1815–1914

See also Aa27,47,75,116,129,b10,42,54,60,67,c17,75,78,87; Ba2,16,38,43,80,88,92, 99,102,108,124,127,139,154–155,170,175,178,188,192,204–205,209,213,b14,19,26–27,34,40,53–54,61,67,76,80,83,118,132,142–143,147,165,188,233–234,258–259, 263,285,c1,14–15,19,63,71,79,106,120,d3,71,73,81; Ca5; Fe65; Ga8,20,b3,8–9,12, 23,f10–11,13,45,g4,6–7,12,i49,60,80; Ib2,5,10,22,42,49,c8–12,d21,e4,f9,24,30,64, 70,88–89,g7–8,12,15,18,h5–7,13,15,30–33,j33,37,55–56,60-61,100,102

(a) *General*

1. Anon. *'The Police Gazette', part 1: issues for 1866–1878, 1882–1897 & 1899–1900 from the Cambridgeshire police archive. A listing and guide to the microfilm collection*. Marlborough; Matthew; 1994. Pp 34.
2. Brendon, Vyvyen. *The age of reform, 1820-1850*. London; Hodder & Stoughton; 1994. Pp 128.
3. Crowther, M.A. 'Response: north of the border' [nineteenth-century Scotland], *Scottish Historical R.* 73 (1994), 100-02.
4. Graham, Angus; Gordon Joanna (ed.). *Skipness: memories of a highland estate on Kintyre Mull*. Edinburgh; Canongate; 1993. Pp 144.
5. Jackson, Ada; Blour, M.M. (ed.). *The diary of Ada Jackson, 1883*. Leicester; Living History Unit; 1993. Pp 156.
6. Jenkins, Terence Andrew. *The liberal ascendancy, 1830-1886*. London; Macmillan; 1994. Pp xi, 292.
7. Jones, Tegwyn. 'Erlid yn Aber' [persecution in Aberystwyth], Bc102, 123–63.
8. Love, John; McMullen, Brenda (ed.). *A salmon for the schoolhouse: a Nairnshire parish in the nineteenth century from the diaries of Robert and Elsie Thomson*. Edinburgh; Canongate; 1994. Pp 160.
9. Morris, R.J.; Morton, Graeme. 'Where was nineteenth-century Scotland?', *Scottish Historical R.* 73 (1994), 89–99.
10. Mulvey, Christopher. *Frank Thomas Morrell: Edwardian entrepreneur*. Winchester; Estcourt; 1994. Pp xi, 181.
11. Popplewell, Joyce. *A Gloucestershire diarist: Lt. Col. A.B. Lloyd-Baker of Hardwicke court: the early years, 1897–1919*. Lydney; Thornhill; 1993. Pp iv, 156.
12. Rock, Joe. 'Relugas and the Dick Lauder family', Bc139, 261–76.
13. Ross, M.S. 'The water supply of Kington Magna', *Dorset Natural History & Arch. Soc. P.* 115 (1994 for 1993), 173–75.
14. Vann, J. Don; VanArsdel, Rosemary T. 'Introduction' [Victorian periodicals and Victorian society], Bc14, 3–8.

15. Wardley, Peter. 'Edwardian Britain: empire, income and political discontent', Bc148, 57–78.
16. Weston, Ron. *Thornsett in 1841: land and people*. Stockport; New Mills Local History Soc.; 1994. Pp 52.

(b) *Politics*

1. Anon. *Fabian economic and social thought, series one: the papers of Edward Carpenter, 1844–1929, from Sheffield archives, Sheffield libraries and information services. A listing and guide to the microfilm edition*. Marlborough; Matthew; 1994. Pp 86.
2. Armbrust, Crys. 'Tennyson's political readers: W.J. Linton's *The National* and the Chartist literary canon', *Victorian Periodicals R.* 26 (1993), 199–202.
3. Belchem, John. '"Freedom and friendship to Ireland": Ribbonism in early nineteenth-century Liverpool', *International R. of Social History* 39 (1994), 33–56.
4. Clayson, Jim; Frow, Edmund; Frow, Ruth. 'John Wade and *The Black Book*', *Labour History R.* 59/2 (1994), 55–57.
5. Collins, Neil. *Politics and elections in nineteenth-century Liverpool*. Aldershot; Scolar; 1994. Pp xii, 237.
6. Conway, Stephen (ed.). *Correspondence of Jeremy Bentham, vol. 10: July 1820 to December 1821*. Oxford; Oxford UP; 1994. Pp 504.
7. Cox, Gary W. 'The development of collective responsibility in the United Kingdom', *Parliamentary History* 13 (1994), 32–47.
8. Cragoe, Matthew. '"A cheaper sort of member"?: Welsh MPs, select committees and the representation of local interests in Parliament, 1852–1865', *Parliaments, Estates & Representations* 14 (1994), 133–48.
9. Dawson, Patricia A. 'Liberalism and the challenge of Labour: the 1906 progressive election in Halifax', *T. of the Halifax Antiquarian Soc.* ns 2 (1994), 107–24.
10. Doherty, Gary; Gray, Tim. 'Herbert Spencer and the relation between economic and political liberty', *History of Political Thought* 14 (1993), 475–90.
11. Dunbabin, J.P.D. 'Some implications of the 1885 British shift towards single-member constituencies: a note', *English Historical R.* 109 (1994), 89–100.
12. Francis, Mark; Morrow, John. *A history of English political thought in the nineteenth century*. London; Duckworth; 1994. Pp viii, 336.
13. Franzmann, Tom L. 'Antislavery and political economy in the early Victorian House of Commons: a research note on "capitalist hegemony"', *J. of Social History* 27 (1994), 579–93.
14. Freeden, Michael. 'Hobson as a political theorist', Bc31, 19–33.
15. Fulcher, Jonathan. 'Gender, politics and class in the early nineteenth-century English reform movement', *Historical Research* 67 (1994), 57–74.
16. Garratt, Morris. *Samuel Bamford: portrait of a radical*. Littleborough; Kelsall; 1992. Pp 101.
17. Green, E.H.H. *The crisis of Conservatism: the politics, economics and*

ideology of the British Conservative party, 1880-1914. London; Routledge; 1994. Pp 412.

18. Haefner, Joel. 'Pressures of the market place: John Hunt's editorial philosophy and strategies, 1805–1831', *Victorian Periodicals R.* 26 (1993), 92–100.
19. Hazlehurst, Cameron. 'Whipping Asquith's Liberals: J.A. Pease's journal as patronage secretary to the Treasury, 1908–1910', *Archives* 21 (1994), 183–99.
20. Hill, C.E. 'Sidney Webb and the common good: 1887–1889', *History of Political Thought* 14 (1993), 591–622.
21. Hilton, Boyd. 'Whiggery, religion and social reform: the case of Lord Morpeth', *Historical J.* 37 (1994), 829–59.
22. Hobhouse, L.T. *Liberalism, and other writings*, ed. James Meadowcroft. Cambridge; Cambridge UP; 1994. Pp xl, 201.
23. Hopkin, Deian. 'The rise of Labour in Wales, 1890-1914', *Llafur* 6/3 (1994), 120-41.
24. Hoppen, K. Theodore. 'Grammars of electoral violence in nineteenth-century England and Ireland', *English Historical R.* 109 (1994), 597–620.
25. Hudson, Katherine. *Royal conflict: Sir John Conroy and the young Victoria*. London; Hodder & Stoughton; 1994. Pp xxiii, 247.
26. Jenkins, Terence Andrew. 'The parliamentary diaries of Sir John Trelawny, 1868–1873', Bc93, 329–513.
27. John, Angela V. '"Run like blazes": the suffragettes and Welshness', *Llafur* 6/3 (1994), 29–43.
28. Jowitt, J.A. 'From Whiggery to Liberalism: Sir Charles Wood and Halifax, 1832–1865', *T. of the Halifax Antiquarian Soc.* ns 2 (1994), 43–54.
29. Kingsford, Peter. 'Policy making in Hertfordshire, 1889–1908: social and economic influence', *Local Historian* 23 (1993), 31–39.
30. Koning, Niek. *The failure of agrarian capitalism: agrarian politics in the United Kingdom, Germany, the Netherlands and the USA, 1846–1919*. London; Routledge; 1994. Pp 292.
31. Lee, Stephen J. *Aspects of British political history, 1815–1914*. London; Routledge; 1994. Pp viii, 375.
32. Lubenow, William C. 'The Liberals and the national question: Irish Home Rule, nationalism, and their relationship to nineteenth-century Liberalism', *Parliamentary History* 13 (1994), 119–42.
33. Machin, Ian. *Disraeli*. London; Longman (Profiles in Power); 1994. Pp 202.
34. Malatesta, Maria. 'Riforma della terra, aristocrazia e liberalismo nell'Inghilterra vittoriana' [Land reform, aristocracy and liberalism in Victorian England], *Ricerche di Storia Politica* 5 (1990), 59–86.
35. Marsh, Peter T. *Joseph Chamberlain: entrepreneur in politics*. London; Yale UP; 1994. Pp xvii, 725.
36. Matthew, H.C.G. (ed.). *The Gladstone diaries with cabinet minutes and prime ministerial correspondence, vol. 12: 1887–1891*. Oxford; Oxford UP; 1994. Pp lxxxvii, 535.
37. Matthew, H.C.G. (ed.). *The Gladstone diaries with cabinet minutes and prime ministerial correspondence, vol. 13: 1892–1896*. Oxford; Oxford UP; 1994. Pp 486.
38. Matthew, H.C.G. (ed.). *The Gladstone diaries with cabinet minutes and*

prime ministerial correspondence, vol. 14: index. Oxford; Oxford UP; 1994. Pp xv, 862.
39. McCord, Norman. 'Taff Vale revisited', *History* 78 (1993), 243–60.
40. O'Brien, Anthony Mór. 'Churchill and the Tonypandy riots', *Welsh History R.* 17 (1994), 67–99.
41. Offer, John (ed.). *Spencer: political writings*. Cambridge; Cambridge UP; 1994. Pp xxxviii, 186.
42. Owen, George. 'The corporation of Doncaster: a self-elected body?', *Yorkshire Arch. J.* 66 (1994), 215–25.
43. Parry, Jonathan. *The rise and fall of Liberal government in Victorian Britain*. London; Yale UP; 1993. Pp viii, 383.
44. Payton, Philip. 'Labour failure and liberal tenacity: radical politics and Cornish political culture, 1880-1939', Bc135, 83–95.
45. Phillips, John A.; Wetherell, Charles. 'Parliamentary parties and municipal politics: 1835 and the party system', *Parliamentary History* 13 (1994), 48–85.
46. Phillips, John A.; Wetherell, Charles. 'Probability and political behaviour: a case study of the Municipal Corporations Act of 1835', *History & Computing* 5 (1993), 135–53.
47. Quinault, Roland. 'The French revolution of 1830 and parliamentary reform', *History* 79 (1994), 377–93.
48. Roberts, Stephen. *Radical politicians and poets in early Victorian Britain: the voices of six Chartist leaders*. Lampeter; Mellen; 1993. Pp viii, 149.
49. Schonhardt-Bailey, Cheryl. 'Linking constituency interests to legislative voting behaviour: the role of district economic and electoral composition in the repeal of the Corn Laws', *Parliamentary History* 13 (1994), 86–118.
50. Smith, Ernest Anthony. *Wellington and the Arbuthnots: a triangular friendship*. Stroud; Sutton; 1994. Pp x, 170.
51. Taylor, Antony. 'Palmerston and radicalism, 1847–1865', *J. of British Studies* 33 (1994), 157–79.
52. Thomas, Kathleen. *Crisis and analysis of John Stuart Mill's life*. Edinburgh; Pentland; 1994. Pp vi, 33.
53. Thompson, Dorothy. *Outsiders: class, gender and nation*. London; Verso; 1993. Pp 186.
54. Turner, Michael J. 'Gas, police and the struggle for mastery in Manchester in the eighteen-twenties', *Historical Research* 67 (1994), 301–17.
55. Turner, Michael J. 'Manchester reformers and the Penryn seats, 1827–1828', *Northern History* 30 (1994), 139–60.
56. Tyler, Paul. 'The origins of labour representation in Woolwich', *Labour History R.* 59/1 (1994), 26–33.
57. Waller, P.J. 'Laughter in the house: a late nineteenth- and early twentieth-century parliamentary survey', *20th-Century British History* 5 (1994), 4–37.
58. Webster, Eric. 'Chartism in the Calder Valley, 1838–1850', *T. of the Halifax Antiquarian Soc.* ns 2 (1994), 55–77.
59. White, Michael V. 'Following strange gods: women in Jevons' political economy', Bc1, 46–78.
60. Wildman, Richard. 'Leaves from the mayor's album: Geoffrey Howard's eventful year of office, 1901–1902', Bc6, 177–92.

(c) *Constitution, Administration and Law*

1. Abraham, David. 'Liberty and property: Lord Bramwell and the political economy of liberal jurisprudence, individualism, freedom and utility', *American J. of Legal History* 38 (1994), 288–321.
2. Behlmer, George. 'Summary justice and working-class marriage in England, 1870-1940', *Law & History R.* 12 (1994), 229–75.
3. Cornwell, S.V.P. 'The nature of a Bristol accountancy practice during the 1820s: some comments', *Accounting, Business & Financial History* 3 (1993), 155–63.
4. Cosgrove, Richard A. 'Victorian periodicals and the professions: law', Bc14, 11–21.
5. Epstein, Richard A. 'For a Bramwell revival', *American J. of Legal History* 38 (1994), 246–87.
6. Epstein, Richard A. 'Introduction: Baron Bramwell at the end of the twentieth century', *American J. of Legal History* 38 (1994), 241–45.
7. Faust, Florian. '[Hadley v. Baxendale]: an understandable miscarriage of justice', *J. of Legal History* 15 (1994), 41–72.
8. Finn, Margot. 'Debt and credit in Bath's court of requests, 1829–1839', *Urban History* 21 (1994), 211–36.
9. Ireland, Richard W. 'Eugene Buckley and the diagnosis of insanity in the early Victorian prison', *Llafur* 6/2 (1993), 5–17.
10. McEldowney, John. 'Miscarriages of justice? The Phoenix Park murders, 1882' [review essay], *Criminal Justice History* 14 (1993), 143–49.
11. McGowen, Randall. 'Civilizing punishment: the end of the public execution in England', *J. of British Studies* 33 (1994), 257–82.
12. Morris, R.J. 'Men, women and property: the reform of the Married Women's Property Act, 1870', Bc55, 171–91.
13. Panton, F.H. 'Finances and government of Canterbury, early to mid-nineteenth century', *Archaeologia Cantiana* 112 (1994 for 1993), 25–54.
14. Petrow, Stefan. *Policing morals: the Metropolitan police and the Home Office, 1870-1914*. Oxford; Oxford UP; 1994. Pp xiii, 343.
15. Petrow, Stefan. 'The rise of the detective in London, 1869–1914', *Criminal Justice History* 14 (1993), 91–108.
16. Philips, David; Storch, Robert D. 'Whigs and coppers: the Grey ministry's national police scheme, 1832', *Historical Research* 67 (1994), 75–90.
17. Polden, Patrick. 'A day at the races: *Wood v. Leadbitter* in context', *J. of Legal History* 14 (1993), 28–38.
18. Pritchard, John Wyn. '"Fit and proper persons": councillors of Denbigh, their status and position, 1835–1894', *Welsh History R.* 17 (1994), 186–204.
19. Ramasastry, Anita. 'The parameters, progressions, and paradoxes of Baron Bramwell', *American J. of Legal History* 38 (1994), 333–73.
20. Rubin, G.R. 'The legal education of British army officers, 1860-1923', *J. of Legal History* 15 (1994), 223–51.
21. Santuari, Alceste. 'The joint stock company in nineteenth-century England and France: *King v. Dodd* and the *Code de Commerce*', *J. of Legal History* 14 (1993), 39–52.
22. Savage, Gail L. '"Intended only for the husband": gender, class and the provision for divorce in England, 1858–1868', Bc63, 11–42.

23. Schofield, Philip (ed.). *Official aptitude maximized, expense minimized* [Jeremy Bentham]. Oxford; Oxford UP; 1993. Pp xlix, 504.
24. Smith, E.A. *A queen on trial: affair of Queen Caroline*. Stroud; Sutton; 1993. Pp xii, 210.
25. Summers, Anne. 'In a few years we shall none of us that now take care of them be here: philanthropy and the state in the thinking of Elizabeth Fry', *Historical Research* 67 (1994), 134–42.
26. Walling, Catherine (ed.). *Hastings gaol records: keeper's commitment book, 1850-1853, and allowance of food to prisoners, 1850-1852*. Eastbourne; PBN; 1993. Pp iv, 51.
27. Weaver, Michael. 'The new science of policing: crime and the Birmingham Police Force, 1839–1842', *Albion* 26 (1994), 289–308.
28. Webb, Adrian James; Parrish, Peter. *The Wilton gaol description register, 1818–1825*. Weston-super-Mare; Galloway; 1994. Pp vii, 53.
29. Weinberger, Barbara. 'Policing juveniles: delinquency in late nineteenth- and early twentieth-century Manchester', *Criminal Justice History* 14 (1993), 43–55.

(d) *External Affairs*

1. Bartlett, C.J. 'After Palmerston: Britain and the Iberian peninsula, 1865–1876', *English Historical R.* 109 (1994), 74–88.
2. Hopkirk, Peter. *On secret service east of Constantinople: the plot to bring down the British empire*. London; Murray; 1994. Pp xii, 431.
3. Hughes, Michael. 'British diplomats in Russia on the eve of war and revolution', *European History Q.* 24 (1994), 341–66.
4. McKay, Alexander. *Scottish Samurai: the life of Thomas Blake Glover, 1838–1911*. Edinburgh; Canongate; 1993. Pp 224.
5. Pennell, C.R. 'Dealing with pirates: British, French and Moroccans, 1834–1856', *J. of Imperial & Commonwealth History* 22 (1994), 54–83.
6. Rosenbach, Harald. *Das Deutsche Reich, Grossbritannien und der Transvaal (1896–1902): Anfange deutsch-britischer Entfremdung* [The German Reich, Great Britain and the Transvaal (1896–1906): the beginnings of Anglo-German estrangement]. Göttingen; Vandenhoeck & Ruprecht (Bayerischen Akademie der Wissenschaften, Historischen Kommission Schriftenreihe 52); 1993. Pp 360.
7. Schaarschmidt, Thomas. *Aussenpolitik und öffentliche Meinung in Grossbritannien während des deutsch-französischen Krieges von 1870/71* [Foreign policy and public opinion in Great Britain during the German-French war of 1870-1871]. Frankfurt-am-Main; Lang; 1993. Pp 757.
8. van der Kraan, Alfons. *George Pocock King: merchant adventurer and catalyst of the Bali war, 1846–1849*. Hull; Centre for South-East Asian Studies, University of Hull; 1992. Pp vii, 46.

(e) *Religion*

1. Ackers, Peter. 'The "Protestant ethic" and the English Labour movement: the case of the Churches of Christ', *Labour History R.* 58/3 (1993), 67–72.

2. Altholz, Josef L. *Anatomy of a controversy: the debate over 'Essays and reviews', 1860-1864*. Aldershot; Scolar; 1994. Pp x, 198.
3. Altholz, Josef L. 'A tale of two controversies: Darwinism in the debate over *Essays and Reviews*', *Church History* 63 (1994), 50-59.
4. Barmann, Lawrence. 'Confronting secularization: origins of the London Society for the Study of Religion', *Church History* 62 (1993), 22–40.
5. Blehl, Vincent Ferrer (ed.). *John Henry Newman: sermons, 1824–1843, vol. 2. Sermons on biblical history, sin and justification, the Christian way of life and biblical theology*. Oxford; Oxford UP; 1994. Pp xx, 472.
6. Boudens, R. 'Newman and the Roman authorities', *Louvain Studies* 15 (1990), 301–17.
7. Briggs, John H.Y. *The English Baptists of the 19th century*. Didcot; Baptist Historical Soc.; 1994. Pp 432.
8. Brown, Roger L. 'Mr Jelf's proprietory chapel at a Welsh watering place', *Merioneth Historical & Record Soc. J.* 12 (1994), 43–51.
9. Brown, Roger L. 'The nineteenth-century parish of Llandudno', *Caernarvonshire Historical Soc. T.* 55 (1994), 7–24.
10. Carwardine, Richard. 'Evangelicals, politics and the coming of the American civil war: a transatlantic perspective', Bc91, 198–218.
11. Cirket, Alan F. 'William Cirket of Elstow', Bc6, 85–94.
12. Conway, Stephen. 'J.F. Colls, M.A. Gathercole, and *Utilitarianism unmasked*: a neglected episode in the Anglican response to Bentham', *J. of Ecclesiastical History* 45 (1994), 435–47.
13. Cox, Jeffrey. 'George Alfred Lefroy, 1854–1919: a bishop in search of a church', Bc20, 54–76.
14. Dulles, Avery. 'Newman on infallibility', *Theological Studies* 51 (1990), 434–49.
15. Fitzgerald-Lombard, Charles. *English and Welsh priests, 1801–1914: a working list*. Bath; Downside; 1993. Pp xxii, 362.
16. Gasquet, F.A. 'Thomas Joseph Brown—bishop of Newport and Menevia', Bc114, 71–81.
17. Gauvreau, Michael. 'The empire of evangelicalism: varieties of common sense in Scotland, Canada and the United States', Bc91, 219–52.
18. Gonzalez Fernandez, P.F. 'John Henry Newman: sua incidenza nella vita della Chiesa' [his coming into the life of the Church], *Euntes* 43 (1990), 477–522.
19. Gordon, Elizabeth B. (ed.). *The diary of the Reverend William Cowan*. Edinburgh; Pentland; 1994. Pp viii, 82.
20. Hamer, Edna. *Elizabeth Prout, 1820-1864: a religious life for industrial England*. Bath; Downside; 1994. Pp vi, 362.
21. Harding, J.A. 'The early bishops of Clifton, 1850-1901', Bc114, 41–56.
22. Harvey, John. 'Spiritual emblems: the visions of the 1904–1905 Welsh revival', *Llafur* 6/2 (1993), 75–93.
23. Houfe, Simon. 'Some unpublished letters of the Rev. Legh Richmond', Bc6, 95–115.
24. Howkins, Alun. 'Politics or quietism: the social history of nonconformity', Bc38, 73–91.
25. Hudson, Timothy P. 'Burton Park, Sussex: a further note', *Recusant History* 22 (1994), 26–28.

26. Jenkins, David. 'Salem, Coedgruffydd, a Siloa, Cwmerfyn' [Salem, Coedgruffydd to Siloa, Cwmerfyn], *Ceredigion* 12/2 (1994), 35–56.
27. Kollar, Rene. 'The 1913 Caldey Island property enquiry: the convert monks and question of ownership', *Revue Benedictine* 102 (1992), 208–26.
28. Linehan, Lawrence. 'The bishop and the editor', *Records of Buckinghamshire* 34 (1994 for 1992), 78–86.
29. Litvack, Leon. *John Mason Neale and the quest for Sobornost*. Oxford; Oxford UP; 1994. Pp xv, 295.
30. McClelland, Maria G. 'The first Hull Mercy nuns: a nineteenth-century case study', *Recusant History* 22 (1994), 199–221.
31. McNaughton, William D. (ed.). *The membership roll of the Congregational church, Kirkcaldy, 1800-1869*. Glasgow; Congregational Union of Scotland; 1994. Pp viii, 40.
32. Meek, Donald E. 'Christopher Anderson, the Scottish highlands and Ireland', Bc80, 17–24.
33. Meek, Donald E. 'Christopher Anderson's missionaries to India', Bc80 38–42.
34. Meek, Donald E. 'The missionary theology of Christopher Anderson', Bc80, 8–16.
35. Merchant, Peter. 'Magnifying martyrdom: Annie Bonus's *Beatrice* and the Victorian vision of the early Church', *Recusant History* 22 (1994), 222–30.
36. Morris, Kevin L. 'Rescuing the scarlet woman: the promotion of Catholicism in English literature, 1829–1850', *Recusant History* 22 (1994), 75–87.
37. Morrish, P.S. 'Leeds and dismemberment of the diocese of Ripon', Bc112, 62–97.
38. Murray, Derek B. 'Anderson and Scotland', Bc80, 3–7.
39. Osborne, Margaret. '"The second spring": Roman Catholicism in Victorian Northamptonshire', *Northamptonshire Past & Present* 9 (1994), 71–79.
40. O'Sullivan, Wayne M. 'Acton, Oxenham and the temporal power', *Recusant History* 22 (1994), 102–11.
41. Palmer, Bernard. *Reverend rebels: five Victorian clerics and their fight against authority*. London; Darton, Longman & Todd; 1993. Pp xviii, 267.
42. Pickering, Paul A.; Tyrrell, Alex. '"In the thickest of the fight": the Reverend James Scholefield (1790-1855) and the Bible Christians of Manchester and Salford', *Albion* 26 (1994), 461–82.
43. Riley-Smith, Jonathan. 'The Order of St John in England, 1827–1858', Bc126, 121–38.
44. Smith, Christopher A. 'Christopher Anderson and "the Serampore Fraternity"', Bc80, 25–37.
45. Smith, Christopher A. 'The early bishops of Plymouth', Bc114, 57–70.
46. Stern, L. 'La Chiesa, il magistero e i teologi secondo J.H. Newman' [The Church, teaching and theology according to J. H. Newman], *Euntes* 43 (1990), 457–76.
47. Stewart, Ian. 'Thomas Mitchell: further notes on his family and church connections', *Innes R.* 44 (1993), 58–68.
48. Walters, Huw. 'Cofnodion Powell, Cross Inn', *National Library of Wales J.* 28 (1994), 315–24.

49. Williams, David H. 'A neglected Radnorshire cleric: the Reverend Jonah Bowen Evans, B.D. ("Cattwg"); vicar of St Harmon: 1845–1876', *J. of Welsh Religious History* 1 (1993), 81–111.
50. Wolffe, John. 'Anti-Catholicism and evangelical identity in Britain and the United States, 1830-1860', Bc91, 179–97.

(f) *Economic Affairs*

1. Anon. *Records of British business and industry, 1760-1914: metal processing and engineering*. London; HMSO; 1994. Pp 202.
2. Anstis, Bess; Anstis, Ralph (ed.). *The diary of a working man, 1872–1873: Bill Williams in the Forest of Dean*. Stroud; Sutton; 1994. Pp xxiv, 136.
3. Ashworth, William J. 'The calculating eye: Baily, Herschel, Babbage and the business of astronomy', *British J. for the History of Science* 27 (1994), 409–41.
4. August, Andrew. 'How separate a sphere?: poor women and paid work in late-Victorian London', *J. of Family History* 19 (1994), 285–309.
5. Backhouse, Roger E. 'Mummery and Hobson's *The physiology of industry*', Bc31, 78–99.
6. Barker, Theo C. 'Workshop of the world, 1870-1914', *History Today* 44/6 (1994), 30-36.
7. Bennett, Norman R. 'The port wine system in the 1890s', *International History R.* 16 (1994), 251–66.
8. Bennison, Brian. 'Concentration in the brewing industry of Northumberland and Durham, 1890-1914', *Northern History* 30 (1994), 161–78.
9. Berghoff, Hartmut; Möller, Roland. 'Tired pioneers and dynamic newcomers?: a comparative essay on English and German entrepreneurial history, 1870-1914', *Economic History R.* 2nd ser. 47 (1994), 262–87.
10. Billington, Edward; Raybould, Trevor; Brown, David. 'The Staffordshire coalfields: north Staffordshire, south Staffordshire, Cannock Chase', Bc79, 8–34.
11. Boyer, George R.; Hatton, Timothy J. 'Did Joseph Arch raise agricultural wages?: rural trade unions and the labour market in late nineteenth-century England', *Economic History R.* 2nd ser. 47 (1994), 310-34.
12. Broadberry, S.N. 'Comparative productivity in British and American manufacturing during the nineteenth century', *Explorations in Economic History* 31 (1994), 521–48.
13. Brooke, Justin (ed.). *The Cunnack manuscript: compiled by Richard John Cunnack (1826–1908) from notes taken between 1845 and 1907*. Pendeen; Trevithick Soc.; 1993. Pp viii, 76.
14. Cain, Peter J. 'Colonies and capital: some aspects of Anglo-colonial financial relations after 1850', Bc55, 213–33.
15. Cant, David. 'Jackson of Coley (1867–1931), master woodcarver', *T. of the Halifax Antiquarian Soc.* ns 2 (1994), 87–105.
16. Casson, Mark. 'Institutional diversity in overseas enterprise: explaining the free-standing company', *Business History* 36/4 (1994), 95–108.
17. Chapman, Stanley. 'The innovating entrepreneurs in the British ready-made clothing industry', *Textile History* 24 (1993), 5–25.

18. Church, Roy A.; Baldwin, Trevor; Berry, Bob. 'Accounting for profitability at the Consett Iron Company before 1914: measurement, sources, and uses', *Economic History R.* 2nd ser. 47 (1994), 703–24.
19. Clark, Gregory. 'Factory discipline', *J. of Economic History* 54 (1994), 128–63.
20. Cook, Bernard A. 'Agriculture' [Victorian periodicals], Bc14, 235–48.
21. Craig, Robin. 'Steamship enterprise in Devon, 1852–1920', Bc87, 91–99.
22. Craig, Robin; Greenhill, Basil; Porter, J.H.; Slade, W.J. 'Some aspects of the business of Devon shipping in the nineteenth century', Bc87, 99–108.
23. Crocker, Alan. 'The tramway at the Chilworth gunpowder works', *Surrey Arch. Collections* 82 (1994), 181–95.
24. Davis, Tracy C. 'Laborers of the nineteenth-century theater: the economies of gender and industrial organization', *J. of British Studies* 33 (1994), 32–53.
25. Dobbin, Frank. *Forging industrial policy: the United States, Britain, and France in the railway age.* Cambridge; Cambridge UP; 1994. Pp 276.
26. Dyke, Gwen. 'The relationship of early market towns and nonconformist centres to nineteenth-century industrial development in Suffolk', Bc38, 143–48.
27. Edgell, Stephen; Tilman, Rick. 'John Hobson: admirer and critic of Thorstein Veblen', Bc31, 211–24.
28. Eve, David C. 'Power stations in Glasgow, 1879–1939', *Scottish Industrial History* 17 (1994), 1–14.
29. Featherstone, Brian. 'John Walter Lukis: the road to Valgodemard, part 2', *Société Guernesiaise Report & T.* 23 (1994 for 1993), 570-98.
30. Fenton, Alexander (ed.). *At Brechin with stirks: farm cash-book from Buskhead, Glenesk, Angus, 1885–1898.* Edinburgh; Canongate; 1994. Pp viii, 95.
31. Gerhold, Dorian. 'A Dorset carrier in 1830', *Dorset Natural History & Arch. Soc. P.* 115 (1994 for 1993), 29–32.
32. Gifford, P. 'Tod and M'Gregor: Atlantic shipbuilders', *Scottish Industrial History* 14–15 (1992), 39–48.
33. Glynn, J.J. 'The development of British railway accounting, 1800-1911', Bc35, 327–42.
34. Greenhill, Robert G. 'Shipping and the refrigerated meat trade from the river Plate, 1900-1930', Bc127, 417–31.
35. Hallas, Christine S. 'Industry in a rural area: employment and wages in the Swaledale and Wensleydale lead industry in the nineteenth century', *Yorkshire Arch. J.* 65 (1993), 153–67.
36. Harley, C. Knick. 'International capital movements and the gold standard: British domestic investment and American railroads in the 1880s', Bc55, 235–56.
37. Hartley, William P. 'Five landed estates in Yorkshire and the development of coalmining during the nineteenth century', *Local Historian* 23 (1993), 188–98.
38. Hartley, William P. 'Involvement in coal mining by the wool textiles industry: some West Yorkshire examples', *Yorkshire Arch. J.* 65 (1993), 169–72.
39. Hathaway, Eileen. *Smuggler: John Rattenbury and his adventures in Devon,*

Dorset and Cornwall, 1778–1844—including 'Memoirs of a smuggler, 1837'. Swanage; Shinglepicker; 1994. Pp 189.

40. Haydon, Edwin S. 'Employment patterns in east Devon parishes in the nineteenth century', *Devonshire Association Report & T.* 126 (1994), 167–80.
41. Hennart, Jean-François. 'Free-standing firms and the internalization of markets for financial capital: a response to Casson', *Business History* 36/4 (1994), 118–31.
42. Horrell, Sara; Humphries, Jane; Weale, Martin. 'An input-output table for 1841', *Economic History R.* 2nd ser. 47 (1994), 545–66.
43. Howkins, Alun. 'Peasants, servants and labourers: the marginal workforce in British agriculture, *c.*1870-1914', *Agricultural History R.* 42 (1994), 49–62.
44. Hudson, Pat. 'Some aspects of nineteenth-century accounting development in the West Riding textile industry', Bc35, 434–49.
45. Huggins, M.J. 'Thoroughbred breeding in the North and East Ridings of Yorkshire in the nineteenth century', *Agricultural History R.* 42 (1994), 115–25.
46. Irving, R.J. 'The branch line problem in British railway history: the financial evidence from north-east England', *J. of Transport History* 3rd ser. 14 (1993), 27–45.
47. Kadish, Alon. 'The non-canonical context of *The physiology of industry*', Bc31, 53–77.
48. Kelly, Henry E. 'The beginning of the pottery industry in Greenock', *Scottish Industrial History* 17 (1994), 57–75.
49. Kent, David A. 'Small businessmen and their credit transactions in early nineteenth-century Britain', *Business History* 36/2 (1994), 47–64.
50. Kim, Dong-Woon. 'The Oldham Limited Company in Scotland: the Glasgow Cotton Spinning Co Ltd', *Scottish Industrial History* 17 (1994), 76–89.
51. Kirby, Maurice. 'Britain in the world economy', Bc148, 21–37.
52. Klovland, Jan Tore. 'Pitfalls in the estimation of the yield on British Consols, 1850-1914', *J. of Economic History* 54 (1994), 164–87.
53. Konvitz, Josef W. 'The crises of Atlantic port cities, 1880 to 1920', *Comparative Studies in Soc. & History* 36 (1994), 293–318.
54. Kostal, R.W. *Law and English railway capitalism, 1825–1875.* Oxford; Oxford UP; 1994. Pp 400.
55. Kynaston, David. *The city of London, vol. 1: a world of its own, 1815–1890.* London; Chatto; 1994. Pp 497.
56. Leach, Belinda. 'Grandma Page's workshop: outwork in Birmingham, 1911–1914', *Oral History* 22 (1994), 35–42.
57. Lee, Clive H. 'Regions and industries in Britain', Bc148, 38–56.
58. Lloyd, Lewis W. 'Early nineteenth-century ships and shareholders of the "Barmouth river"' [part 2], *Merioneth Historical & Record Soc. J.* 11 (1992), 284–302.
59. Lovell, John. 'Why the railway missed Northampton: some further thoughts', *Northamptonshire Past & Present* 9 (1994), 80-84.
60. MacHaffie, Fraser G. 'The Clyde Screw Steam Packet Company: an 1850s venture, part 1', *Scottish Industrial History* 17 (1994), 15–33.

61. Macleod, Christine. 'The peculiarities of Yorkshire inventors: a reply', *Economic History R.* 2nd ser. 47 (1994), 754–59.
62. Magnusson, Lars. 'Hobson and imperialism: an appraisal', Bc31, 143–62.
63. Marshall, J.D. 'Industrial colonies and the local historian', *Local Historian* 23 (1993), 146–54.
64. McConnell, Anita. 'Thomas Cooke's order book: analysis of an optical business, 1856–1868', Bc36, 431–42.
65. Milward, Robert.; Ward, Robert. 'From private to public ownership of gas undertakings in England and Wales, 1851–1947; chronology, incidence and causes', *Business History* 35/3 (1993), 1–21.
66. Morgan, Gerald. 'The Aberystwyth brig *Renown*, 1816–1841', *National Library of Wales J.* 28 (1994), 291–98.
67. Morris, Richard D. 'Distributable profit in nineteenth-century British regulated industries', *Accounting, Business & Financial History* 3 (1993), 165–95.
68. Moss, David J.; Hosgood, Chris. 'Financial and trade press' [Victorian periodicals], Bc14, 199–218.
69. Moss, Michael S. 'Shipbuilding in Ireland in the nineteenth century', Bc120, 177–95.
70. Nabb, Harold. *The Bristol Gas Light Company: the Breillat dynasty of engineers*. Bristol; Historical Association, Bristol Branch; 1993. Pp 38.
71. Napier, C.J. 'Fixed asset accounting in the shipping industry: P & O, 1840-1914', Bc35, 343–73.
72. Neal, Frank. 'Shipbuilding in the north-west of England in the nineteenth century', Bc120, 111–52.
73. Owen, Trefor M. 'Cyflog y Groes yng Nghlwyd, ["Wages of the Cross in Clwyd"]', *Llafur* 6/2 (1993), 18–31.
74. Palmer, John E.C.; Paar, Harold W. 'Transport' [Victorian periodicals], Bc14, 179–98.
75. Palmer, Sarah. 'Shipbuilding in south-east England, 1800-1913', Bc120, 45–74.
76. Parry, Ann. 'Theories of formation: *Macmillan's Magazine*, vol. 1, November 1859', *Victorian Periodicals R.* 26 (1993), 100-04.
77. Patrick, Amber. 'Tamar traffic in the nineteenth century', Bc87, 60-68.
78. Rawding, Charles; Short, Brian. 'Binbrook in 1910: the use of the Finance (1909–1910) Act records', *Lincolnshire History & Archaeology* 28 (1993), 58–65.
79. Roberts, Elizabeth. 'Women and the domestic economy, 1890-1970: the oral evidence', Bc29, 129–43.
80. Robertson, Una A. 'The supply and sale of milk in 19th-century Edinburgh', Bc30, 64–70.
81. Roper, Robert Stephenson. *The Co-op chapel of Rochdale*. Rochdale; Roper; 1993. Pp 47.
82. Rose, Sonya O. 'Widowhood and poverty in nineteenth-century Nottinghamshire', Bc151, 269–91.
83. Rossler, Hörst. 'Constantine stonemasons in search of work abroad, 1870-1900', Bc135, 48–82.
84. Saad-Filho, Alfredo. 'Labor, money and "labour-money": a review of Marx's critique of John Gray's monetary analysis', *History of Political Economy* 25 (1993), 65–84.

85. Sanger, Chesley W. '"We are now in a splendid position for whales": environmental factors affecting nineteenth-century whaling in Baffin Bay', *Mariner's Mirror* 80 (1994), 159–77.
86. Schmidt, Albert J. 'The Smith-Kelham-Langdale nexus: country attorneys, family connections, and London business in the early nineteenth century', *Lincolnshire History & Archaeology* 29 (1994), 17–27.
87. Slaven, Anthony. 'Shipbuilding in nineteenth-century Scotland', Bc120, 153–76.
88. Sockwell, W.D. *Popularizing classical economics: Henry Brougham and William Ellis*. London; Macmillan; 1994. Pp viii, 224.
89. Springhall, John. '"Disseminating impure literature": the "penny dreadful" publishing business since 1860', *Economic History R.* 2nd ser. 47 (1994), 567–84.
90. Starkey, David J. 'The shipbuilding industry of south-west England, 1790-1913', Bc120, 75–110.
91. Streeter, Patrick. *Streeter of Bond Street: a biography of a Victorian jeweller*. Harlow; Matching; 1993. Pp 174.
92. Thompson, F.M.L. 'Business and landed élites in the nineteenth century', Bc55, 139–70.
93. Turner, Michael J. 'Before the Manchester school: economic theory in early nineteenth-century Manchester', *History* 79 (1994), 216–41.
94. Ville, Simon P. 'Shipbuilding in the north-east of England in the nineteenth century', Bc120, 1–43.
95. Welskopp, Thomas. '"Defensive élitism" and early craft unions in the wrought iron industry after 1850: Britain, the USA and Germany in comparative perspective', *Labour History R.* 58/3 (1993), 13–18.
96. Williams, David M. 'Growth, market shifts and technological change: the citrus trade to Britain in the nineteenth century', Bc127, 325–47.
97. Williams, J. Llewellyn; Jenkins, D.A. 'Dwr a llechi ym mhywyf Llanllechid, Bethesda: agweddau ar ddat blygiad diwydiant yn Nyffryn Ogwen' [Water and slates in the parish of Llanllechid, Bethesda: aspects of the development of industry in the Ogwen Valley], *Caernarvonshire Historical Soc. T.* 54 (1993), 29–62.
98. Williams, J. Llewellyn; Jenkins, D.A. 'Tair chwarel ym mhlwyf Llanllechid, Bethesda' [Slate quarries around Llanllechid, Bethesda], *Caernarvonshire Historical Soc. T.* 55 (1994), 47–70.
99. Williams, Sydna Ann. 'Women's employment in nineteenth-century Anglesey', *Llafur* 6/2 (1993), 32–49.
100. Young, Craig. 'The economic characteristics of small craft businesses in rural lowland Perthshire, *c.*1830-*c.*1900', *Business History* 36/4 (1994), 33–52.

(g) *Social Structure and Population*

1. Anderson, Michael; Morse, Donald J. 'High fertility high emigration, low nuptiality: adjustment processes in Scotland's demographic experience, 1861–1914', *Population Studies* 47 (1993), 5–25; 319–43.
2. Anon. *Working women in Victorian Britain, 1850-1910: the diaries and letters of Arthur J. Munby (1828–1910) and Hannah Cullwick (1833–1909)*

from Trinity College, Cambridge. A listing and guide to the microfilm collection. Marlborough; Matthew; 1994. Pp 88.

3. Armstrong, W.A. 'The population of Victorian and Edwardian Kent', *Archaeologia Cantiana* 112 (1994 for 1993), 1–16.
4. Belcher, Sherry. 'Home ownership and miners' housing on Cannock Chase in the late-nineteenth century', Bc79, 54–73.
5. Bell, E.T. 'Family structure in 19th-century Oakham: a study of the 1851 census', *Rutland Record* (1993), 108–17.
6. Benson, John. 'Introduction' [miners of Staffordshire], Bc79, 3–7.
7. Berghoff, Hartmut; Möller, Roland. 'Wirtschaftsbürger in Bremen und Bristol, 1870-1914' [the middle class in Bremen and Bristol, 1870-1914], Bc132, 156–77.
8. Berrol, Selma C. *East Side/East End: Eastern European Jews in London and New York, 1870-1920.* New York; Praeger; 1994. Pp xiv, 159.
9. Billington, Edward. 'Mining settlements in north Staffordshire in the mid-nineteenth century', Bc79, 35–53.
10. Bourke, Joanna. 'Housewifery in working-class England, 1860-1914', *Past & Present* 143 (1994), 167–97.
11. Bourke, Joanna. '"I was always fond of my pillow": the handmade lace industry in the United Kingdom, 1870-1914', *Rural History* 5 (1994), 155–69.
12. Bragg, William. *A general directory of the county of Somerset, 1840.* Weston-super-Mare; Galloway; 1994. Microfiche edn.
13. Brayshay, Mark. 'The emigration trade in nineteenth-century Devon', Bc87, 108–19.
14. Chase, Malcolm. 'The Teesside Irish in the nineteenth century', Bc140, 47–58.
15. Cole, Jean A. (ed.). *Coroners' inquisitions for the borough of Malmesbury, Wiltshire, 1830-1854.* Trowbridge; Wiltshire Family History Soc.; 1994. Pp 34.
16. Dennis, Richard; Daniels, Stephen. '"Community" and the social geography of Victorian cities', Bc29, 201–24.
17. Donnachie, Ian. 'Work and community: changing occupational profiles', Bc117, 68–88.
18. Drake, Michael; Golby, John; Finnegan, Ruth. 'Work case studies: some occupations and roles', Bc117, 38–67.
19. Dreher, Nan H. 'Redundancy and emigration: the "woman question" in mid-Victorian Britain', *Victorian Periodicals R.* 26 (1993), 3–7.
20. Duncan, Shirley; Cooper, Janet; Studd, Pamela; Board, Beryl. 'Modern Colchester', Bc86, 176–418.
21. Dunn, Edgar. 'Poor children in the diocese of Southwark', *Catholic Ancestor* 5 (1994), 65–73.
22. Edgar, Michael J.D. 'Occupational diversity in seven rural parishes in Dorset, 1851', *Local Population Studies* 52 (1994), 48–54.
23. Englander, David (ed.). *A documentary history of Jewish immigrants in Britain, 1840-1920.* London; Leicester UP; 1993. Pp xv, 380.
24. Erickson, Charlotte. 'Agrarian myths of English immigrants', Bc106, 34–59.
25. Erickson, Charlotte. 'British immigrants in the old northwest, 1815–1860', Bc106, 60-86.

26. Erickson, Charlotte. 'Depression emigrants: who went where from the British Isles in 1841?', Bc106, 167–206.
27. Erickson, Charlotte. 'Emigration from the British Isles to the United States of America in 1831', Bc106, 126–66.
28. Erickson, Charlotte. 'Englishwomen in America in the nineteenth century: expectations and reality', Bc106, 239–65.
29. Erickson, Charlotte. 'Introduction' [leaving England], Bc106, 1–33.
30. Erickson, Charlotte. 'Was the American west a safety valve for Lancashire?', Bc106, 207–38.
31. Erickson, Charlotte. 'Who were the English and the Scots emigrants to the United States in the late nineteenth century?', Bc106, 87–125.
32. Feldman, David. 'Nationality and ethnicity', Bc148, 127–48.
33. Flint, Kate. *The woman reader, 1837–1914*. Oxford; Oxford UP; 1993. Pp xii, 366.
34. Forrester, Richard. 'Owners and occupiers of Treasurer's House, York, 1815–1900', *York Historian* 10 (1992), 52–63.
35. Garrett, Eilidh; Reid, Alice. 'Satanic mills, pleasant lands: spatial variation in women's work, fertility and infant mortality as viewed from the 1911 census', *Historical Research* 67 (1994), 156–77.
36. Gerard, Jessica. *Country house life: family and servants, 1815–1914*. Oxford; Blackwell; 1994. Pp 352.
37. Gerrish, Margaret M. 'The dock builders: dock navvies at Grimsby in 1851', *J. of Transport History* 3rd ser. 15 (1994), 45–58.
38. Giffard, Ann. 'Emigration—the human dimension', Bc87, 119–26.
39. Groenewegen, Peter. 'Alfred Marshall—women and economic development: labour, family and race', Bc1, 79–109.
40. Groenewegen, Peter. 'Introduction: women in political economy and women as political economists in Victorian England', Bc1, 1–24.
41. Hennock, E.P. 'Poverty and social reforms', Bc148, 79–93.
42. Higginbotham, Ann R. '"Sin of the age": infanticide and illegitimacy in Victorian London', Bc63, 257–88.
43. Hinde, P.R. Andrew; Davies, H. Rhodri; Kirby, Darron M. 'Hampshire village populations in the nineteenth century', *Southern History* 15 (1993), 140-61.
44. Hopkins, Eric. *Childhood transformed: working-class children in nineteenth-century England*. Manchester; Manchester UP; 1994. Pp viii, 343.
45. Ittmann, Karl. *Work, gender and family in Victorian England*. London; Macmillan; 1994. Pp 341.
46. Killingray, David. 'Rights, "riot" and ritual: the Knole Park access dispute, Sevenoaks, Kent, 1883–1885', *Rural History* 5 (1994), 63–79.
47. Knelman, Judith. 'Class and gender bias in Victorian newspapers', *Victorian Periodicals R.* 26 (1993), 29–35.
48. Lilly, Derek B.; Lilly, Jane. *The Kenn hangings: the full story of 1830*. Weston-super-Mare; Galloway; 1994. Pp 50.
49. Lowe, Jeremy. 'Survey of thematic-based research: an industrial "housing-stock"', Bc144, 59–86.
50. Lummis, Trevor. *The labour aristocracy, 1851–1914*. Aldershot; Scolar; 1994. Pp xii, 190.

51. Mills, Mary. 'Women and community on Cannock Chase in the late-nineteenth century', Bc79, 74–96.
52. Mooney, Graham. 'Still-births and the measurement of urban infant mortality rates c.1880-1930', *Local Population Studies* 53 (1994), 42–52.
53. Neal, Frank. 'English-Irish conflict in the north-east of England', Bc140, 59–85.
54. Nicholas, Stephen J.; Nicholas, Jacqueline M. 'Male literacy, "deskilling" and the industrial revolution', *J. Interdisciplinary History* 23 (1993), 1–18.
55. Oppenheim, Janet. 'Victorian and Edwardian women: the halves of modern English social history' [review article], *J. of Modern History* 66 (1994), 79–91.
56. Page, Stephen J. 'Lodging and poverty in late Victorian Leicester: a socio-geographic perspective', *Leicestershire Arch. & Historical Soc. T.* 68 (1994), 121–44.
57. Pooley, Colin G. 'The mobility of criminals in north-west England, *c.*1880-1910', *Local Population Studies* 53 (1994), 14–28.
58. Pryce, W.T.R. 'Using quantitive and cartographic techniques: two examples', Bc116, 106–20.
59. Pryce, W.T.R.; Donnachie, Ian. 'Aspects of English, Welsh, Scottish and Irish migrations', Bc116, 32–69.
60. Rawding, Charles. 'Village type and employment structure: an analysis in the nineteenth-century Lincolnshire wolds', *Local Population Studies* 53 (1994), 53–68.
61. Ray, Michael. 'Domestic servants in a superior suburb: Brunswick Town, Hove', *Sussex Arch. Collections* 131 (1993), 172–84.
62. Razzell, Peter. 'Further evaluation of Anglican baptism as a form of birth registration through cross-matching census, parish and civil register data', Bc88, 119–49.
63. Reay, Barry. 'Before the transition: fertility in English villages, 1800-1880', *Continuity & Change* 9 (1994), 91–120.
64. Riley, James C. 'Height, nutrition, and mortality risk reconsidered', *J. of Interdisciplinary History* 24 (1994), 465–92.
65. Shelley, Monica; Braham, Peter; Finnegan, Ruth. 'Movements into the British Isles: three complementary perspectives', Bc116, 70-105.
66. Shelley, Monica; Golby, John; Emsley, Clive. 'Work case studies: individual examples', Bc117, 11–37.
67. Smith, D. Ian. 'Worker migration into Workington, Cumbria, based upon the census enumerators' books for 1861, 1881 and 1891', *T. of the Cumberland & Westmorland Antiquarian & Arch. Soc.* 94 (1994), 249–76.
68. Somers, Margaret R. 'The "mistress" of property: relationality, rural-industrialization, and community in Chartist narratives of political rights', Bc150, 62–94.
69. Stoney, Benita; Weltzein, Heinrich (ed.). *My mistress the queen: letters of Frieda Arnold, dresser to Queen Victoria.* London; Weidenfeld & Nicolson; 1994. Pp 252.
70. Sullivan, Kate. '"The biggest room in Merthyr": working-class housing in Dowlais, 1850-1914', *Welsh History R.* 17 (1994), 155–85.
71. Thane, Pat. 'The social, economic and political status of women', Bc148, 94–110.

72. Townsend, Camilla. '"I am the woman for spirit": a working woman's gender transgression in Victorian London', *Victorian Studies* 36 (1993), 293–314.
73. Wahrman, Dror. '"Middle class" domesticity goes public: gender, class and politics from Queen Caroline to Queen Victoria', *J. of British Studies* 32 (1993), 396–432.
74. Werth, Paul W. 'Through the prism of prostitution: state, society and power', *Social History* 19 (1994), 1–15.
75. Wilson, Richard. 'The autobiography of Elizabeth Oakley, 1831–1900', Bc4, 113–50.

(h) *Social Policy*

1. Allott, Stephen. *John Wilhelm Rowntree, 1868–1905*. York; Sessions; 1994. Pp xiv, 138.
2. Arnot, Margaret L. 'Infant death, child care and the state: the baby-farming scandal and the first infant life protection legislation of 1872', *Continuity & Change* 9 (1994), 271–311.
3. Atkins, P.J. 'How the West End was won: the struggle to remove street barriers in Victorian London', *J. of Historical Geography* 19 (1993), 265–77.
4. Bartrip, P.W.J. 'How green was my valance: environmental arsenic poisoning and the Victorian domestic ideal', *English Historical R.* 109 (1994), 891–913.
5. Caine, Barbara. 'Feminism and political economy in Victorian England—or John Stuart Mill, Henry Fawcett and Henry Sidgwick ponder the "woman question"', Bc1, 25–45.
6. Davidson, Roger. 'The state and social investigation in Britain, 1880-1914', Bc27, 242–75.
7. Feldman, David. *Englishmen and Jews: social relations and political culture, 1840-1914*. London; Yale UP; 1994. Pp xii, 401.
8. Frost, Ginger S. '"I shall not sit down and crie": women, class and breach of promise of marriage plaintiffs in England, 1850-1900', *Gender & History* 6 (1994), 224–45.
9. Gorsky, Martin. 'James Tuckfield's "ride": combination and social drama in early nineteenth-century Bristol', *Social History* 19 (1994), 320-38.
10. Gould, Jack. 'In search of Sydney Ward', *Northamptonshire Past & Present* 8 (1993/4), 393–406.
11. Greenhill, Basil (comp.). 'The seaman's life in his own words', Bc87, 25–32.
12. Groenewegen, Peter. 'A neglected daughter of Adam Smith: Clara Elizabeth Collett, 1860-1948', Bc1, 147–73.
13. Guha, Sumit. 'The importance of social intervention in England's mortality decline: the evidence reviewed', *Social History of Medicine* 7 (1994), 89–113.
14. Hartwell, R.M. 'Entrepreneurship and public inquiry: the growth of government in nineteenth-century Britain', Bc55, 193–211.
15. Johnston, W. Ross. 'The Welsh diaspora: emigrating around the world in the late nineteenth century', *Llafur* 6/2 (1993), 50-74.
16. Jones, Aled. 'Sir John Gibson and the *Cambrian news*', *Ceredigion* 12/2 (1994), 57–83.

17. Jones, J. Gordon. 'The social consequences of industrialization in the Ffestiniog district, 1800-1914', *Merioneth Historical & Record Soc. J.* 12 (1994), 29–42.
18. Jones, J. Graham. 'Select Committee or Royal Commission?: Wales and "the land question", 1892', *Welsh History R.* 17 (1994), 205–29.
19. Koven, Seth. 'Henrietta Barnett, 1851–1936: the (auto)biography of a late Victorian marriage', Bc20, 30-55.
20. Levine, Philippa. 'Consistent contradictions: prostitution and protective labour legislation in nineteenth-century England', *Social History* 19 (1994), 17–35.
21. Martin, Mary Clare. 'Women and philanthropy in Walthamstow and Leyton, 1740-1870', *London J.* 19 (1994), 119–50.
22. Mayne, Alan. *The imagined slum: newspaper representation in three cities, 1870-1914.* Leicester; Leicester UP; 1993. Pp x, 228.
23. Nurse, Bernard. 'Planning a London suburban estate: Dulwich, 1882–1920', *London J.* 19 (1994), 54–70.
24. Nyland, Chris; Ramia, Gaby. 'The Webbs and the rights of women', Bc1, 110-46.
25. Paris, Michael. 'The rise of the airmen: the origins of airforce élitism, *c.*1890-1918', *J. of Contemporary History* 28 (1993), 123–41.
26. Rendel, Margherita. 'Women in Torquay in the first half of the nineteenth century', *Devonshire Association Report & T.* 126 (1994), 17–39.
27. Schneer, Jonathan. 'London's docks in 1900: nexus of empire', *Labour History R.* 59/3 (1994), 20-33.
28. Sheppard, June A. 'Housing the agricultural worker in nineteenth-century Sussex: a case study', *Sussex Arch. Collections* 131 (1993), 185–92.
29. Waddington, Keir. 'Bastard benevolence: centralization, voluntarism and the Sunday Fund, 1873–1898', *London J.* 19 (1994), 151–67.
30. Weiner, Deborah. *Architecture and social reform in late-Victorian London.* Manchester; Manchester UP; 1994. Pp 256.
31. Whittick, Margaret. 'A three-headed Cerberus: Brighton and the Health of Towns Bill, 1847', *Sussex Arch. Collections* 131 (1993), 159–71.

(i) *Education*

1. Aldrich, Richard. 'Pioneers of female education in Victorian Britain', *History of Education Soc. B.* 54 (1994), 56–61.
2. Aspinwall, Bernard. 'Catholic teachers for Scotland: the Liverpool connection', *Innes R.* 45 (1994), 47–70.
3. Booth, Rosemary. 'Crisis at the Ladies' college, 1899–1900, or the dovecote fluttered', *Société Guernesiaise Report & T.* 23 (1994 for 1993), 473–92.
4. Davies, Brynmor Lewis. *Henry Richard: a radical dissenter and education in Wales.* Aberystwyth; Centre for Educational Studies; 1993. Pp 122.
5. Davies, Brynmor Lewis. 'The university college of Wales, Aberystwyth: appointing the first principal', *National Library of Wales J.* 28 (1994), 299–313.
6. Edwards, J.B. 'Llandudno, a school and John Bright: an unexpected link?', *Caernarvonshire Historical Soc. T.* 54 (1993), 63–78.

7. Evans, W. Gareth. 'T.E. Ellis, MP, and the Free Elementary Education Act, 1891', *Merioneth Historical & Record Soc. J.* 11 (1993), 446–58.
8. Evans, W. Gareth. 'The "bilingual difficulty": the inspectorate and the failure of a Welsh language teacher-training experiment in Victorian Wales', *National Library of Wales J.* 28 (1994), 325–33.
9. Fitzpatrick, T.A. 'Scottish Catholic teacher education: the wider context', *Innes R.* 45 (1994), 147–70.
10. Foster, Stewart. 'Monsignor Lord William Joseph Petre (1847–1893): a pillar of Downside', *Recusant History* 22 (1994), 88–101.
11. Girvin, Stephen D. 'Nineteenth-century reforms in Scottish legal education: the universities and the bar', *J. of Legal History* 14 (1993), 127–41.
12. Guagnini, Anna. 'Worlds apart: academic instruction and professional qualifications in the training of mechanical engineers in England, 1850-1914', Bc47, 16–41.
13. Hawley, John C. 'The muscular Christian as schoolmarm', Bc63, 134–56.
14. Jenkins, E.W. 'Leeds grammar school, the charity commissioners and the education of girls, 1870-1901', Bc112, 98–132.
15. Jennings, Ruth. *Lofty aims and lowly duties: three Victorian schoolmasters.* Sheffield; Sheffield Academic; 1994. Pp 286.
16. Jones, John Ellis. 'Crimean medals—a museum note: a Bangor veteran and the nineteenth-century British army', *Caernarvonshire Historical Soc. T.* 55 (1994), 71–107.
17. Jons, Christopher J.M. '"Doer of the hard dull duty": Rosamond Davenport-Hill at the London School Board, 1879–1897', *History of Education Soc. B.* 54 (1994), 8–20.
18. Leach, J.H.C. *Sparks of reform: the career of Francis Jeune, 1806–1868.* Oxford; Leach; 1994. Pp xii, 116.
19. Phillips, Trevor R. 'Elementary school teachers in business, 1870-1882', *History of Education Soc. B.* 54 (1994), 52–55.
20. Roderick, Gordon. 'An English champion of Welsh education', *National Library of Wales J.* 28 (1994), 285–90.
21. Rose, Jonathan. 'Willingly to school: the working-class response to elementary education in Britain, 1875–1918', *J. of British Studies* 32 (1993), 114–38.
22. Savage, David W. 'Evangelical education policy in Britain and India, 1857–1860', *J. of Imperial & Commonwealth History* 22 (1994), 432–61.
23. Selleck, R.J.W. *James Kay-Shuttleworth: journey of an outsider.* London; Woburn; 1993. Pp xiv, 494.
24. Spittal, C.J. 'Three generations of the Exley family in Bristol education and Methodism, 1800-1901', *T. of the Bristol & Gloucestershire Arch. Soc.* 111 (1993), 201–13.
25. Stewart, John. 'The political economy of agrarian education: England in the late nineteenth century', *Agricultural History R.* 42 (1994), 126–39.
26. Tozer, Malcolm. 'Genial solvents: German influences at Thring's Uppingham', *History of Education Soc. B.* 54 (1994), 21–28.
27. VanArsdel, Rosemary T.; North, John S. 'Student journals', Bc14, 311–31.
28. Williams, J. Gwynn. 'Prifysgol Cymru 1893' [Welsh language, 1893], *T. of the Honourable Soc. of Cymmrodorion* (1993), 83–97.

29. Woods, Sylvia. 'A remarkable school: Bedford county school, 1869–1916', Bc6, 219–27.

(j) *Naval and Military*

1. Brown, D.K. *Paddle warships: the earliest naval steamers, 1815–1850*. London; Conway; 1993. Pp 94.
2. Burroughs, Peter. 'An unreformed army? 1815–1868', Bc109, 160-89.
3. Caldwell, George. *Rifle green in the Crimea: an account of the Rifle Brigade in the Crimean war of 1854–1856, with a full medal and casualty roll and details of weapons, clothes and equipment used in the campaign*. Leicester; Bugle-Horn; 1994. Pp 360.
4. Fraser, John (ed.). 'Field surgeon at the battle of Aliwal: the letters of Dr John Murray during the Sikh war', *J. of the Soc. for Army Historical Research* 72 (1994), 35–48.
5. Greenhill, Basil; Giffard, Ann. *Steam, politics and patronage: the transformation of the Royal Navy, 1815–1854*. London; Conway; 1994. Pp 257.
6. Haraszti, Eva H. *Kossuth as an English journalist*. Budapest; Akademiai Kiado; 1990. Pp 404.
7. Hobson, John M. 'The military-extraction gap and the wary Titan: the fiscal-sociology of British defence policy, 1870-1913', *J. of European Economic History* 22 (1993), 461–506.
8. Lambert, Andrew D.; Badsey, Stephen. *War correspondents: the Crimean war*. Stroud; Sutton; 1994. Pp 345.
9. Lloyd, William Glyn. *John Williams, VC: a biography*. Cwmbran; Lloyd; 1993. Pp 120.
10. Maxwell, Robert (ed.). *Jimmie Stewart—frontiersman: the edited memoirs of Major General Sir J.M. Stewart*. Durham; Pentland; 1992. Pp xiv, 113.
11. McBride, Keith. 'Queen Victoria's battlecruisers: the *Canopus* and *Duncan* class battleships', *Mariner's Mirror* 80 (1994), 431–49.
12. Partridge, Michael Stephen. 'The battle of Braye Bay, 1901', *B. of the Alderney Soc.* 27 (1992), 77–81.
13. Powell, Geoffrey. *Buller: scapegoat?—a history of General Sir Redvers Buller, VC*. London; Cooper; 1994. Pp 220.
14. Raugh, Harold E., Jr. 'Training ground for a future field marshal: Wavell in the Boer war and Edwardian India, 1901–1908', *J. of the Soc. for Army Historical Research* 72 (1994), 8–18.
15. Spiers, Edward M. 'The late-Victorian army, 1868–1914', Bc109, 189–215
16. Trebilcock, Clive. 'Science, technology and the armaments industry in the UK and Europe, with special reference to the period 1880-1914', *J. of European Economic History* 22 (1993), 565–80.
17. Tucker, Albert. 'Victorian periodicals and the professions: military', Bc14, 62–80.
18. Vale, Brian. 'British sailors and the Brazilian navy, 1822–1850', *Mariner's Mirror* 80 (1994), 312–25.
19. Welch, David A. *Justice and the genesis of war*. Cambridge; Cambridge UP; 1993. Pp xvi, 335.

20. Westlake, Ray. *The volunteer infantry, 1880-1908*. London; Military Historical Soc.; 1992. Pp i, 139.
21. Worth, George J. '*Macmillan's magazine* and the American civil war: a reconsideration', *Victorian Periodicals R.* 26 (1993), 193–98.
22. Yorke, Edmund. 'Isandlwana, 1879: further reflections of the ammunition controversy', *J. of the Soc. for Army Historical Research* 72 (1994), 205–18.

(k) *Science and Medicine*

1. Bearn, Alexander G. *Archibald Garrod and the individuality of man*. Oxford; Oxford UP; 1993. Pp xvi, 227.
2. Brock, William H. 'Victorian periodicals and the professions: science', Bc14, 81–96.
3. Burbridge, David. 'Galton's 100: an exploration of Francis Galton's imagery studies', *British J. for the History of Science* 27 (1994), 443–63.
4. Burkhardt, Frederick; Porter, Duncan M.; Browne, Janet; Richmond, Marsha. *The correspondence of Charles Darwin, vol. 9: 1861*. Cambridge; Cambridge UP; 1994. Pp 700.
5. Burnett, John. 'William Prout and the urinometer: some interpretations', Bc36, 242–54.
6. Cashman, Bernard. 'The Bedford medical community in the nineteenth century', Bc6, 193–204.
7. Caudill, Edward. 'The bishop-eaters: the publicity campaign for Darwin and *On the origin of species*', *J. of the History of Ideas 55* (1994), 441–60.
8. Chauvin, Michael E. 'Astronomy in the Sandwich Islands: the 1874 transit of Venus', *Hawaiian J. of History* 27 (1993), 185–225.
9. Creese, Mary R.S.; Creese, Thomas M. 'British women who contributed to research in the geological sciences in the nineteenth century', *British J. for the History of Science* 27 (1994), 23–54.
10. Croft, L.R. 'Edmund Gosse and the "new and fantastic cure" for breast cancer', *Medical History* 38 (1994), 143–59.
11. Crook, Paul. *Darwinism, war and history: the debate over the biology of war from the 'Origin of species' to the first world war*. Cambridge; Cambridge UP; 1994. Pp xii, 306.
12. Dawes, Howard A.L. 'Wheatstone's wave machine: a physical model of light', Bc36, 127–38.
13. Fisch, Menachem. '"The emergency which has arrived": the problematic history of the nineteenth-century British algebra—a programmatic outline', *British J. for the History of Science* 27 (1994), 247–76.
14. Greenaway, Frank. 'More than "a mere gazing place": the special loan exhibition and the science conferences of 1876', Bc36, 139–45.
15. Hackmann, Willem. 'Electricity from steam: Armstrong's hydroelectric machine in the 1840s', Bc36, 146–72.
16. Hardy, Anne. *The epidemic streets: infectious diseases and the rise of preventive medicine, 1856–1900*. Oxford; Oxford UP; 1993. Pp xii, 325.
17. Harrison, Hilda M. *Voyager in time and space: the life of John Couch Adams, Cambridge astronomer*. Lewes; Book Guild; 1994. Pp xii, 282.

18. Kurata, Marilyn J. 'Wrongful confinement: the betrayal of women by men, medicine and law', Bc63, 43–68.
19. Lomax, Elizabeth. 'The control of contagious disease in nineteenth-century British paediatric hospitals', *Social History of Medicine* 7 (1994), 383–400.
20. MacNeill, Niall. 'An early Dumbarton helicopter', *Scottish Industrial History* 17 (1994), 34–43.
21. McConnell, Anita. *R.B. Bate of the Poultry, 1782–1847: the life and times of a scientific instrument maker*. London; Scientific Instrument Soc.; 1993. Pp vii, 74.
22. Miller, Norman; Miller, Margaret. 'Improvement to the Glasgow-Carlisle road, 1815: the metal bridge', *Dumfriesshire & Galloway Natural History & Antiquarian Soc. T.* 3rd ser. 67 (1992), 67–73.
23. Milner, Richard. *Charles Darwin: evolution of a naturalist*. New York; Facts on File; 1994. Pp 160.
24. Morris, Susan W. 'Fleeming Jenkin and *The origin of species*: a reassessment', *British J. for the History of Science* 27 (1994), 313–43.
25. Pennington, T.H. 'Osteotomy as an indicator of antiseptic surgical practice', *Medical History* 38 (1994), 178–88.
26. Peterson, M. Jeanne. 'Victorian periodicals and the professions: medicine', Bc14, 22–44.
27. Preston, James. 'Rochester bridge, 1825–1950', Bc12, 221–73.
28. Roberts, Glynne. '"Closing the stable door after the horse has bolted": preventing the spread of smallpox and cholera in Caernarfonshire, 1870-1910', *Caernarvonshire Historical Soc. T.* 55 (1994), 109–28.
29. Schaaf, Larry J. (ed.). *Selected correspondence of William Henry Fox Talbot, 1823–1874*. London; Science Museum; 1994. Pp 92.
30. Second, Anne. 'Corresponding interests: artisans and gentlemen in nineteenth-century natural history', *British J. for the History of Science* 27 (1994), 383–408.
31. Sigsworth, Michael; Warboys, Michael. 'The public's view of public health in mid-Victorian Britain', *Urban History* 21 (1994), 237–50.
32. Simpson, A.D.C. 'The pendulum as the British length standard: a nineteenth-century legal aberration', Bc36, 174–90.
33. Taylor, L.J. 'William Froude and the model-testing tank at Torquay', Bc87, 188–91.
34. Twort, Antony. *In focus, out of step: a biography of Frederick William Twort FRS, 1877–1950*. Stroud; Sutton; 1993. Pp xi, 340.
35. Williams, Mari E.W. 'Training for specialists: the precision instruments industry in Britain and France, 1890-1925', Bc47, 227–50.

(l) *Intellectual and Cultural*

1. Anderson, Mary R. 'Cultural dissonance and the ideology of transition in late Victorian England', *Victorian Periodicals R.* 26 (1993), 108–14.
2. Anderson, Patricia. 'Victorian periodicals and the arts: illustration', Bc14, 127–42.
3. Andrews, Malcolm. 'The metropolitan picturesque', Bc41, 282–98.

4. Arnold, Godwin H.; Gold, Sidney M. 'Morris of Reading: a supplement', *T. of the Ancient Monuments Soc.* 38 (1994), 147–68.
5. Bailey, Peter. 'Conspiracies of meaning: music-hall and the knowingness of popular culture', *Past & Present* 144 (1994), 138–70.
6. Barlow, Stuart. 'The diffusion of "rugby" football in the industrialized context of Rochdale, 1868–1890: a conflict of ethical values', *International J. of the History of Sport* 10 (1993), 49–67.
7. Bennison, Brian. 'Drunkenness in turn-of-the-century Newcastle upon Tyne', *Local Population Studies* 52 (1994), 14–22.
8. Binfield, Clyde. 'An excursion into architectural cousinhood: the East Anglian connexion', Bc38, 92–140.
9. Birch, Christopher. 'Miners' leisure in nineteenth-century Staffordshire', Bc79, 97–111.
10. Booth, Bradford A.; Mehew, Ernest (ed.). *Letters of Robert Louis Stevenson* [August 1879–September 1882]. London; Yale UP; 1994. Vol. 3. Pp 372.
11. Booth, Bradford A.; Mehew, Ernest (ed.). *Letters of Robert Louis Stevenson* [October 1882–June 1884]. London; Yale UP; 1994. Vol. 4. Pp 326.
12. Born, Anne. 'William Pengelly and his circle, 12 January 1812–16 March 1894', *Devonshire Association Report & T.* 126 (1994), 41–55.
13. Boyd, Kelly. 'Examplars and ingrates: imperialism and the boys' story paper, 1880-1930', *Historical Research* 67 (1994), 143–55.
14. Brandwood, Geoff. '"A disgrace to the town": Aylesbury church and its restoration by George Gilbert Scott', *Records of Buckinghamshire* 34 (1994 for 1992), 1–10.
15. Brogden, William A. 'John Smith and Craigievar castle', Bc139, 229–36.
16. Brown, P.S.; Brown, Dorothy N. 'Victorian church tiles in Somerset, part 1: floor tiles', *Somerset & Dorset Notes & Queries* 33 (1994), 275–82.
17. Brown, P.S.; Brown, Dorothy N. 'Victorian church tiles in Somerset, part 2: architects and pavements', *Somerset & Dorset Notes & Queries* 33 (1994), 338–43.
18. Charity, Kate; Watkins, Malcom J. *A many sided man, John Bellows of Gloucester, 1831–1902: Quaker, printer, lexicographer and archaeologist.* York; Sessions; 1993. Pp x, 130.
19. Charnley, Robert (comp.). *A voyage to St Kilda*. Portree; Maclean; 1993. Pp 72.
20. Colby, Robert A. 'Victorian periodicals and the arts: authorship and the book trade', Bc14, 143–61.
21. Coombes, Annie E. *Reinventing Africa: museums, material culture, and popular imagination in late Victorian and Edwardian England.* London/Newhaven; Yale UP; 1994. Pp vii, 280.
22. Cranston, Maurice. *The Romantic movement*. Oxford; Blackwell; 1994. Pp 169.
23. Crook, J. Mordaunt. 'The new square style: Robert Smirke's Scottish houses', Bc139, 206–16.
24. Davies, Hazel Walford (ed.). 'Edward Thomas: twelve unpublished letters to O.M. Edwards', *National Library of Wales J.* 28 (1994), 335–45.
25. Davis, Tracy C. 'The actress in Victorian pornography', Bc63, 99–133.
26. Day, John C. 'Coffee houses and book clubs in Northumberland', *Archaeologia Aeliana* 5th ser. 22 (1994), 255–62.

27. Dowling, Linda. *Hellenism and homosexuality in Victorian Oxford.* Ithaca (NY); Cornell UP; 1994. Pp xvi, 173.
28. Fischer, Steven Roger. 'The calling of HMS *Seringapatam* at Rapanui (Easter Island) on 6 March 1830', *Pacific Studies* 16/1 (1993), 67–84.
29. Frank, Katherine. *Lucie Duff Gordon: a passage to Egypt.* London; Hamilton; 1994. Pp xi, 398.
30. Fryckstedt, Monica Correa. 'Retrieving a synchronic perspective of Victorian culture: the *Athenaeum* as a research tool', *Victorian Periodicals R.* 26 (1993), 87–92.
31. Gagnier, Regenia. 'On the insatiability of human wants: economic and aesthetic man', *Victorian Studies* 36 (1993), 125–53.
32. Garratt, G.R.M. *The early history of radio: from Faraday to Marconi.* Stevenage; Inst. of Electrical Engineers; 1993. Pp 96.
33. Garrigan, Kristine Ottesen. 'Decorum, scandal and the press', Bc63, 1–8.
34. Gash, Norman. *Robert Surtees and early Victorian society.* Oxford; Oxford UP; 1993. Pp 407.
35. Gifford, John. 'The duchess-countess of Sutherland and Dunrobin castle', Bc139, 249–60.
36. Goldman, Paul. *Victorian illustrated books, 1850-1870: the heyday of wood-engraving.* London; British Museum; 1994. Pp 144.
37. Gow, Ian. 'David (or John?) Henderson's designs for Caprington castle', Bc139, 181–92.
38. Gutzke, David W. 'Gentrifying the British public house, 1896–1914', *International Labor & Working-Class-History* 45 (1993), 29–43.
39. Harrington, Ralph. 'The neuroses of the railway', *History Today* 44/7 (1994), 15–21.
40. Harris, Eileen. 'The parent style or the original sin? Adam revived', *Architectural History* 4 (1993), 45–57.
41. Herstein, Sheila. 'The Langham Place circle and feminist periodicals of the 1860s', *Victorian Periodicals R.* 26 (1993), 24–27.
42. Hinxman, Neville. 'Emmeline Fisher, a forgotten Wiltshire poet: her links with William Wordsworth and the national anthem', *Hatcher R.* 37 (1994), 16–30.
43. Itzkowitz, David C. 'The (other) great evil: gambling, scandal and the national anti-gambling league', Bc63, 235–56.
44. Janowitz, Anne. 'The Chartist picturesque', Bc41, 261–81.
45. Jones, Dewi. 'The old Snowdon guides', *Caernarvonshire Historical Soc. T.* 54 (1993), 7–28.
46. Joyce, Patrick. *Democratic subjects: self and the social in nineteenth-century England.* Cambridge; Cambridge UP; 1994. Pp 242.
47. Kaye, David; Scorer, Sam; Robinson, David N. (ed.). *Fowler of Louth: the life and works of James Fowler, Louth architect, 1828–1892.* Louth; Louth Naturalists', Antiquarian & Literary Soc.; 1992. Pp 80.
48. Kelley, Philip; Lewis, Scott (ed.). *The Brownings' correspondence, vol. 12: January-May 1846.* London; Athlone; 1994. Pp xii, 426.
49. Koropeckyj, Roman. 'Unexpected affinities: Slavic literature in the *Illustrated Polytechnic Review*', *Victorian Periodicals R.* 26 (1993), 213–20.
50. Langley, Leanne. 'Victorian periodicals and the arts: music', Bc14, 99–126.

51. Laver, F.J.M. 'William Kingdon Clifford, F.R.S., 1845–1879', *Devonshire Association Report & T.* 126 (1994), 145–66.
52. Levitt, Sarah. 'From Mrs Bloomer to the bloomer: the social significance of the nineteenth-century English dress reform movement', *Textile History* 24 (1993), 27–37.
53. Linnell, David. *Blake, Palmer, Linnell & Co: the life of John Linnell.* Lewes; Book Guild; 1994. Pp 413.
54. Lukitsh, Joanne. 'Julia Margaret Cameron and the "enoblement" of photographic portraiture', Bc63, 207–32.
55. Macaulay, James. 'J.P. Neale and the *Views* of Scotland', Bc139, 217–28.
56. Mancoff, Debra N. 'Reluctant redactor: William Dyce reads the legend', Bc100, 254–73.
57. Mangum, Teresa. 'George Eliot and the journalists: making the mistress moral', Bc63, 159–79.
58. Mason, Michael. *The making of Victorian sexuality.* Oxford; Oxford UP; 1994. Pp 338.
59. Mason, Tony. 'Sport and recreation', Bc148, 111–26.
60. Mason, Tony. 'Sport' [Victorian periodicals], Bc14, 291–97.
61. McCrimmon, Barbara. 'Public relations, Panizzi-style', *British Library J.* 20 (1994), 214–21.
62. McGarvie, Michael (ed.). *Sir Stephen Glynne's church notes for Somerset.* Taunton; Somerset Record Soc. vol. 82; 1994. Pp xxxi, 400.
63. Moore, Helena. 'Touring pictures exhibited in York, 1840-1860', *York Historian* 10 (1992), 64–81.
64. Morgan, Thäis E. 'Victorian scandals, Victorian strategies', Bc63, 289–319.
65. Morse, David. *High Victorian culture.* Basingstoke; Macmillan; 1992. Pp viii, 553.
66. Mullen, Richard. *Birds of passage: five Englishwomen in search of America.* London; Duckworth; 1994. Pp 246.
67. Murray, Hugh. 'Queen Victoria, York and the royal myths', *York Historian* 11 (1994), 56–68.
68. Nevett, Terence. 'Advertising' [Victorian periodicals], Bc14, 219–34.
69. Niessen, Olwen C. 'Temperance' [Victorian periodicals], Bc14, 251–77.
70. Pearce, Colin D. 'Lord Brougham's neo-paganism', *J. of the History of Ideas* 55 (1994), 651–70.
71. Port, M.H. *Imperial London: civil government building in London, 1851–1915.* London; Yale UP; 1994. Pp 352.
72. Richardson, Ruth; Thorne, Robert. 'Victorian periodicals and the professions: architecture', Bc14, 45–61.
73. Roberts, Brynley F. 'Syr John a'r Eidalwr' [Sir John and the Italian], Bc102, 179–94.
74. Rose, Jonathan. '*Everyman*: an experiment in culture for the masses', *Victorian Periodicals R.* 26 (1993), 79–87.
75. Rose, Jonathan. 'Workers' journals' [Victorian periodicals], Bc14, 301–10.
76. Rowan, Alistair. 'A neo-baronial Monymusk not built by William Burn', Bc139, 237–48.
77. Rupke, Nicolaas A. *Richard Owen: a Victorian naturalist.* London; Yale UP; 1994. Pp 480.

78. Sainsbury, A.B. 'The origins of the Society for Nautical Research', *Mariner's Mirror* 80 (1994), 450-58.
79. Sanderson, Margaret H.B. 'The disagreeable business: John Paterson against the earl of Eglinton', Bc139, 193–205.
80. Sheeran, George. *Brass castles: west Yorkshire new rich and their houses, 1800-1914*. Halifax; Ryburn; 1992. Pp 159.
81. Siberry, Elizabeth. 'Victorian perceptions of the military Orders', Bc126, 365–72.
82. Smith, Jonathan. 'De Quincey's revisions to "The system of the heavens"', *Victorian Periodicals R.* 26 (1993), 203–12.
83. Stansky, Peter. 'E.M. Forster, 1879–1970: connecting the prose and the passion in 1910', Bc20, 126–46.
84. Stedman, Jane W. '"Come substantial damages!", W.S. Gilbert, *Trial by jury*', Bc63, 69–96.
85. Stedman, Jane W. 'Victorian periodicals and the arts: theatre', Bc14, 162–76.
86. Telfer, Hamish. 'Play, customs and popular culture of west coast communities, 1840-1900', Bc3, 113–24.
87. Tranter, Neil L. 'Women and sport in nineteenth-century Scotland', Bc3, 27–42.
88. Trela, D.J. 'Froude on the Carlyles: the Victorian debate over biography', Bc63, 180-206.
89. Tyson, Nicola. 'Candie cemetery and its monuments: death and self-presentation in nineteenth-century St Peter Port, Guernsey', *Société Guernesiaise Report & T.* 23 (1994 for 1993), 599–626.
90. Umland, Rebecca. 'The snake in the woodpile: Tennyson's Vivien as Victorian prostitute', Bc100, 274–87.
91. Vaisey, David. 'Connections: Gloucestershire, the Bodleian Library, and the adventures of Roger Plowman', *T. of the Bristol & Gloucestershire Arch. Soc.* 111 (1993), 7–20.
92. van Biervliet, Lori. *Leven en werk van W.H'. James Weale: een Engels kunsthistoricus in Vlaanderen in de 19de eeuw* [Life and work of W.H. James Weale: an English church historian in Flanders in the nineteenth century]. Brussells; Royal Academy of Sciences; 1991. Pp 243.
93. Vann, J. Don. 'Comic periodicals' [Victorian periodicals], Bc14, 278–90.
94. Waterfield, Giles; Smith, Nicola. 'Art for the people', *History Today* 44/6 (1994), 55–58.
95. Weber, Gary. 'Henry Labouchère, *Truth* and the new journalism of late Victorian Britain', *Victorian Periodicals R.* 26 (1993), 36–43.
96. Weinbren, Dan. 'Against *all* cruelty: the Humanitarian League, 1891–1919', *History Workshop* 38 (1994), 86–105.
97. Weston, Stephen J. 'A Northamptonshire composer: Thomas Jarman of Clipston', *Northamptonshire Past & Present* 8 (1993/4), 385–90.
98. Wilton, Andrew. 'Turner and Shakespeare's Jessica', *History Today* 44/4 (1994), 14–19.
99. Wood, Marcus. *Radical satire and print culture, 1790-1822*. Oxford; Oxford UP; 1994. Pp 318.

100. Worth, George J. 'Alexander Macmillan and his magazine' [part 1], *Victorian Periodicals R.* 26 (1993), 105–07.
101. Wright, Alan. *Tom Sayers: the last great bare-knuckle champion.* Lewes; Book Guild; 1994. Pp 304.
102. Yarrington, Alison; Leiberman, Ilene D.; Potts, Alex; Baker, Malcolm (ed.). 'An edition of the ledger of Sir Francis Chantrey, R.A., at the Royal Academy, 1809–1841', *Walpole Soc.* 56 (1994 for 1991/2), 1–343.

I. BRITAIN SINCE 1914

See also Aa3,63,73,84,90,92,138,b60–61; Ba16,43,88,99,108,124,155,175,188,205, 209,b14,19,34,53,61,67,80,83,118,132,147,165,171,233,263,c8,21,26,33–34,37,43, 120,148; Hb1,c20,f15,28,79,h19,i21,k27,35,l13,83

(a) *General*

1. Anon (ed.). *When peace broke out: Britain, 1945.* London; HMSO; 1994. Pp vii, 148.
2. Bonnard, Brian. 'The Alderney's "Domesday book"', *Guernsey Soc. R.* 48 (1992), 69–75.
3. Ceadel, Martin. 'Attitudes to war: pacifism and collective security', Bc148, 221–41.
4. Channon, John; Gill, Crispin; Alston, Kennerley; Redfern, Tony; Starkey, David J. 'Towards the twenty-first century', Bc87, 258–67.
5. Cook, Chris; Leonard, Jane; Leese, Peter. *The Longman guide to sources in contemporary British history, vol. 2: individuals.* Harlow; Longman; 1994. Pp 383.
6. Forty, George. *Frontline Dorset: the county at war, 1939–1945.* Wincanton; Dorset Books; 1994. Pp 168.
7. Gilbert, Martin. *In search of Churchill: a historian's journey.* London; HarperCollins; 1994. Pp xiii, 338.
8. Harris, José. 'Great Britain: the people's war?', Bc26, 233–59.
9. Hichens, Mark. *Troubled century: British and world history, 1914–1993.* Edinburgh; Pentland; 1994. Pp viii, 394.
10. Ivey, Allen E. 'Towards a Cornish identity theory', Bc135, 151–63.
11. Johnson, Paul. 'The role of the state in twentieth-century Britain', Bc148, 476–91.
12. Kellas, James. 'The party in Scotland' [Conservative century], Bc141, 671–94.
13. Lawrence, Jon. 'The first world war and its aftermath', Bc148, 151–68.
14. Macintyre, Lorn. *Sir David Russell: a biography.* Edinburgh; Canongate; 1994. Pp 259.
15. Muchitsch, Wolfgang. *Mit Spaten, Waffen un Worten: die Einbindung österreichischer Flüchtlinge in die britischen Kriegsanstrengungen, 1939–1945* [With spade, weapons and words: the involvement of Austrian refugees in the British war effort, 1939–1945.] Vienna; Europa; 1992. Pp 280.

16. Sherman, A.J. *Island refuge: Britain and refugees from the third Reich, 1933–1939*. London; Cass; 1994. Pp 291.
17. Walker, Brian. *Spilsby, 1914–1939*. Spilsby; Spilsby Branch Workers' Educational Association; 1994. Pp vii, 202.
18. Wasley, Gerald D. *Devon at war, 1939–1945*. Wincanton; Devon Books; 1994. Pp 208.

(b) *Politics*

1. Alderman, Keith. 'Legislating on Maastricht', *Contemporary Record* 7 (1993), 499–521.
2. Bagwell, Philip S.; Lawley, Joan. *From prison cell to council chamber: the life of Philip William Bagwell, 1885–1958*. York; Sessions; 1994. Pp ix, 185.
3. Ball, Stuart. 'Local Conservatism and the evolution of the party organization', Bc141, 261–314.
4. Ball, Stuart. 'The national and regional party structure' [Conservative century], Bc141, 169–220.
5. Barker, Rodney. *Politics, peoples and government: themes in British political thought since the nineteenth century*. Basingstoke; Macmillan; 1994. Pp 157.
6. Barnes, John. 'Ideology and factions' [Conservative century], Bc141, 315–46.
7. Barnes, John; Cockett, Richard. 'The making of party policy' [Conservative century], Bc141, 347–83.
8. Barrett, Neil. 'The Communist party in Manchester and Salford, 1935–1936' [documents], *Labour History R.* 58/3 (1993), 51–66.
9. Beckett, Francis. 'The rebel who lost his cause', *History Today* 44/5 (1994), 36–42.
10. Berger, Stefan. *The British Labour party and the German SPD, 1900-1931: a comparative study*. Oxford; Oxford UP; 1994. Pp xiii, 302.
11. Bogdanor, Vernon. 'The rise and fall of the post-war settlement', *Parliamentary History* 13 (1994), 351–62.
12. Bogdanor, Vernon. 'The selection of the party leader' [Conservative century], Bc141, 69–96.
13. Byrne, Paul. 'Pressure groups and popular campaigns', Bc148, 442–59.
14. Carter, Bob; Harris, Clive; Joshi, Shirley. 'The 1951–1955 Conservative government and the racialization of black immigration', Bc65, 55–71.
15. Cassar, George H. *Asquith as war leader*. London; Hambledon; 1994. Pp 288.
16. Catterall, Peter (ed.). 'The battle of Cable Street', *Contemporary Record* 8 (1994), 105–32.
17. Cockett, Richard. 'The party, publicity and the media' [Conservative century], Bc141, 547–78.
18. Criddle, Byron. 'Members of parliament' [Conservative century], Bc141, 145–68.
19. Cullen, Stephen M. 'Another nationalism: the British Union of Fascists in Glamorgan, 1932–1940', *Welsh History R.* 17 (1994), 101–14.
20. Dean, Dennis. 'The dilemmas of an academic liberal historian in Lloyd

George's government: H.A.L. Fisher at the Board of Education, 1916–1922', *History* 79 (1994), 57–81.

21. Donoughue of Ashton, Lord. 'The 1975 Chrysler rescue: a political view from number ten', *Contemporary Record* 8 (1994), 77–83.
22. Doyle, Barry M. 'Who paid the price of patriotism?: the funding of Charles Stanton during the Merthyr Boroughs by-election of 1915', *English Historical R.* 109 (1994), 1215–22.
23. Dutton, David. 'Anticipating Maastricht: the Conservative party and Britain's first application to join the European Community', *Contemporary Record* 7 (1993), 522–40.
24. Fielding, Steven. 'Neil Kinnock: an overview of the Labour party', *Contemporary Record* 8 (1994), 589–99.
25. Finlay, Richard J. *Independent and free: Scottish politics and the origins of the Scottish National Party, 1918–1945*. Edinburgh; Donald; 1994. Pp 259.
26. Finlay, Richard J. 'National identity in crisis: politicians, intellectuals and the "end of Scotland", 1920-1939', *History* 79 (1994), 242–59.
27. Finlay, Richard J. 'Scotland in the twentieth century: in defence of oligarchy?', *Scottish Historical R.* 73 (1994), 103–12.
28. Geddes, Andrew. 'Labour and the European community, 1973–1993: pro-Europeanism, "Europeanization" and their implications', *Contemporary Record* 8 (1994), 370-80.
29. George, A.D. 'Was Amy Johnson a feminist?', *Yorkshire Arch. J.* 66 (1994), 238–41.
30. Girvin, Brian. 'The party in comparative and international context' [Conservative century], Bc141, 695–726.
31. Goodlad, Graham D. 'Lord Sankey and labour: the radicalization of a conservative', *Labour History R.* 59/1 (1994), 16–26.
32. Graves, Pamela M. *Labour women: women in British working class politics, 1918–1939*. Cambridge; Cambridge UP; 1994. Pp xvi, 271.
33. Green, E.H.H. 'Churchill reappraized', *Parliamentary History* 13 (1994), 338–50.
34. Grieves, Keith. 'Sir Eric Geddes, Lloyd George and the transport problem, 1918–1921', *J. of Transport History* 3rd ser. 13 (1992), 23–42.
35. Halkier, Henrik. 'Den "engelske" syge: konsensus og konflikt i Storbritannien, 1951–1979' [The "English" disease: consensus and conflict in Great Britain, 1951–1979], *Den Jyske Historiker* 54/55 (1991), 199–231.
36. Harrison, Brian. 'Mrs Thatcher and the intellectuals', *20th-Century British History* 5 (1994), 206–45.
37. Harrison, Brian. 'Politics' [Oxford], Bc8, 377–412.
38. Hinton, James. 'Militant housewives: the British Housewives' League and the Attlee government', *History Workshop* 38 (1994), 128–56.
39. Hutchison, I.G.C. 'Beyond the oligarchs' [twentieth-century Scotland], *Scottish Historical R.* 73 (1994), 113–16.
40. Jarvis, David. 'Mrs Maggs and Betty: the Conservative appeal to women voters in the 1920s', *20th-Century British History* 5 (1994), 129–52.
41. Jefferys, Kevin. *War and reform: British politics during the second world war*. Manchester; Manchester UP; 1994. Pp xii, 170.
42. Johnson, Paul. 'Introduction: Britain, 1900-1990', Bc148, 1–18.

43. Jones, J. Graham. 'The Cardiganshire election of 1959', *Ceredigion* 12/2 (1994), 84–105.
44. Jones, J. Graham. 'The Merioneth election of 1959', *Merioneth Historical & Record Soc. J.* 11 (1992), 331–41.
45. Jones, Tudor. 'Neil Kinnock's socialist journey: from clause four to the policy review', *Contemporary Record* 8 (1994), 567–88.
46. Kaiser, Wolfram. 'Using Europe and the abusing the Europeans: the Conservatives and the European Community', *Contemporary Record* 8 (1994), 381–99.
47. Kandiah, Michael David. 'Yes minister and Yes prime minister (1): Sir Antony Jay, CVO', *Contemporary Record* 8 (1994), 506–20.
48. Kandiah, Michael David. 'Yes minister and Yes prime minister (2): Jonathan Lynn', *Contemporary Record* 8 (1994), 521–34.
49. Keating, Joan. 'The making of the Catholic Labour activist: the Catholic Social Guild and Catholic Workers' College, 1909–1939', *Labour History R.* 59/3 (1994), 44–56.
50. Kelly, Richard. 'The party conferences' [Conservative century], Bc141, 221–60.
51. Kinnock, Neil. 'Reforming the Labour party', *Contemporary Record* 8 (1994), 535–54.
52. Koerner, Steve. 'The history of the Labour party' [conference report], *Labour History R.* 59/3 (1994), 11–16.
53. Lawlor, Sheila. *Churchill and the politics of war, 1940-1941*. Cambridge; Cambridge UP; 1994. Pp xvi, 270.
54. Lineham, Thomas P. *East London for Mosley: the British Union of Fascists in east London and south-west Essex, 1933–1940*. London; Cass; 1994. Pp 256.
55. Lovenduski, Joni; Norris, Pippa; Burness, Catriona. 'The party and women' [Conservative century], Bc141, 611–36.
56. Mandolesi, Ilaria. 'Il partito laburista inglese negli anni '30: tra grande crisi e crisi del socialismo' [The British Labour party in the 1930s: between the great crisis and the crisis of socialism], *Annali della Fondazione Luigi Einaudi* 24 (1990), 379–472.
57. Marr, Andrew. 'The fading vision: Scottish politics in the twentieth century', Bc2, 373–84.
58. McIlroy, John. 'Labour legislation: anatomy of the counter-revolution', *Labour History R.* 59/2 (1994), 68–71.
59. Middlemas, Keith. 'The party, industry, and the city' [Conservative century], Bc141, 445–98.
60. Morgan, Kenneth O. 'Leaders and led in the Labour movement: the Welsh experience', *Llafur* 6/3 (1994), 109–19.
61. Murray, David. 'Local elections in the 1920s and 1930s: Northampton', Bc117, 131–54.
62. Norton, Philip. 'The parliamentary party and party committees' [Conservative century], Bc141, 97–144.
63. Ponting, Clive. *Winston Churchill*. London; Sinclair-Stevenson; 1994. Pp xii, 900.
64. Roberts, Andrew. *Eminent Churchillians*. London; Weidenfeld & Nicolson; 1994. Pp 356.

65. Rose, Norman. *Churchill: an unruly life*. Hemel Hempstead; Simon & Schuster; 1994. Pp 435.
66. Scammell, Margaret. 'The phenomenon of political marketing: the Thatcher contribution', *Contemporary Record* 8 (1994), 23–43.
67. Seldon, Anthony. 'Conservative century', Bc141, 17–68.
68. Seldon, Anthony; Ball, Stuart. 'Introduction' [Conservative century], Bc141, 1–16.
69. Shepherd, Robert. *Iain Macleod: a biography*. London; Hutchinson; 1994. Pp 320.
70. Shorrocks, Derek. 'The Somerset election of 1966', *Somerset & Dorset Notes & Queries* 33 (1994), 267–73.
71. Smith, Martin J. 'Neil Kinnock and the modernization of the Labour party', *Contemporary Record* 8 (1994), 555–66.
72. Sørensen, Nils Arne. 'Nationalismer og nationale identiteter i Storbritannien' [Nationalism and national identity in Great Britain], *Historisk Tidskrift* 94 (1994), 271–91.
73. Srebrnik, Henry Felix. *London Jews and British communism, 1935–1945*. London; Mitchell; 1994. Pp 260.
74. Stevens, Richard. '"Disruptive elements"?: the influence of the Communist party in Nottingham and District Trades Council, 1929–1951', *Labour History R.* 58/3 (1993), 22–37.
75. Tanner, Duncan. 'The strange death of Liberal England', *Historical J.* 37 (1994), 971–94.
76. Taylor, Andrew. 'The party and the trade unions' [Conservative century], Bc141, 499–546.
77. Theakston, Kevin; Fry, Geoffrey. 'The party and the civil service' [Conservative century], Bc141, 383–402.
78. Thomas, Alys. 'Cornwall's territorial dilemma: European region or "Westcountry" sub-region?', Bc135, 138–50.
79. Thomas, David A. *Churchill, the member for Woodford*. London; Cass; 1994. Pp 224.
80. Thompson, Noel W. *John Strachey: an intellectual biography*. Basingstoke; Macmillan; 1993. Pp 275.
81. Turner, John. *Macmillan*. London; Longman; 1994. Pp 302.
82. Veldman, Meredith. *Fantasy, the bomb and the greening of Britain: romantic protest, 1945–1980*. Cambridge; Cambridge UP; 1994. Pp xii, 325.
83. Waller, Robert. 'Conservative electoral support and social class', Bc141, 579–610.
84. White, Stephen. 'British labour in Soviet Russia, 1920', *English Historical R.* 109 (1994), 621–40.
85. Williams, Claudette. 'We are a natural part of many different struggles: black women organizing', Bc65, 153–63.
86. Young, Ken. 'The party and English local government' [Conservative century], Bc141, 403–44.
87. Zweiniger-Bargielowska, Ina. 'Rationing, austerity and the Conservative party recovery after 1945', *Historical J.* 37 (1994), 173–97.
88. Zweiniger-Bargielowska, Ina. 'South Wales miners' attitudes towards nationalization: an essay in oral history', *Llafur* 6/3 (1994), 70-84.

(c) *Constitution, Administration and Law*

1. Dell, Edmund; Hunt, John. 'The failings of Cabinet government in the mid-to-late 1970s', *Contemporary Record* 8 (1994), 453–71.
2. Grant, Mariel. *Propaganda and the role of the state in inter-war Britain.* Oxford; Oxford UP; 1994. Pp 360.
3. Harris, Clive. 'Configurations of racism: the civil service, 1945–1960', *Race & Class* 33 (1991), 1–30.
4. Hennessy, Peter. 'Cabinet government: a commentary', *Contemporary Record* 8 (1994), 484–94.
5. James, Simon. 'Cabinet government: a commentary', *Contemporary Record* 8 (1994), 495–505.
6. Lawson, Nigel; Armstrong, Robert. 'Cabinet government in the Thatcher years', *Contemporary Record* 8 (1994), 440-52.
7. Levitt, Ian (ed.). *Scottish Office: depression and reconstruction, 1919–1959.* Edinburgh; Scottish History Soc. 5th ser. vol. 5; 1993. Pp 250.
8. Morgan, Richard I. 'The introduction of civil legal aid in England and Wales, 1914–1949', *20th-Century British History* 5 (1994), 38–76.
9. Poole, K.; Keith-Lucas, B. *Parish government, 1894–1994.* London; National Association of Local Councils; 1994. Pp 281.
10. Porter, Valerie. *The village parliaments: the centenary of West Sussex parish councils, 1994.* Chichester; Phillimore; 1994. Pp xi, 106.
11. Rubin, G.R. *Private property, government requisition and the constitution, 1914–1927.* London; Hambledon; 1994. Pp 288.
12. Stein, Alex. 'From Blackstone to Woolmington: on the development of a legal doctrine', *J. of Legal History* 14 (1993), 14–27.
13. Stevens, Robert. 'Judicial salaries: financial independence in the age of equality', *J. of Legal History* 13 (1992), 155–77.
14. Wakeham, John. 'Cabinet government', *Contemporary Record* 8 (1994), 473–83.
15. Warne, E.J.D. *A review of the Architects (Registration) Acts, 1931–1969.* London; HMSO; 1993. Pp 51.

(d) *External Affairs*

1. Adams, R.J.Q. *British politics and foreign policy in the age of appeasment, 1935–1939.* Basingstoke; Macmillan; 1993. Pp 208.
2. Aldrich, Richard J. *The key to the south: Britain, the United States and Thailand during the approach of the Pacific war, 1929–1942.* Oxford; Oxford UP; 1993. Pp xxii, 416.
3. Aldrich, Richard J.; Hopkins, Michael F. (ed.). *Intelligence, defence, and diplomacy: British policy in the post-war world.* London; Cass; 1994. Pp x, 273.
4. Almog, Orna. 'An end of an era: the crisis of 1958 and the Anglo-Israeli relationship', *Contemporary Record* 8 (1994), 49–76.
5. Atherton, Louise. 'Lord Lloyd at the British Council and the Balkan front, 1937–1940', *International History R.* 16 (1994), 25–48.
6. Becker, Bert. *Die DDR und Grossbritannien 1945/49 bis 1973: politische,*

wirtschaftliche und kulturelle Kontakte im Zeichen der Nichtanerkennungs-politik [The German Democratic Republic and Great Britain 1945–49 to 1973: political, economic and cultural contacts under non-recognition]. Bochum; Brockmeyer; 1991. Pp 366.

7. Bennett, G.H. 'Britain's relations with France after Versailles: the problem of Tangier, 1919–1922', *European History Q.* 24 (1994), 53–84.
8. Beyersdorf-Zimeray, Karine. 'L'impact de la réunification allemande sur les rélations Franco-Britanniques', *Revue d'Histoire Diplomatique* 108 (1994), 223–56.
9. Bickers, Robert A. 'History, legend and treaty port ideology, 1925–1931', Bc44, 81–92.
10. Brown, Anthony Cave. *Treason in the blood: H. St John Philby, Kim Philby and the spy case of the century*. London; Hale; 1994. Pp 640.
11. Burk, Kathleen. 'The Americans, the Germans, and the British: the 1976 IMF crisis', *20th-Century British History* 5 (1994), 351–69.
12. Cain, Frank. 'Exporting the cold war: British responses to the USA's establishment of COCOM, 1947–1951', *J. of Contemporary History* 29 (1994), 501–22.
13. Callahan, Michael D. 'NOMANSLAND: the British Colonial Office and the League of Nations mandate for German East Africa, 1916–1920', *Albion* 25 (1993), 443–64.
14. Carley, Michael Jabara. 'Down a blind-alley: Anglo-Franco-Soviet relations, 1920-1939', *Canadian J. of History* 29 (1994), 147–72.
15. Catterall, Peter (ed.). 'The east of Suez decision', *Contemporary Record* 7 (1993), 612–53.
16. Chandler, Andrew. 'Munich and morality: the bishops of the Church of England and appeasement', *20th-Century British History* 5 (1994), 77–99.
17. Clark, Ian. *Nuclear diplomacy and the special relationship: Britain's deterrent and America, 1957–1962*. Oxford; Oxford UP; 1994. Pp ix, 461.
18. Cohen, Stuart A. 'Imperial policing against illegal immigration: the Royal Navy and Palestine, 1945–1948', *J. of Imperial & Commonwealth History* 22 (1994), 275–93.
19. Croft, Stuart. *The end of superpower: British Foreign Office conceptions of a changing world, 1945–1951*. Aldershot; Dartmouth; 1994. Pp xv, 225.
20. Danchev, Alex. 'Great Britain: the indirect strategy', Bc26, 1–26.
21. Deakin, F.W. 'Great Britain and Italian claims in Africa, 1915–1919', Bc128, 231–41.
22. Dell, Edmund. 'Britain and the origins of the European monetary system', *Contemporary European History* 3 (1994), 1–60.
23. Doherty, Martin. 'Black propaganda by radio: the German Concordia broadcasts to Britain, 1940-1941', *Historical J. of Film, Radio & Television* 14 (1994), 167–97.
24. Fitzgerald, Edward Peter. 'France's middle eastern ambitions, the Sykes-Picot negotiations, and the oil fields of Mosul, 1915–1918', *J. of Modern History* 66 (1994), 697–725.
25. Gerke, Gerwin. 'The Iraq Development Board and British policy, 1945–1950', *Middle Eastern Studies* 27 (1991), 231–55.

26. Goda, Norman J.W. 'Hitler's demand for Casablanca in 1940: incident or policy?', *International History R.* 16 (1994), 491–510.
27. Goldsworthy, David (ed.). *The Conservative government and the end of empire, 1951–1957.* London; HMSO; 1994. 3 vol. Pp 551; 437; 450.
28. Gordon, G.A.H. 'The Admiralty and appeasement', *Naval History* 5 (1991), 44–48.
29. Heiss, Mary Ann. 'The United States, Great Britain, and the creation of the Iranian oil consortium, 1953–1954', *International History R.* 16 (1994), 522–35.
30. Hodgkin, E.C. (ed.). *Letters from Tehran: a British ambassador in world war two Persia.* London; Tauris; 1991. Pp xvi, 301.
31. Howard, Michael. 'A thirty years war?: the two world wars in historical perspective', *T. of the Royal Historical Soc.* 6th ser. 3 (1993), 171–84.
32. Huan, Claude. 'Les negoçiations franco-brittaniques de l'automne 1940', *Guerres Mondiales et Conflits Contemporains* 176 (1994), 139–54.
33. Jablonsky, David (ed.). 'Introduction' [the Clausewitzian trinity], Bc122, 1–60.
34. Jamet, Catherine M. 'Les frères ennemis: guerres mondiales et mémoire des étudiants allemands dans les collèges d'Oxford et de Cambridge', *Guerres Mondiales et Conflits Contemporains* 175 (1994), 79–86.
35. Kennedy, Greg C. 'Great Britain's maritime strength and the British merchant marine, 1922–1935', *Mariner's Mirror* 80 (1994), 66–76.
36. Kitchen, Martin. 'British policy towards the Soviet Union, 1945–1948', Bc37, 111–34.
37. Krieger, Wolfgang. 'Foundation and history of the treaty of Brussels, 1948–1950', Bc33, 229–49.
38. Larres, Klaus. 'Preserving law and order: Britain, the United States, and the East German uprising of 1953', *20th-Century British History* 5 (1994), 320-50.
39. Louis, William Roger. 'Britain at the crossroads in Palestine, 1952–1954', *Jerusalem J. of International Relations* 12/3 (1990), 59–82.
40. Maisel, Ephraim. *The Foreign Office and foreign policy, 1919–1926.* Brighton; Sussex Academic; 1994. Pp 323.
41. Narinsky, Mikhail N.; Pozdeeva, Lydia V. 'Mutual perceptions: images, ideals and illusions', Bc26, 307–32.
42. Ovendale, Ritchie. 'Great Britain and the Anglo-American invasion of Jordan and Lebanon in 1958', *International History R.* 16 (1994), 284–303.
43. Overy, Richard J. 'Co-operation: trade, aid and technology', Bc26, 201–32.
44. Overy, Richard J. 'Great Britain: cyclops', Bc26, 113–44.
45. Pappe, Ilan. 'Overt conflict to tacit alliance: Anglo-Israeli relations, 1948–1951', *Middle Eastern Studies* 26 (1990), 561–81.
46. Pfeiffer, Rolf. 'Die Grenzen imperialer Loyalität: Britisch-Niederländische Interessengegensätze im Umfeld des ersten Weltkrieges 1914 bis 1919' [The limits of imperial loyalty: Anglo-Dutch conflicts of interest in the context of the first world war, 1914–1919], *Zeitschrift für Geschichtswissenschaft* 39 (1991), 554–76.
47. Pundik, Ron. *Struggle for sovereignty: relations between Great Britain and Jordan, 1946–1951.* Oxford; Blackwell; 1994. Pp xi, 363.

48. Reynolds, David. 'Great Britain: imperial diplomacy', Bc26, 333–53.
49. Reynolds, David. 'Great Britain' [the cold war], Bc34, 77–95; 247–50.
50. Reynolds, David. 'Introduction' [cold war], Bc34, 1–21.
51. Reynolds, David. 'Legacies: allies, enemies and posterity', Bc26, 417–40.
52. Robertson, Edwin H. 'George Bell's peace initiatives', *Kirchliche Zeitgeschichte* 4 (1991), 153–76.
53. Roi, Michael L. '"A completely immoral and cowardly attitude": the British Foreign Office, American neutrality and the Hoare-Laval plan', *Canadian J. of History* 29 (1994), 333–51.
54. Ruane, Kevin. 'Anthony Eden, British diplomacy and the origins of the Geneva conference of 1954', *Historical J.* 37 (1994), 153–72.
55. Saul, Samir. 'Bienveillance et méprise: les reáctions françaises au coup de force britannique du 4 fevrier 1942 en Egypte', *Guerres Mondiales et Conflits Contemporains* 173 (1994), 107–30.
56. Schwabe, Klaus. 'Efforts towards cooperation and integration in Europe, 1948–1950', Bc33, 29–42.
57. Seton-Watson, Christopher. 'Britain and Italy, 1914–1915: the role of Sir Rennel Rodd', Bc128, 215–29.
58. Seton-Watson, Christopher. 'Il trattato di pace italiano: la prospettiva inglese' [The Italian peace treaty: the British perspective], *Italia Contemporanea* 182 (1991), 27–50.
59. Sfikas, Thanais D. 'Britain, the United States and the Soviet Union in the United Nations commission of investigation in Greece, January-May 1947', *Contemporary European History* 2 (1993), 243–63.
60. Shaw, Tony. 'Government manipulation of the press during the 1956 Suez crisis', *Contemporary Record* 8 (1994), 274–88.
61. Skogmar, Gunnar. *Nuclear triangle: relations between the United States, Great Britain and France in the atomic energy field, 1939–1950*. Copenhagen; Copenhagen Political Studies; 1993. Pp 237.
62. Sluga, Glenda. 'Trieste: ethnicity and the cold war, 1945–1954', *J. of Contemporary History* 29 (1994), 285–303.
63. Stone, Glyn. *Oldest ally: Britain and the Portuguese connection, 1936–1941*. London; Royal Historical Soc. (studies in history 69); 1994. Pp x, 182.
64. Thomas, Neil. *Compromised policy: Britain, Germany and the Soviet threat, 1945–1946*. London; Minerva; 1994. Pp v, 298.
65. Walton, J.K. 'British perceptions of Spain and their impact on attitudes to the Spanish civil war', *20th-Century British History* 5 (1994), 283–99.
66. Warner, Geoffrey. 'The British Labour government and the Atlantic alliance, 1949–1951', Bc33, 149–71.
67. Whyte, Neil; Gummett, Philip. 'The military and early United Kingdom space policy', *Contemporary Record* 8 (1994), 343–69.
68. Wilson, Keith M. 'Understanding the "misunderstanding" of 1 August 1914', *Historical J.* 37 (1994), 885–89.
69. Wilson, Theodore A. 'Coalition: structure, strategy and statecraft', Bc26, 79–111.
70. Woyke, Wichard. 'Foundation and history of NATO, 1948–1950', Bc33, 251–71.

71. Young, John W. *Britain and European unity, 1945–1992*. Basingstoke; Macmillan; 1993. Pp 217.
72. Young, John W. *The Longman companion to cold war and detente, 1941–1991*. London; Longman; 1993. Pp xv, 355.
73. Zhai, Qiang. *The dragon, the lion and the eagle: Chinese-British-American relations, 1949–1958*. Kent (Oh); Kent State UP; 1994. Pp 304.

(e) *Religion*

1. Bagshaw, Paul. *The Church beyond the Church: the Sheffield industrial mission, 1944–1994*. Sheffield; Industrial Mission in South Yorkshire; 1994. Pp 128.
2. Brown, Roger L. 'A matter of principle: the great row of Dolgellan church, 1933–1934', *Merioneth Historical & Record Soc. J.* 11 (1993), 459–69.
3. Brown, Stewart J. '"A solemn purification by fire": responses to the great war in the Scottish Presbyterian churches, 1914–1919', *J. of Ecclesiastical History* 45 (1994), 82–104.
4. Clwyd, Hafina. 'Robert Hughes (1889–1917): un o'r bechgyn a gollwyd' one of the lads who were lost, *Merioneth Historical & Record Soc. J.* 12 (1994), 52–68.
5. Lofthouse, Mark. *The Church colleges, 1918–1939: the struggle for survival.* Rearsby; Billings; 1993. Pp xviii, 116.
6. McCluskey, Raymond. 'Robert Mocogni and the Scottish Catholic Film Institute', *Innes R.* 44 (1993), 76–84.
7. Sherwood, Marika. *Pastor Danies Ekarte and the African Churches Mission, Liverpool, 1932–1964*. Savannah; 1994. Pp 104.
8. Turner, F.M. 'Religion' [Oxford], Bc8, 293–316.
9. Wolffe, John. 'Religion and "secularization"', Bc148, 427–41.

(f) *Economic Affairs*

1. Ackers, Peter. 'Colliery deputies in the British coal industry before nationalization', *International R. of Social History* 39 (1994), 383–414.
2. Ainley, Pat; Vickerstaff, Sarah. 'Transitions from corporatism: the privatization of policy failure', *Contemporary Record* 7 (1993), 541–56.
3. Alexander, Carol O. 'The changing relationship between productivity, wages and unemployment in the UK', *Oxford B. of Economics and Statistics* 55 (1993), 87–102.
4. Ally, Russell. 'The South African pound comes of age: sterling, the Bank of England and South Africa's monetary policy, 1914–1925', *J. of Imperial & Commonwealth History* 22 (1994), 109–26.
5. Attard, Bernard. 'The jobbers of the London stock exchange: an oral history', *Oral History* 22 (1994), 43–48.
6. Baines, Dudley. 'Recovery from unemployment', Bc148, 188–203.
7. Baines, Dudley. 'The onset of depression', Bc148, 169–87.
8. Bamberg, J.H. *The history of the British Petroleum Company, vol. 2: the Anglo-Iranian years, 1928–1954*. Cambridge; Cambridge UP; 1994. Pp xxvi, 639.

9. Barber, Norman. *A century of British brewers, 1890-1990*. New Ash Green; Brewery History Soc.; 1994. Pp 147.
10. Barkai, H. 'Productivity patterns, exchange rates and the gold standard restoration debate of the 1920s', *History of Political Economy* 25 (1993), 1–37.
11. Black, Margaret. 'Clerical workers in the 1950s and 1960s: the use of oral evidence', *Oral History* 22 (1994), 49–55.
12. Blackman, Simon; Matthews, Kent. 'Wages, unemployment and benefits in inter-war Britain: an exercise in robust estimation', *B. of Economic Research* 45 (1993), 119–33.
13. Boyns, T. Kate. 'Jigging and shaking: technical choice in the south Wales coal industry between the wars', *Welsh History R.* 17 (1994), 230-51.
14. Broadberry, S.N. 'Why was unemployment in inter-war Britain so low?', *B. of Economic Research* 46 (1994), 241–62.
15. Brown, Keith D. 'Through a glass darkly: cost control in British industry: a case study', *Accounting, Business & Financial History* 3 (1993), 291–302.
16. Burchell, S.; Clubb, C.; Hopwood, A.G. 'Accounting in its social context: towards a history of value added in the United Kingdom', Bc35, 539–89.
17. Burge, Alun. 'In search of Harry Blount: scabbing between the wars in one south Wales community', *Llafur* 6/3 (1994), 58–69.
18. Burness, Catriona. 'Ardrossan as a fishing port', *Scottish Industrial History* 14–15 (1992), 19–22.
19. Channon, John. 'Pollution and resort development: Teignmouth in the twentieth century, and especially in recent years', Bc136, 180-98.
20. Church, Roy A. 'How did business think?', *Scottish Economic & Social History* 14 (1994), 5–18.
21. Clapson, Mark. 'The revival of the state lottery in Britain', *Contemporary Record* 8 (1994), 321–42.
22. Clegg, Hugh Armstrong. *A history of British trade unions since 1889, vol. 3: 1934–1951*. Oxford; Oxford UP; 1994. Pp iv, 458.
23. Cooper, P.F.; Jones, Glyn C. 'Rochester bridge, 1950-1993: an epilogue', Bc12, 275–85.
24. Corley, T.A.B. 'Britain's overseas investments in 1914 revisited', *Business History* 36/1 (1994), 71–88.
25. Davies, Sam. '"Three on the hook and three on the book": dock labourers and unemployment insurance between the wars', *Labour History R.* 59/3 (1994), 34–43.
26. de la Haye, A. 'The dissemination of design from haute couture to fashionable ready-to-wear during the 1920s', *Textile History* 24 (1993), 39–48.
27. Dickenson, A.P. 'The cyclicality of British strike frequency', *Oxford B. of Economics and Statistics* 56 (1994), 285–304.
28. Dobson, Alan P. 'Regulation or competition?: negotiating the Anglo-American air service agreement of 1977', *J. of Transport History* 3rd ser. 15 (1994), 144–64.
29. Earnshaw, Neil. 'The relocation of south Ayrshire mining communities after 1946: a consequence of coal nationalisation?', *Scottish Industrial History* 17 (1994), 44–56

30. Edgerton, D.E.H. 'British industrial R and D, 1900-1970', *J. of European Economic History* 23 (1994), 49–67.
31. Edgerton, D.E.H. 'The rise and fall of British technology', *History Today* 44/6 (1994), 43–48.
32. Edgerton, D.E.H.; Horrocks, S.M. 'British industrial research and development before 1945', *Economic History R.* 2nd ser. 47 (1994), 213–38.
33. Evans, G.W. 'Sectoral imbalance and unemployment in the United Kingdom', *Oxford Economic Papers* 45 (1993), 440-56.
34. French, Michael. 'Manufacturing and models in the commercial vehicle industry: the Albion Motor Company, 1920-1956', *J. of Transport History* 3rd ser. 15 (1994), 59–77.
35. French, Michael. 'Organization and profitability in exporting: Albion Motors overseas, 1920-1956', *Scottish Industrial History* 14–15 (1992), 1–12.
36. Gazeley, Ian. 'Prices in interwar Britain', *Explorations in Economic History* 31 (1994), 195–209.
37. Glynn, Sean; Gospel, Howard. 'Britain's low skill equilibrium: problem of demand?', *Industrial Relations J.* 24 (1993), 112–25.
38. Hannah, Leslie. 'Crisis and turnaround? 1973–1993', Bc148, 340-55.
39. Harris, Clive. 'Post-war migration and the industrial reserve army', Bc65, 9–54.
40. Haskel, J.; Szymanski, S. 'Privatization, liberalization, wages and employment: theory and evidence for the UK', *Economica* 60 (1993), 161–81.
41. Hilditch, Peter. 'The dockyard in the local economy' [Devon], Bc87, 215–26.
42. Hinde, K.S.G. *The City of London Solicitors' Company, 1908–1994.* London; City of London Solicitors' Company; 1994. Pp x, 118.
43. Holmes, Harry. *Avro: history of an aeronautical company, 1910-1963.* Shrewsbury; Airlife; 1994. Pp 180.
44. Horn, Pamela. 'Women workers in the 1920s: Cadbury's of Bournville', *Genealogists' Magazine* 24 (1994), 399–402.
45. Howlett, Peter. 'Resource allocation in wartime Britain: the case of steel, 1939–1945', *J. of Contemporary History* 29 (1994), 523–44.
46. Howlett, Peter. 'The "golden age", 1955–1973', Bc148, 320-39.
47. Howlett, Peter. 'The war economy', Bc148, 283–99.
48. Hughes, Geoffrey. 'The board of directors of the London and North Eastern Railways', *J. of Transport History* 3rd ser. 13 (1992), 163–79.
49. Jenkins, David. *Shipping in Cardiff: photographs from the Hansen collection, 1920-1975.* Cardiff; Wales UP; 1994. Pp xx, 98.
50. Johnson, Paul; Webb, S. 'Explaining the growth in UK income inequality: 1979–1988', *Economic J.* 103 (1993), 429–35.
51. Knapp, Manfred. 'Economic aspects of the creation of the north American-western European alliance system, 1948–1950', Bc33, 343–73.
52. Lee, T.A. 'The early debate on financial and physical capital', Bc35, 493–520.
53. Lewis, Gail. 'Black women's employment and the British economy', Bc65, 73–96.
54. Lowerson, John. 'Celtic tourism—some recent magnets', Bc135, 128–38.

55. Lyddon, Dave. '"Trade union traditions", the Oxford Welsh, and the 1934 Pressed Steel strike', *Llafur* 6/2 (1993), 106–14.
56. Marriner, Sheila. 'The Ministry of Munitions, 1915–1919, and government accounting procedures', Bc35, 450-72.
57. Melling, Joe; Johannson, Alf. 'Employers, craft workers and the effort bargain: the institutional context of engineering firms' labour policies in Britain and Sweden, *c.*1920-1940', *Scandinavian Economic History R.* 42 (1994), 200-19.
58. Millward, Robert. 'British industry since the second world war', *History Today* 44/6 (1994), 49–54.
59. Morrice, Andrew A.G. '"The medical pundits": doctors and indirect advertising in the lay press, 1922–1927', *Medical History* 38 (1994), 255–80.
60. Mort, Frank; Thompson, Peter. 'Retailing, commercial culture and masculinity in 1950s Britain: the case of Montague Burton, the "tailor of taste"', *History Workshop* 38 (1994), 106–27.
61. Mortimer, J.E. *A history of the Boilermakers' Society, vol. 3: 1939–1989.* London; Verso; 1994. Pp 474.
62. Mullins, Samuel; Stockdale, David. *Talking shop: retailing in the Harborough area in the twentieth century.* Stroud; Sutton; 1994. Pp 192.
63. Mumford, M.J. 'The end of a familiar inflation accounting cycle', Bc35, 521–35.
64. Newman, Oksana; Foster, Allan (ed.). *Value of a pound: prices and incomes in Britain, 1900-1993.* Andover; Gale Research; 1994. Pp 306.
65. O'Grady, Joseph P. 'A troubled triangle: Great Britain, Ireland, the United States and civil aviation, 1944–1945', *J. of Transport History* 3rd ser. 15 (1994), 179–96.
66. Oliver, George. *Motor trials and tribulations: a history of Scottish motor vehicle manufacture.* London; HMSO; 1993. Pp 96.
67. Oulton, Nicholas; O'Mahony, Mary. *Productivity and growth: a study of British industry, 1954–1986.* Cambridge; Cambridge UP; 1994. Pp xvii, 315.
68. Palmer, Sarah. 'From London to Tilbury—the port of London since 1945', Bc121, 185–200.
69. Phillips, Andrew. 'Women on the shop floor: the Colchester rag trade, 1918–1950', *Oral History* 22 (1994), 56–65.
70. Porter, Mark. 'Devon's fishing industry, 1880-1990', Bc87, 243–50.
71. Porter, Mark. 'Devon's port industry since 1914', Bc87, 235–43.
72. Ritchie, Sebastian. 'A new audit of war: the productivity of Britain's wartime aircraft industry reconsidered', *War & Soc.* 12 (1994), 125–47.
73. Rollings, Neil. '"Poor Mr Butskell: a short life, wrecked by schizophrenia"?', *20th-Century British History* 5 (1994), 183–205.
74. Rudder, Ben. *Builders of the borough: a century of achievement by Battersea and Wandsworth Trades Union Council from 1894 to 1994.* London; Battersea & Wandsworth Trades Union Council; 1993. Pp 166.
75. Sapsford, David; Turnbull, Peter. 'Strikes and industrial conflicts in Britain's docks: balloon or iceberg? ', *Oxford B. of Economics and Statistics* 56 (1994), 249–66.
76. Schenk, Catherine R. 'Austerity and boom', Bc148, 300-19.

77. Scott, Peter. 'Learning to multiply: the property market and the growth of multiple retailing in Britain', *Business History* 36/3 (1994), 1–28.
78. Self, Robert. 'Treasury control and the Empire Marketing Board: the rise and fall of non-tariff preference in Britain', *20th-Century British History* 5 (1994), 153–82.
79. Singleton, John. 'The cotton industry and the British war effort, 1914–1918', *Economic History R.* 2nd ser. 47 (1994), 601–18.
80. Solomou, S.; Weale, M. 'Balanced estimates of national accounts when measurement errors are auto-correlated: the UK, 1920-1938', *J. of the Royal Statistical Soc.* ser. A 156 (1993), 89–105.
81. Steward, Robert. '"To keep the highlander at home"—the story of A.I. Welders', *Scottish Industrial History* 14–15 (1992), 13–17.
82. Sumner, Scott. 'The role of the gold standard in the Gibson paradox', *B. of Economic Research* 45 (1993), 215–28.
83. Supple, Barry. 'Fear of failing: economic history and the decline of Britain', *Economic History R.* 2nd ser. 47 (1994), 441–58.
84. Thornton, Paul. 'Tourism in Cornwall', Bc135, 108–27.
85. Tiratsoo, Nick; Tomlinson, Jim. 'Restrictive practices on the shopfloor in Britain, 1945–1960: myth and reality', *Business History* 36/2 (1994), 65–82.
86. Tomlinson, Jim. *Government and the enterprise since 1900.* Oxford; Oxford UP; 1994. Pp xv, 455.
87. Tsokhas, Kosmas. 'The Australian role in Britain's return to the gold standard', *Economic History R.* 2nd ser. 47 (1994), 129–46.
88. Waddington, Jeremy. 'Managing trade union membership concentration, 1892–1987: the relative effects of mergers and membership change', *British J. of Industrial Relations* 32 (1994), 449–64.
89. Waddington, Jeremy. 'Trade union membership concentration, 1892–1987: development and causation', *British J. of Industrial Relations* 31 (1993), 433–57.
90. Wale, Judith. 'Influences on structural change in the British coal industry, 1914–1946: evidence from the Northumberland coalfield', *Business Archives* 68 (1994), 15–36.
91. Whisler, Timothy R. 'The outstanding potential market: the British motor industry and Europe, 1945–1975', *J. of Transport History* 3rd ser. 15 (1994), 1–19.
92. Worswick, David. 'Has mass unemployment come to stay?', *Contemporary Record* 8 (1994), 1–22.

(g) *Social Structure and Population*

1. Bowden, Sue. 'The new consumerism', Bc148, 242–62.
2. Bowden, Sue; Offer, Avner. 'Household appliances and the use of time: the United States and Britain since the 1920s', *Economic History R.* 2nd ser. 47 (1994), 725–48.
3. Boyce, Frank. 'Irish Catholicism in Liverpool: the 1920s and 1930s', Bc140, 86–101.
4. Cage, R.A. 'Infant mortality rates and housing: twentieth-century Glasgow', *Scottish Economic & Social History* 14 (1994), 77–92.

5. Cannadine, David. *Aspects of aristocracy: grandeur and decline in modern Britain*. London; Yale UP; 1994. Pp x, 320.
6. Catterall, Peter. 'The party and religion' [Conservative century], Bc141, 637–70.
7. Cross, Joy; Staple, Margaret. *Memoirs of a Loughborough man: A.E. Shepherd, 1872–1962*. Nottingham; Department of Adult Education, University of Nottingham; 1994. Pp x, 46.
8. Graham, David. 'Female employment and infant mortality: some evidence from British towns, 1911, 1931 and 1951', *Continuity & Change* 9 (1994), 313–46.
9. Greenslade, Liam. '*An Tonn Gheal*: historical reflections on Irish migrants in Britain in the 1980s', Bc140, 29–46.
10. James, Winston. 'Migration, racism and identity formation: the Caribbean experience in Britain', Bc65, 231–87.
11. James, Winston; Harris, Clive. 'Introduction' [inside Babylon], Bc65, 1–8.
12. Lomas, Janis. '"So I married again": letters from British widows of the first and second world wars', *History Workshop* 38 (1994), 218–27.
13. Lyth, Peter J.; Dierikx, Marc L.J. 'From privilege to popularity: the growth of leisure air travel since 1945', *J. of Transport History* 3rd ser. 15 (1994), 97–116.
14. Mant, Joan. *All muck, no medals: landgirls by landgirls*. Lewes; Book Guild; 1994. Pp 280.
15. Morgan, N.J. 'Devon seaside tourism since 1900', Bc87, 250-58.
16. Murphy, M. 'The contraceptive pill and women's employment as factor in fertility change in Britain, 1963–1980: a challenge to the conventional view', *Population Studies* 47 (1993), 221–43.
17. Petter, Martin. '"Temporary gentlemen" in the aftermath of the great war: rank, status and the ex-officer problem', *Historical J.* 37 (1994), 127–52.
18. Read, Brian Ahier. *Jersey in London: a history of the Jersey Society in London, 1896–1989*. St Helier; Seaflower; 1994. Pp xii, 178.
19. Roper, Michael. *Masculinity and the British organization man since 1945*. Oxford; Oxford UP; 1994. Pp xi, 259.
20. Street, John. 'Youth culture', Bc148, 460-75.
21. Supple, Barry. 'The state and social investigation in Britain between the world wars', Bc27, 365–80.
22. Tabili, Laura. 'The construction of racial difference in twentieth-century Britain: the Special Restriction (Coloured Alien Seamen) Order, 1925', *J. of British Studies* 33 (1994), 54–98.
23. Thane, Pat. 'Women since 1945', Bc148, 392–410.
24. Thompson, F.M.L. 'English landed society in the twentieth century, part IV: prestige without power', *T. of the Royal Historical Soc.* 6th ser. 3 (1993), 1–22.
25. Vertovec, Steven. 'Indo-Caribbean experience in Britain: overlooked, mis-categorized, misunderstood', Bc65, 165–78.
26. Walton, J.K.; O'Neill, C. 'Numbering the holidaymakers: the problems and possibilities of the June census of 1921 for historians of resorts', *Local Historian* 23 (1993), 205–16.

27. Wardley, Peter; Woollard, Matthew. 'Retrieving the past: a reclamation and reconstruction of the social survey of Bristol, 1937', *History & Computing* 6 (1994), 85–105.
28. Watts, Martin; Rich, Judith. 'Occupational sex aggregation in Britain, 1979–1989: the persistence of sexual stereotyping', *Cambridge J. of Economics* 17 (1993), 159–77.

(h) *Social Policy*

1. Abel-Smith, Brian. 'The Beveridge report: its origins and outcomes', Bc13, 10-22.
2. Baldwin, Peter. 'Beveridge in the "longue dureé"', Bc13, 37–55.
3. Barber, Michael. *The making of the 1944 Education Act.* London; Cassell; 1994. Pp 144.
4. Cresswell, Yvonne M. *Living with the wire: civilian internment in the Isle of Man during two world wars.* Douglas (IoM); Manx National Heritage; 1994.
5. Englander, David. 'Soldiers and social reform in the first and second world wars', *Historical Research* 67 (1994), 318–26.
6. Evans, Neil; Jones, Dot. '"A blessing for the miner's wife": the campaign for pithead baths in the South Wales coalfield, 1908–1950', *Llafur* 6/3 (1994), 5–28.
7. Fox, Enid. 'District nursing in England and Wales before the National Health Service: the neglected evidence', *Medical History* 38 (1994), 303–21.
8. Francis, Errol. 'Psychiatric racism and social police: black people and the psychiatric services', Bc65, 179–205.
9. Glennerster, Howard; Evans, Martin. 'Beveridge and his assumptive worlds: the incompatibilites of a flawed design', Bc13, 56–72.
10. Harris, Bernard. 'Unemployment and the dole in interwar Britain', Bc148, 203–20.
11. Hills, John; Ditch, John; Glennerster, Howard. 'Introduction' [Beveridge and social security], Bc13, 1–9.
12. Hinchcliffe, Tanis. '"This rather foolish piece of panic administration": the government's flat conversion programme in London, 1919', *London J.* 19 (1994), 168–82.
13. Hogg, Gordon (ed.). *The professional education of teachers at the university of Newcastle upon Tyne, 1890-1990.* Newcastle; University of Newcastle upon Tyne; 1990. Pp 202.
14. Ingleton, Roy D. *The gentlemen at war: policing Britain, 1939–1945.* Maidstone; Cranborne; 1994. Pp viii, 392.
15. Jones, Helen. *Health and society in twentieth-century Britain.* London; Longman; 1994. Pp x, 204.
16. Jones, Peter Ellis. 'Elementary education in a Welsh rural community in the 1930s: a case-study of Carrog council school', *Merioneth Historical & Record Soc. J.* 12 (1994), 69–82.
17. Jones, Peter Ellis. 'The grammar school in rural Wales: a study of the boys' grammar school, Bala, 1939–1946', *Merioneth Historical & Record Soc. J.* 11 (1992), 303–30.

18. Kushner, Tony. 'Immigration and "race relations" in postwar British society', Bc148, 411–26.
19. Leneman, Leah. *In the service of life: the story of Elsie Inglis and the Scottish women's hospitals*. Edinburgh; Mercat; 1994. Pp 288.
20. Leneman, Leah. 'Medical women at war, 1914–1918', *Medical History* 38 (1994), 160-77.
21. Levine, Philippa. '"Walking the streets in a way no decent women should": women police in world war one', *J. of Modern History* 66 (1994), 34–78.
22. Lewis, Jane. 'Gender, the family and women's agency in the building of "welfare states": the British case', *Social History* 19 (1994), 37–55.
23. Lindsay, Jean. 'The impact of world war one on education in Caernarfonshire', *Caernarvonshire Historical Soc. T.* 54 (1993), 79–90.
24. Lowe, Rodney. 'A prophet dishonoured in his own country?: the rejection of Beveridge in Britain, 1945–1970', Bc13, 118–33.
25. Lowe, Rodney. 'Postwar welfare', Bc148, 356–73.
26. Lunn, Kenneth. '"Good for a few hundreds at least": Irish labour recruitment into Britain during the second world war', Bc140, 102–114.
27. Macnicol, John. 'Beveridge and old age', Bc13, 73–96.
28. Mama, Amina. 'Black women and the police: a place where the law is not upheld', Bc65, 135–51.
29. Mama, Amina. 'Woman abuse in London's black communities', Bc65, 97–134.
30. Meacham, Standish. 'Raymond Unwin, 1863–1940: designing for democracy in Edwardian England', Bc20, 78–102.
31. Parkin, Bernard. *St Ignatius college, 1894–1994*. London; St Ignatius College; 1994. Pp xii, 180.
32. Pedersen, Susan. 'Eleanor Rathbone, 1872–1946: the Victorian family under the daughter's eye', Bc20, 104–25.
33. Sanderson, Michael. *The missing stratum: technical school education in England, 1900-1990s*. London; Athlone; 1994. Pp ix, 235.
34. Sanderson, Michael. 'Education and social mobility', Bc148, 374–91.
35. Sheail, John. 'Sewering the English suburbs: an inter-war perspective', *J. of Historical Geography* 19 (1993), 433–47.
36. Simmons, Colin; Caruana, Viv. 'Neighbourhood issues in the development of Manchester airport, 1934–1982', *J. of Transport History* 3rd ser. 15 (1994), 117–43.
37. Stanton, Jennifer. 'What shapes vaccine policy?: the case of hepatitis B in the UK', *Social History of Medicine* 7 (1994), 427–46.
38. Tansey, E.M. 'Protection against dog distemper and Dogs Protection Bills: the Medical Research Council and anti-vivisectionist protest, 1911–1933', *Medical History* 38 (1994), 1–26.
39. Veit-Wilson, John. 'Condemned to deprivation?: Beveridge's responsibility for the invisibility of poverty', Bc13, 97–117.
40. Woollacott, Angela. '"Khaki fever" and its control: gender, class, age and sexual morality on the British home front in the first world war', *J. of Contemporary History* 29 (1994), 325–47.

(i) *Naval and Military*

1. Alban, J.R. *The three nights' blitz: select contemporary reports relating to Swansea's air raids of February 1941*. Swansea; Swansea City Council; 1994. Pp xvi, 179.
2. Allan, Jim. *In the trade of war: XIV Army in Burma—2nd Bn Green Howards*. Tunbridge Wells; Parapress; 1994. Pp 208.
3. Ambrose, Stephen E. *D-Day: 6 June 1944—the climactic battle of world war two*. Hemel Hempstead; Simon & Schuster; 1994. Pp 704.
4. Anderson, David G. 'British rearmament and the "merchants of death": the 1935–1936 Royal Commission on the manufacture of and trade in armaments', *J. of Contemporary History* 29 (1994), 5–37.
5. Anon. *Invasion Europe: the D-Day landings*. London; HMSO; 1994. 3 vol. Pp 96; 104; 368.
6. Anon. *The battle of the Atlantic: an anthology of personal memories from those involved with the battle of the Atlantic*. Birkenhead; Picton; 1993. Pp xii, 113.
7. Bangert-Larsen, Chris. '50-aret for den anden verdenskrig og de britiske militaerhistoriske udgivelser' [50 years from the second world war and British military history publications], *Historisk Tidskrift* 94 (1994), 74–81.
8. Barber, Laurie. 'Freyberg and Crete: the Australasian perspective', *J. of the Soc. for Army Historical Research* 72 (1994), 247–54.
9. Barnes, John. *The diary of a national serviceman in the Royal Navy*. Edinburgh; Pentland; 1994. Pp x, 85.
10. Baylis, John. 'The development of Britain's thermonuclear capability, 1954–1961: myth or reality?', *Contemporary Record* 8 (1994), 159–74.
11. Baynes, John. *Urquhart of Arnhem: the life of Major General R.E. Urquhart, CB, DSO*. London; Brassey's; 1993. Pp 285.
12. Beavan, Stanley. *Aegean journey: RAF wireless operations in the Mediterranean*. Studley; Brewin; 1994. Pp 112.
13. Bennett, Ralph. *Behind the battle: intelligence in the war with Germany, 1939–1945*. London; Sinclair-Stevenson; 1994. Pp xxiv, 328.
14. Beresford, Christine; Newbould, Christopher (ed.). *The Fifth Gloster gazette, 1915–1919: a trench magazine of the first world war*. Stroud; Sutton; 1993. Pp 280.
15. Bet-El, Ilana R. 'A soldier's pilgrimage: Jerusalem, 1918', *Mediterranean Historical R.* 8 (1993), 218–35.
16. Bloch, Gilbert. 'Les manoeuvres d'intoxication allieés en Mediterranée en 1944 et le débarquement de Provence', *Guerres Mondiales et Conflits Contemporains* 174 (1994), 59–68.
17. Bond, Brian. 'The army between the two world wars, 1918–1939', Bc109, 263–79.
18. Bowyer, Michael J.F. *Force for freedom: the USAF in the UK, 1948–1993*. Yeovil; Stephens; 1994. Pp 232.
19. Brown, O. Arthur. 'Les Jedburghs: un coup de maître ou une occasion manquée', *Guerres Mondiales et Conflits Contemporains* 174 (1994), 127–42.
20. Brown, Robin A. *Shark squadron: the history of 112 squadron, 1917–1975*. Bristol; Crecy; 1994. Pp 304.

21. Buckley, John. 'Contradictions in British defence policy, 1937–1939: the RAFs and the defence of trade', *20th-Century British History* 5 (1994), 100-13.
22. Buffetaut, Yves. *D-Day ships: the allied invasion fleet, June 1944*. London; Conway; 1994. Pp 162.
23. Clutton-Brock, Oliver. *Massacre over the Marne: RAF bombing raids on Révigny, July 1944*. Yeovil; Stephens; 1994. Pp 264.
24. Cooper, Anthony (ed.). *We who knew: the journal of an infantry subaltern during the great war*. Lewes; Book Guild; 1994. Pp 102.
25. Croft, John. 'A company commander at Imphal, 1944', *J. of the Soc. for Army Historical Research* 72 (1994), 19–34.
26. Cronin, Dick. *Royal Navy shipboard aircraft developments, 1912–1931*. Tonbridge; Air Britain; 1990. Pp 384.
27. Crook, Paul. 'Science and war: radical scientists and the Tizard-Cherwell area bombing debate in Britain', *War & Soc.* 12 (1994), 69–101.
28. Danchev, Alex. 'The army and the home front, 1939–1945', Bc109, 307–329.
29. David, Saul. *Churchill's sacrifice of the Highland division: France, 1940*. London; Brassey's; 1994. Pp 276.
30. Delaforce, Patrick. *The black bull: from Normandy to the Baltic with the 11th armoured division*. Stroud; Sutton; 1993. Pp 252.
31. d'Este, Carlo. 'The army and the challenge of war, 1939–1945', Bc109, 279–307.
32. Doherty, Richard. *Only the enemy in front: history of the Reconnaissance Corps, 1941–1946*. London; Donovan; 1994. Pp 294.
33. Farrar-Hockley, Sir Anthony. *The army in the air: the history of the Army Air Corps*. Stroud; Sutton; 1994. Pp ix, 268.
34. Farrar-Hockley, Sir Anthony. 'The post-war army, 1945–1963', Bc109, 329–57.
35. Fletcher, David. *The universal tank: British armour in the second world war*, part 2. London; HMSO; 1993. Pp 130.
36. Fletcher, Ian (ed.). *Letters from the front: the great war correspondence of Lieutenant Brian Lawrence, Grenadier Guards, 1916–1917*. Tunbridge Wells; Parapress; 1994. Pp 136.
37. Foreman, John. *Over the beaches: the air war of June 1944*. Walton on Thames; Air Research; 1994. Pp 300.
38. Funk, Arthur L. 'Les Américains et les britanniques dans la libération de la Corse', *Guerres Mondiales et Conflits Contemporains* 174 (1994), 7–21.
39. Gardiner, Juliet. *D-Day*. London; Collins & Brown; 1994. Pp 224.
40. Gelb, Norman. *Ike and Monty: generals at war*. London; Constable; 1994. Pp 480.
41. Giddon, Gerald. *VCs of the first world war: 1914*. Stroud; Sutton; 1994. Pp 224.
42. Godfrey, E.G. *The duke of Cornwall's light infantry, 1939–1945*. Upton-upon-Severn; Images; 1994. Pp 416.
43. Graham, Dominick; Bidwell, Shelford. *Coalitions, politicians and generals: some aspects of command in two world wars*. London; Brassey's; 1993. Pp 260.
44. Griffith, Paddy. *Battle tactics of the western front: the British army's art of attack, 1916–1918*. London; Yale UP; 1994. Pp xvi, 286.

45. Grounds, Tom. *Some letters from Burma: the story of the 25th Dragoons at war*. Tunbridge Wells; Parapress; 1994. Pp 208.
46. Grove, Eric J. 'A war fleet built for peace: British naval rearmament in the 1930s and the dilemma of deterrence versus defence', *Naval War College R.* 44 (1991), 82–92.
47. Gurley, Franklin L. 'Le débarquement en Provence: le 15 août 1944', *Guerres Mondiales et Conflits Contemporains* 174 (1994), 23–45.
48. Hamilton, Tim. *Identification friend or foe: being the story of aircraft recognition*. London; HMSO; 1994. Pp 137.
49. Hampton, James. *Selected for aircrew: bomber command in the second world war*. Walton on Thames; Air Research; 1993. Pp 376.
50. Harris, J.P. 'British military intelligence and the rise of German mechanized forces, 1929–1940', *Intelligence & National Security* 6 (1991), 395–417.
51. Harris, Sir Alan. 'Gathering Mulberries', *History Today* 44/5 (1994), 15–22.
52. Hausen, G. 'Englands "schwarzer Tag": die Seeschlacht bei Coronel' [England's "black day": the sea battle at Coronel], *Kohlers Flottenkalender* (1994), 20-26.
53. Henshaw, Peter J. 'The British chief of staff committee and the preparation of the Dieppe raid, March-August 1942: did Mountbatten really evade the committee's authority?', *War in History* 1 (1994), 197–214.
54. Hinsley, F.H.; Stripp, Alan (ed.). *Codebreakers: the inside story of Bletchley Park*. Oxford; Oxford UP; 1993. Pp xxi, 321.
55. Hoe, Alan; Morris, Eric. *Re-enter the SAS: the Special Air Service and the Malaysian emergency*. London; Cooper; 1994. Pp 230.
56. Horne, Alistair; Montgomery, David. *Lonely leader: Monty, 1944–1945*. Basingstoke; Macmillan; 1994. Pp xxvii, 381.
57. Howarth, Stephen; Law, Derek G. (ed.). *The battle of the Atlantic, 1939–1945: 50th anniversary international naval conference*. London; Greenhill; 1994. Pp 639.
58. Kavanagh, Gaynor. *Museums and the first world war: a social history*. Leicester; Leicester UP; 1994. Pp 210.
59. Kemp, Anthony. *The SAS: the savage wars of peace, 1947 to the present*. London; Murray; 1994. Pp xviii, 222.
60. Kennerley, Alston. 'Occupational conditioning: first voyage in *Passat*', Bc136, 210-32.
61. Kerr, Sheila. 'Roger Hollis and the dangers of the Anglo-Soviet treaty of 1942', *Intelligence & National Security* 5 (1990), 148–57.
62. Kingsley, F.A. (ed.). *Applications of radar and other electronic systems in the Royal Navy in world war II*. London; Macmillan; 1994. Pp 294.
63. Kingsley, F.A. (ed.). *The development of radar equipment for the Royal Navy, 1935–1945*. London; Macmillan; 1994. Pp 310.
64. Koven, Seth. 'Remembering and dismemberment: crippled children, wounded soldiers, and the great war in Great Britain', *American Historical R.* 99 (1994), 1167–1202.
65. Latter, Edwin. 'The Indian army in Mesopotamia, 1914–1918, part I', *J. of the Soc. for Army Historical Research* 72 (1994), 92–102.
66. Latter, Edwin. 'The Indian army in Mesopotamia, 1914–1918, part II', *J. of the Soc. for Army Historical Research* 72 (1994), 160-79.

67. Love, Robert William; Major, John (ed.). *The year of D-Day: the 1944 diary of Admiral Sir Bertram Ramsay*. Hull; Hull UP; 1994. Pp xxxiii, 208.
68. Loyd, William. *Challengers and chargers: a history of the Life Guards, 1945–1992*. London; Cooper; 1991. Pp 168.
69. Mann, John. *The atlas of D-Day and the Normandy landings*. London; Viking; 1994. Pp 144.
70. Mason, Francis Kenneth. *The British fighter since 1912*. London; Putnam; 1992. Pp 448.
71. Mason, Michael. 'Killing time: the British army and its antagonists in Egypt, 1945–1954', *War & Soc.* 12 (1994), 103–26.
72. McIntyre, M.J. 'These for remembrance: the "biscuit boys" at the battle of Loos', *Hatcher R.* 37 (1994), 2–15.
73. Meisel, Joseph S. 'Air raid shelter policy and its critics in Britain before the second world war', *20th-Century British History* 5 (1994), 300-19.
74. Millman, Brock. 'Toward war with Russia: British naval and air planning for conflict in the near east, 1939–1940', *J. of Contemporary History* 29 (1994), 261–83.
75. Morrison, David E. 'Journalists and the social construction of war', *Contemporary Record* 8 (1994), 305–20.
76. Paget, George Charles Henry Victor. *A history of the British cavalry, 1816–1919, vol. 5: Egypt, Palestine and Syria, 1914–1918*. London; Cooper; 1994. Pp 480.
77. Payton, Philip; Oakley, Derek. 'Naval education and training in Devon', Bc87, 191–204.
78. Phillips, Gervase. 'Dai bach y soldiwr [Little Dai, the soldier]: Welsh soldiers in the British army, 1914–1918', *Llafur* 6/2 (1993), 94–105.
79. Picot, Geoffrey. *Accidental warrior: in the front line from Normandy to victory*. Lewes; Book Guild; 1993. Pp 283.
80. Raugh, Harold E., Jr. *Wavell in the middle east, 1939–1941: a study in generalship*. London; Brassey's; 1993. Pp xxiv, 323.
81. Ray, John. *The battle of Britain: new perspectives—behind the scenes of the great air war*. London; Arms & Armour; 1994. Pp 240.
82. Reusch, Ulrich. 'Das Porträt [The portrait]: Sir Brian Robertson (1896–1974)' , *Geschichte im Westen* 5 (1990), 69–80.
83. Richards, John (ed.). *Wales on the western front*. Cardiff; Wales UP; 1994. Pp xiv, 225.
84. Rohwer, Jurgen. 'The wireless world at war, 1939–1945', *International History R.* 16 (1994), 536–48.
85. Rommetveit, Karl (ed.). *Narvik 1940: five-nation war in the high north*. Oslo; Institutt for Forsvarsstudier; 1991. Pp 174.
86. Rooney, David. *Wingate and the Chindits: redressing the balance*. London; Arms & Armour; 1994. Pp 272.
87. Routledge, N.W. *A history of the Royal Regiment of Artillery, vol. 4: anti-aircraft artillery, 1914–1955*. London; Brassey's; 1994. Pp 474.
88. Simkins, Peter. 'The four armies, 1914–1918', Bc109, 241–63.
89. Skinner, Ian. 'Devon and the navy in the second world war', Bc87, 204–15.
90. Skinner, Ian. 'The naval threat on the western flank of Operation Neptune, June 1944', *Mariner's Mirror* 80 (1994), 178–90.

91. Sladen, John. *The fourth service: merchantmen at war, 1939–1945*. London; Hale; 1994. Pp 347.
92. Smithers, A.J. *The fighting nation: Lord Kitchener and his armies*. London; Cooper; 1994. Pp 224.
93. Stedman, Michael. *Manchester Pals: 16th, 17th, 18th, 19th, 20th, 21st, 22nd and 23rd Battalions of the Manchester Regiment*. London; Cooper; 1994. Pp 264.
94. Strawson, John. 'The thirty years peace', Bc109, 357–76.
95. Sumida, Jon T. 'British naval operational logistics, 1914–1918', *J. of Military History* 57 (1993), 447–80.
96. Taylor, Gordon. 'We shelled Alderney: the bombardment of 12 August 1944', *B. of the Alderney Soc.* 27 (1992), 54–57.
97. Townshend-Bickers, Richard. *Air war: Normandy*. London; Cooper; 1994. Pp 192.
98. Travers, Tim. 'Command and leadership styles in the British army: the 1915 Gallipoli model', *J. of Contemporary History* 29 (1994), 403–42.
99. Travers, Tim. 'The army and the challenge of war, 1914–1918', Bc109, 215–41.
100. Travers, Tim. 'The evolution of British strategy and tactics on the western front in 1918: GHQ, manpower, and technology', *J. of Military History* 54 (1990), 173–200.
101. Trevelyan, Raleigh. *Clear premonition: letters of Lt. Tim Lloyd to his mother: Italy and north Africa, 1943–1944*. London; Cooper; 1994. Pp 224.
102. Villa, Brian L. 'Mountbatten, the British chiefs of staff, and approval of the Dieppe raid', *J. of Military History* 54 (1990), 201–06.
103. Walker, Adrian. *Barren place: national servicemen in Korea, 1950-1954*. London; Cooper; 1994. Pp 176.
104. Whitby, Michael. 'The seaward defence of the British assault area, 6–14 June 1944', *Mariner's Mirror* 80 (1994), 191–207.
105. Winter, J.M. (ed.). *First world war: documentary record series 1: European war, 1914–1919, the war reserve collection from Cambridge university library, part 3–4: Allied and German propaganda of the first world war*. Marlborough; Matthew; 1993. Pp 100.
106. Wollocombe, R.H. (ed.). *In the trenches with the 9th Devons: Frank Wollocombe's war diary*. Bath; Wollocombe; 1994. Pp ii, 126.
107. Wynn, Humphrey. *The RAF strategic nuclear deterrent forces: their origins, roles and deployment, 1946–1969: a documentary history*. London; HMSO; 1994. Pp xiv, 653.

(j) *Intellectual and Cultural*

1. Adamthwaite, Anthony. '"Nation shall speak peace unto nation": the BBC's response to peace and defence issues, 1945–1958', *Contemporary Record* 7 (1993), 557–77.
2. Addison, Paul. 'Oxford and the second world war', Bc8, 167–88.
3. Agar, Jon. 'Making a meal of the big dish: the construction of the Jodrell Bank mark I radio telescope as a stable edifice, 1946–1957', *British J. for the History of Science* 27 (1994), 3–21.

4. Allett, John. 'The moral philosophy of J.A. Hobson', Bc31, 1–18.
5. Anon. 'Chronological list of houses in Scotland designed by William Leiper', Bc139, 337–43.
6. Bailey, John. *Theatre for all seasons: the Nottingham Playhouse—the first thirty years, 1948–1978*. Stroud; Sutton; 1994. Pp 210.
7. Balmer, Judith (ed.). *Thomson's China: travels and adventures of a nineteenth-century photographer*. Oxford; Oxford UP; 1994. Pp xxiii, 225.
8. Barber, Giles. 'Libraries' [Oxford], Bc8, 471–84.
9. Bartley, Mary M. 'Conflicts in human progress: sexual selection and the Fisherian "runaway"', *British J. for the History of Science* 27 (1994), 177–96.
10. Beasley, John D. *Peckham Rye park, 1894–1994*. London; South Riding; 1994. Pp 80.
11. Bennett, S. *A history of control engineering, 1930-1955*. Stevenage; Inst. of Electrical Engineers; 1993. Pp x, 250.
12. Berry, David. *Wales and cinema: the first hundred years*. Cardiff; Wales UP; 1994. Pp xii, 567.
13. Betjeman, John. *Letters, vol. 1: 1926–1951*, ed. Candida Lycett-Green. London; Methuen; 1994. Pp xxiii, 584.
14. Boddington, Sue. *A source of pride: a brief history of Calne library, written and published for National Library Week*. Wiltshire; Wiltshire County Council Library & Museum Service; 1993. Pp 41.
15. Bracco, Rosa Maria. *Merchants of hope: British middlebrow writers and the first world war, 1919–1939*. Oxford; Berg; 1993. Pp 208.
16. Brivati, Brian. 'The historian as history: an interview with Adam Sisman, biographer of A.J.P. Taylor', *Contemporary Record* 7 (1993), 594–611.
17. Brock, Michael. 'The university since 1970' [Oxford], Bc8, 739–74.
18. Burgess, Keith. 'British employers and education policy, 1935–1945: a decade of "missed opportunities"?', *Business History* 36/3 (1994), 29–61.
19. Burton-Page, Piers. *Philharmonic concerto: the life and music of Malcolm Arnold*. London; Lime Tree; 1994. Pp 194.
20. Cantor, Geoffrey. 'The making of a British theoretical physicist—E.C. Stoner's early career', *British J. for the History of Science* 27 (1994), 277–90.
21. Clarke, Peter. 'J.M. Keynes, 1883–1946: "the best of both worlds"', Bc20, 170-87.
22. Cochrane, Claire. *Shakespeare and the Birmingham repertory theatre, 1913–1929*. London; Soc. for Theatre Research; 1993. Pp 181.
23. Colvin, Howard. 'Introducing Alexander Roos', Bc139, 277–84.
24. Cruickshank, Stewart. 'Scottish rock music, 1955–1993', Bc2, 191–94.
25. Cunningham, Valentine. 'Literary culture' [Oxford], Bc8, 413–50.
26. Currie, Robert. 'The arts and social studies, 1914–1939' [Oxford], Bc8, 109–38.
27. Darwin, J.G. 'A world university' [Oxford], Bc8, 607–36.
28. Davies, Andrew. 'Cinema and broadcasting', Bc148, 263–81.
29. Davies, John. *Broadcasting and the BBC in Wales*. Cardiff; Wales UP; 1994. Pp xii, 445.
30. Davies, John. '"L'art du possible": the Board of Education, the Catholic Church and negotiations over the White Paper and the Education Bill, 1943–1944', *Recusant History* 22 (1994), 231–50.

31. Deacon, Margaret. 'British governmental attitudes to marine science', Bc136, 11–35.
32. Denniston, R.A. 'Publishing and bookselling' [Oxford], Bc8, 451–70.
33. Driver, Mary; Scott, Audrey (ed.). *Queen Anne's school, Caversham: the first hundred years*. Chichester; Phillimore; 1994. Pp x, 134.
34. Dunbabin, J.P.D. 'Finance since 1914' [Oxford], Bc8, 639–82.
35. Foulkes, Richard. 'The Northampton history pageant of 1925', *Northamptonshire Past & Present* 8 (1993/4), 410-15.
36. Greenberg, Morris. 'Knowledge of the health hazard of asbestos prior to the Merewether and Price report of 1930', *Social History of Medicine* 7 (1994), 493–516.
37. Greenstein, Daniel I. 'The junior members, 1900-1990: a profile' [Oxford], Bc8, 45–77.
38. Gutzmore, Cecil. 'Carnival, the state and the black masses in the United Kingdom', Bc65, 207–30.
39. Halsey, A.H. 'Oxford and the British universities', Bc8, 577–606.
40. Halsey, A.H. 'The Franks commission' [Oxford], Bc8, 721–36.
41. Harcourt, G.C. 'Kahn, Keynes and the making of the *General theory*', *Cambridge J. of Economics* 18 (1994), 11–24.
42. Harris, José. 'Beveridge's social and political thought', Bc13, 23–36.
43. Harris, José. 'The arts and social sciences, 1939–1970' [Oxford], Bc8, 217–49.
44. Harrison, Brian. 'College life, 1918–1939' [Oxford], Bc8, 81–108.
45. Harrison, Brian. 'Government and administration, 1914–1964' [Oxford], Bc8, 683–719.
46. Hasted, Rachel. 'Tales of the city: women in a community history exhibition', *Gender & History* 6 (1994), 397–409.
47. Hayes, Nick. 'Municipal subsidy and Tory minimalism: building the Nottingham Playhouse, 1942–1963', *Midland History* 19 (1994), 128–46.
48. Holt, Richard. 'King across the border: Denis Law and Scottish football', Bc3, 58–74.
49. Howarth, Janet. 'Women' [Oxford], Bc8, 345–75.
50. Hume, John R. 'The Scottish houses of William Leiper', Bc139, 285–97.
51. Ireson, Tony. 'Archibald Spencer Ireson: an appreciation', *Northamptonshire Past & Present* 8 (1993/4), 416–22.
52. Kay, Diane. 'Architecture' [Oxford], Bc8, 499–518.
53. King, John. 'J.A. Hobson's macroeconomics: the last ten years, 1930-1940', Bc31, 124–42.
54. Klosse, Saskia; Jaspers, Teun; Westerveld, Mies. 'Beveridge fifty years on: second youth or early retirement?', Bc13, 171–88.
55. LeMahieu, D.L. 'John Reith, 1889–1971: entrepreneur of collectivism', Bc20, 188–206.
56. Leventhal, F.M. 'Leonard Woolf, 1880-1969: the conscience of a Bloomsbury socialist', Bc20, 148–68.
57. Long, David. 'Three modes of internationalism in the work of J.A. Hobson', Bc31, 163–87.
58. Lucas, A.M.; Lucas, P.J.; Darragh, Thomas A.; Maroske, Sara. 'Colonial

pride and metropolitan expectations: the British Museum and Melbourne's meteorites', *British J. for the History of Science* 27 (1994), 65–87.

59. MacLeod, Roy. '"All for each and each for all": reflections on Anglo-American and Commonwealth scientific cooperation, 1940-1945', *Albion* 26 (1994), 79–112.
60. Mandler, Peter. 'John Summerson, 1904–1992: the architectural critic and the quest for the modern', Bc20, 228–45.
61. Mandler, Peter; Pedersen, Susan. 'Introduction: the British intelligentsia after the Victorians', Bc20, 1–28.
62. Mathews, Tom Dewe. *Censored!: the story of film censorship in Britain*. London; Chatto; 1994. Pp 298.
63. Merricks, Linda. 'An invisible man: on writing biography', *History Workshop* 37 (1994), 194–204.
64. Millward, Roy. 'William George Hoskins, landscape historian (1908–1992)', *Landscape History* 14 (1992), 65–70.
65. Mini, Piero V. *John Maynard Keynes: study in the psychology of original work*. London; Macmillan; 1994. Pp 254.
66. Morrell, J.B. 'The non-medical sciences, 1914–1939' [Oxford], Bc8, 139–63.
67. Morse, Elizabeth J. 'English civic universities and the myth of decline', *History of Universities* 11 (1992), 177–204.
68. Neale, Walter C.; Mayhew, Anne. 'Hobson, Veblen and American institutionalism', Bc31, 225–37.
69. Nehring, Neil. *Flowers in the dustbin: culture, anarchy and postwar England*. London; Michigan UP; 1993. Pp 330.
70. Nenadic, Stana. 'Museums, gender and cultural identity in Scotland', *Gender & History* 6 (1994), 426–34.
71. Paget, George Charles Henry Victor; Harvey, John H.; Saunders, Matthew; Bowles, John. 'Ivor Bulmer-Thomas, 1905–1993: in memoriam', *T. of the Ancient Monuments Soc.* 38 (1994), 1–14.
72. Penglase, Charles. 'Authenticity in the revival of Cornish', Bc135, 96–107.
73. Pollock, Stephen. *In a class of its own: BBC education, 1924–1994*. London; BBC; 1994.
74. Porter, Bernard. '"Though not an historian myself ...": Margaret Thatcher and the historians', *20th-Century British History* 5 (1994), 246–56.
75. Prest, John. 'The Asquith commission, 1919–1922' [Oxford], Bc8, 27–43.
76. Prior, Robin; Wilson, Trevor. 'Paul Fussell at war' [review article], *War in History* 1 (1994), 63–80.
77. Read, Donald. 'Reuters: news agency of the British empire', *Contemporary Record* 8 (1994), 195–212.
78. Rees, Jenny. *Mr Nobody: secret life of Goronwy Rees*. London; Weidenfeld & Nicolson; 1994. Pp 320.
79. Richards, Huw. 'The ragged man of Fleet Street: the *Daily Herald* in the 1920s', *Contemporary Record* 8 (1994), 242–57.
80. Richards, Jeffrey. 'Cinema-going in worktown: regional film audiences in 1930s Britain', *Historical J. of Film, Radio & Television* 14 (1994), 147–66.
81. Richards, Pamela Spence. *Scientific information in wartime: Allied-German rivalry, 1939–1945*. London; Greenwood; 1994. Pp 192.
82. Roche, John. 'The non-medical sciences, 1939–1970' [Oxford], Bc8, 251–89.

83. Rowse, A.L. *All Souls in my time*. London; Duckworth; 1993. Pp 214.
84. Rutherford, Malcolm. 'J.A. Hobson and American institutionalism: underconsumption and technological change', Bc31, 188–210.
85. Schama, Simon. 'John Clive, 1924–1990: TBM and John', Bc20, 246–60.
86. Scheiber, Harry N. 'Modern U.S. Pacific oceanography and the legacy of British and northern European science', Bc136, 36–75.
87. Schneider, Michael. 'Modelling Hobson's underconsumption theory', Bc31, 100-23.
88. Sedgwick, John. 'The market for feature films in Britain, 1934: a viable national cinema', *Historical J. of Film, Radio & Television* 14 (1994), 15–36.
89. Seldon, Anthony. 'The Heathman: an interview with John Campbell', *Contemporary Record* 7 (1993), 578–93.
90. Seymour-Ure, Colin. 'Who owns the national press?', *Contemporary Record* 8 (1994), 258–73.
91. Shone, Richard. 'Beaverbrook—Sickert's "political portrait"', *History Today* 44/9 (1994), 41–46.
92. Smith, David. 'Churchill's wartime radio rival', *The Historian* 44 (1994), 12–15.
93. Tansey, E.M.; Catterall, P.P. 'Monoclonal antibodies: a seminar in contemporary medical history', *Medical History* 38 (1994), 322–27.
94. Thomas, Keith. 'College life, 1945–1970' [Oxford], Bc8, 189–215.
95. Townshend, Jules. 'Hobson and the socialist tradition', Bc31, 34–52.
96. Trentmann, Frank. 'Civilization and its discontents: English neo-romanticism and the transformation of anti-modernism in twentieth-century western culture', *J. of Contemporary History* 29 (1994), 583–625.
97. van Splunter, Jacob M. 'Nuclear fission across the North Sea: Anglo-Dutch co-operation on the peaceful use of atomic energy, 1950-1963', *J. of Contemporary History* 29 (1994), 663–709.
98. Walker, David M. 'The country house, larger villas and related hotels designs of Sir Sir John James Burnet', Bc139, 298–324.
99. Walker, Stephen P. *Accountancy at the university of Edinburgh, 1919–1994: the emergence of a 'viable academic department'*. Edinburgh; Inst. of Chartered Accountants of Scotland; 1994. Pp 119.
100. Waters, Chris. 'J.B. Priestley, 1894–1984: Englishness and the politics of nostalgia', Bc20, 208–26.
101. Watson, Don. '"Where do we go from here?": education, theatre and politics in the British army, 1945–1945', *Labour History R.* 59/3 (1994), 57–67.
102. Watson, Nigel. *In hortis reginae: story of Queenswood school, 1894–1994*. London; James & James; 1994. Pp 240.
103. Webster, Charles. 'Medicine' [Oxford], Bc8, 317–43.
104. Weight, Richard. 'Return to Albion: intellectuals in wartime Britain', *History Today* 44/12 (1994), 37–43.
105. Wells, Harwell. 'The philosophical Michael Oakeshott', *J. of the History of Ideas* 55 (1994), 129–45.
106. Wenden, D.J. 'Sport' [Oxford], Bc8, 519–39.
107. White, Christopher. 'Museums and art galleries' [Oxford], Bc8, 485–98.
108. Whiting, Richard. 'University and locality' [Oxford], Bc8, 543–76.

109. Winter, J.M. 'Oxford and the first world war', Bc8, 3–25.
110. Wood, Joanne A. 'Lighthouse bodies: the neutral monism of Virginia Woolf and Bertrand Russell', *J. of the History of Ideas* 55 (1994), 483–502.

J. MEDIEVAL WALES

See also Aa54,132,b51,53; Ba184,c24; Fe60

(a) *General*

1. Bartrum, Peter C. 'Corrections to *Early Welsh genealogical tracts*', *B. of the Board of Celtic Studies* 40 (1993), 171–72.
2. Coplestone-Crow, Bruce. 'Early Rhwng Gwy a Hafren', *Herefordshire Arch. News* 60 (1993), 12–16.
3. Dark, Kenneth R. *Discovery by design: the identification of secular élite settlements in western Britain, AD 400-700*. Oxford; BAR (British ser. 237); 1994. Pp vii, 190.
4. Harrison, Julian. 'A note on Gerald of Wales and *Annales Cambriae*', *Welsh History R.* 17 (1994), 252–55.
5. Siddons, Michael. 'Welsh heraldry', *T. of the Honourable Soc. of Cymmrodorion* (1993), 27–46.
6. Sims-Williams, Patrick. 'Historical need and literary narrative: a caveat from ninth-century Wales', *Welsh History R.* 17 (1994), 1–40.
7. Thomas, Charles. *And shall these mute stones speak? Post-Roman inscriptions in western Britain*. Cardiff; Wales UP; 1994. Pp xxiv, 353.
8. Trotter, D.A. 'L'anglo-français au Pays de Galles: une enquête préliminaire', *Revue de la linguistique romane* 58 (1994), 461–87.

(b) *Politics*

1. Babcock, Robert S. 'Rhys ap Tewdwr, king of Deheubarth', Bc69, 21–35.
2. Kirby, D.P. 'The political development of Ceredigion, *c.*400-1081', Bc24, 318–42.
3. Robinson, W.R.B. 'A letter from Sir Richard Croft to Sir Gilbert Talbot in 1486 concerning Sir James Tyrell's offices in Wales', *Historical Research* 67 (1994), 178–89.
4. Walker, R.F. 'The supporters of Richard Marshall, earl of Pembroke, in the rebellion of 1233–1234', *Welsh History R.* 17 (1994), 41–65.

(c) *Constitution, Administration and Law*

1. James, Christine. 'Tradition and innovation in some later medieval Welsh lawbooks', *B. of the Board of Celtic Studies* 40 (1993), 148–56.
2. Jenkins, Dafydd. 'Ail olwg ar lawysgrif colan' [A second look at the Colan manuscript], Bc102, 63–78.

(d) *External Affairs*

1. Flanagan, Marie Therese. '*Historia Gruffud vab Kenan* and the origins of Balrothery, county Dublin', *Cambrian Medieval Celtic Studies* 28 (1994), 71–94.

(e) *Religion*

1. Anon. *An architectural study: the cathedral church of St John the Evangelist Brecon* Brecon; Friends of Brecon Cathedral; 1994. Pp 88.
2. Bacon, Olive L. 'Masons' marks in Brecon cathedral', *Brycheiniog* 25 (1992–3), 27–29.
3. Britnell, W.J. 'Excavation and recording at Pennant Melangell church', *Montgomeryshire Collections* 82 (1994), 41–102.
4. Britnell, W.J. 'The boundaries of the parish of Pennant Melangell', *Montgomeryshire Collections* 82 (1994), 1–8.
5. Britnell, W.J.; Watson, K. 'Saint Melangell's shrine, Pennant Melangell', *Montgomeryshire Collections* 82 (1994), 147–66.
6. Campbell, Ewan; MacDonald, Philip. 'Excavations at Caerwent vicarage orchard garden, 1973: an extra-mural post-Roman cemetery', *Archaeologia Cambrensis* 142 (1993), 74–98.
7. Cole, J.R.; Pratt, Derrick. 'Capel Spon, Buckley, Clwyd', *Archaeology in Wales* 33 (1993), 25–30.
8. Edwards, Nancy; Hulse, Tristan Gray. 'A fragment of a reliquary casket from Gwytherin, north Wales', *Antiquaries J.* 72 (1994 for 1992), 91–101.
9. Evans, J. Wyn. 'Meidrum: some sidelights on the church and parish', *The Carmarthenshire Antiquary* 29 (1993), 13–22.
10. Greatrex, Joan. 'Medieval converts from Judaism in the Welsh borders', *The Monmouthshire Antiquary* 10 (1994), 19–20.
11. Haycock, Marged. *Blodeugerdd Barddas o ganu crefyddol cynnar* [The 'Barddas' anthology of early religious poetry]. Llandybïe; Cyhoeddiadau Barddas; 1994. Pp xxvii, 385.
12. Heaton, R.B.; Britnell, W.J. 'A structural history of Pennant Melangell church', *Montgomeryshire Collections* 82 (1994), 103–26.
13. James, Heather. 'The archaeology of early Christianity in Cardiganshire', Bc24, 397–407.
14. James, Terrence. 'Carmarthen priory gatehouse', *The Carmarthenshire Antiquary* 29 (1993), 23–26.
15. Kirby, D.P. 'The coming of Christianity' [Cardiganshire], Bc24, 365–77.
16. Ó Riain, Pádraig. 'The saints of Cardiganshire', Bc24, 378–96.
17. Parkinson, A.J. 'Paintings and inscriptions in Pennant Melangell church', *Montgomeryshire Collections* 82 (1994), 139–46.
18. Pryce, Huw. 'A new edition of the *Historia divae Monacellae*', *Montgomeryshire Collections* 82 (1994), 23–40.
19. Redknap, Mark. 'A pre-Norman cross from Caerwent and its context', *The Monmouthshire Antiquary* 10 (1994), 1–6.
20. Ridgway, Maurice H. 'Furnishings and fittings in Pennant Melangell church', *Montgomeryshire Collections* 82 (1994), 127–38.

21. Salter, Mike. *The old parish churches of south-west Wales*. Malvern; Folly; 1994. Pp 80.
22. Sell, S.H. 'Excavations at St Mary's, Swansea, 1976–1977', *Archaeology in Wales* 33 (1993), 1–16.
23. Thomas, W. Gwyn. 'The early Christian monuments' [Cardiganshire], Bc24, 407–20.
24. Williams, David H. 'Medieval monasticism in Monmouthshire', *The Monmouthshire Antiquary* 10 (1994), 7–17.

(f) *Economic Affairs*

none

(g) *Social Structure and Population*

1. Alban, J.R.; Thomas, W.S.K. 'Charters of the borough of Brecon, 1276–1517', *Brycheiniog* 25 (1992–3), 31–55.
2. Berry, André Q. 'The parks and forests of the lordship of Dyffryn Clwyd', *Denbighshire Historical Soc. T.* 43 (1994) 7–25.
3. Campbell, Ewan; Lane, Alan. 'Excavations at Longbury Bank, Dyfed, and early medieval settlement in south Wales', *Medieval Archaeology* 37 (1993), 15–77.
4. Coplestone-Crow, Bruce. 'Trewalkin and its Anglo-Norman lords', *Brycheiniog* 26 (1992–3), 43–51.
5. Davies, R.R. 'Dinbych yn oes Owain Glyndŵr' [Denbigh in the age of Owain Glyndŵr], *Denbighshire Historical Soc. T.* 43 (1994), 27–44.
6. Dodgshon, R.A. 'Early society and economy' [Cardiganshire], Bc24, 343–64.
7. Hughes, R. Elfyn. 'Land, agricultural resources and population in parts of Penllyn in 1318' [part 1], *Merioneth Historical & Record Soc. J.* 11 (1993), 355–78.
8. Hughes, R. Elfyn. 'Land, agricultural resources and population in parts of Penllyn in 1318: part 2', *Merioneth Historical & Record Soc. J.* 12 (1994), 1–16.
9. Kissock, J.A. 'Some examples of co-axial field systems in Pembrokeshire', *B. of the Board of Celtic Studies* 40 (1993), 190-97.
10. Mytum, Harold C.; Webster, C.J. 'Survey and excavation at Berry Hill Camp, Nevern, Dyfed', *B. of the Board of Celtic Studies* 40 (1993), 198–211.
11. Powell, R.F. Peter. 'Early place-names in the former borough of Brecknock', *Brycheiniog* 25 (1992–3), 15–26.
12. Pratt, Derrick. 'A Chirk charter, 1334', *Denbighshire Historical Soc. T.* 43 (1994), 109–15.
13. Pratt, Derrick. 'Parkers of Clocaenog', *Denbighshire Historical Soc. T.* 43 (1994), 116–20.
14. Thomas, Colin. 'A cultural-ecological model of agrarian colonization in upland Wales', *Landscape History* 14 (1992), 37–50.
15. Walker, David. 'Brecon: the early borough', *Brycheiniog* 25 (1992–3), 7–14.

(h) *Naval and Military*

1. Avent, Richard. *Dolwyddelan castle, Dolbadarn castle.* Cardiff; Cadw; 1994. Pp 70.
2. Avent, Richard; Jones, Martin. 'Laugharne castle', *Archaeology in Wales* 33 (1993), 70-72.
3. Butler, Lawrence A.S. 'Dolforwyn castle', *Archaeology in Wales* 33 (1993), 82.
4. Hannaford, Hugh R. 'Flint castle, Clwyd: excavations to the west of the castle, 1988', *Archaeology in Wales* 33 (1993), 30-33.
5. Knight, Jeremy K. 'Excavations at Montgomery castle, part II, the finds: metalwork', *Archaeologia Cambrensis* 142 (1993), 182–242.
6. Lewis, J.M. 'Excavations at Loughor castle, West Glamorgan, 1969–1973', *Archaeologia Cambrensis* 142 (1993), 99–181.
7. Mein, A.G. 'Trostrey castle, Trostrey', *Archaeology in Wales* 33 (1993), 79–80.
8. Quinell, Henrietta; Blockley, Marion R. *Excavations at Rhuddlan, Clwyd, 1969–1973: mesolithic to medieval.* York; CBA; 1994. Pp 248.
9. Remfry, Paul M. 'The last campaign of Rhys ap Gruffydd, prince of Deheubarth', *Herefordshire Arch. News* 62 (1994), 35–38.
10. Rowlands, M.L.J. 'Monnow bridge and gate, Monmouth', *Archaeologia Cambrensis* 142 (1993), 243–87.
11. Spurgeon, C.J. 'St Donat's castle: a recent revised interpretation by the R.C.A.H.M. Wales', *Arch. J.* 150 (1994 for 1993), 498–503.

(i) *Intellectual and Cultural*

1. Breeze, Andrew. 'Master John of St David's, a new twelfth-century poet?', *B. of the Board of Celtic Studies* 40 (1993), 73–82.
2. Charles-Edwards, G.; Charles-Edwards, T.M. 'The continuation of "Brut y Tywysogion" in Peniarth MS 20', Bc102, 293–305.
3. Daniel, R. Iestyn (ed.). *Gwaith Bleddyn Ddu* [The works of Bleddyn Ddu]. Aberystwyth; Canolfan Uwchefrydiau Cymreig a Cheltaidd; 1994. Pp xv, 81.
4. Davies, Sioned. *The four branches of the Mabinogi: Pedeir Keinc y Mabinogi.* Llandysul; Gomer; 1993. Pp 86.
5. Diverres, Armel. 'Arthur in *Culhwch and Olwen* and in the romances of Chrétien de Troyes', Bc100, 54–69.
6. Haycock, Marged. 'Taliesin's "lesser song of the world"', Bc102, 229–50.
7. Thomas, Peter Wynn. 'Middle Welsh dialects: problems and perspectives', *B. of the Board of Celtic Studies* 40 (1993), 17–50.
8. Williams, David H. 'Gwent seals: VI', *The Monmouthshire Antiquary* 10 (1994), 21–24.

K. SCOTLAND BEFORE THE UNION

See also Aa9,68,79,139; Ba184,b71–72,127,168–169,181,208,245,c1,23,83,111,119, 133,39; Ea1–2,53,k16; Fb27,48,g17,45,h13,j43; Gf16

(a) *General*

1. Ballantyne, John; Smith, Brian (ed.). *Shetland documents, 1580-1611.* Lerwick; Shetland Times; 1994. Pp 352.
2. Brooke, Daphne. *Wild men and holy places: St Ninian, Whithorn and the medieval realm of Galloway.* Edinburgh; Canongate; 1994. Pp 192.
3. Devine, Thomas M. 'Whither Scottish history?', *Scottish Historical R.* 73 (1994), 1–3.
4. Grant, Alexander. 'To the medieval foundations', *Scottish Historical R.* 73 (1994), 4–24.
5. Hall, Ursula. *St Andrew and Scotland.* St Andrews; University of St Andrews Library; 1994. Pp 192.
6. Hudson, Benjamin T. *Kings of Celtic Scotland.* London; Greenwood; 1994. Pp 216.
7. Lee, Maurice, Jr. 'Scotland, the Union and the idea of a "general crisis"', Bc119, 41–58.
8. Lynch, Michael. 'Old games and new' [early modern Scotland], *Scottish Historical R.* 73 (1994), 47–63.
9. Macdougall, Norman. 'At the medieval bedrock', *Scottish Historical R.* 73 (1994), 25–29.
10. Macinnes, Allan J. 'Early modern Scotland: the current state of play', *Scottish Historical R.* 73 (1994), 30-46.
11. Omand, Donald. 'The landscape of Caithness and Orkney', Bc23, 102–10.
12. Sutherland, Elizabeth. *Picts.* London; Constable; 1994. Pp 256.
13. Williamson, Arthur H. 'Scotland and the British revolutions' [review article], *Scottish Historical R.* 73 (1994), 117–27.

(b) *Politics*

1. Brooke, Daphne. 'Fergus of Galloway: miscellaneous notes for a revised portrait', *Dumfriesshire & Galloway Natural History & Antiquarian Soc. T.* 3rd ser. 66 (1991), 47–58.
2. Brown, Michael. *James I.* Edinburgh; Canongate; 1994. Pp 256.
3. Brown, Michael. 'Scotland tamed?: kings and magnates in late medieval Scotland: a review of recent work', *Innes R.* 45 (1994), 120-46.
4. Burns, J.H. 'George Buchanan and the anti-monarchomacs', Bc119, 138–61.
5. Crawford, Barbara E. 'Norse earls and Scottish bishops in Caithness', Bc23, 129–47.
6. Hill, J. Michael. *Fire and sword: Sorley Boy MacDonnell and the rise of Clan Ian Mor, 1538–1590.* London; Athlone; 1993. Pp xiv, 321.

7. Knox, John. *On rebellion*, ed. Roger A. Mason. Cambridge; Cambridge UP; 1994. Pp lxviii, 219.
8. Loughlin, Mark. '"The dialogue of the twa wyfeis": Maitland, Machiavelli and the propaganda of the Scottish civil war', Bc83, 226–45.
9. Mason, Roger A. 'George Buchanan, James VI and the presbyterians', Bc119, 112–38.
10. Mason, Roger A. 'Imagining Scotland: Scottish political thought and the problem of Britain, 1560-1650', Bc119, 3–17.
11. Oram, Richard D. 'Bruce, Balliol and the lordship of Galloway: south-west Scotland and the wars of independence', *Dumfriesshire & Galloway Natural History & Antiquarian Soc. T.* 3rd ser. 67 (1992), 29–47.
12. Watt, Diane. 'Nationalism in Barbour's *Bruce*', *Parergon* 12 (1994), 89–107.
13. Wormald, Jenny. 'The union of 1603', Bc119, 17–41.

(c) *Constitution, Administration and Law*

1. Levack, Brian P. 'Law, sovereignty and the union', Bc119, 213–41.
2. MacQueen, Hector L. '*Glanvill* resarcinate: Sir John Skene and *Regiam Majestam*', Bc83, 385–403.
3. Maley, Teresa; Elliot, Walter (ed.). *The protocol book of Robert Wedderop, notary, Lauder, 1543–1553*. Selkirk; Walter Mason Trust; 1993. Pp 44.
4. Smith, David B. 'The spiritual jurisdiction, 1560-1564', *Records of the Scottish Church History Soc.* xxv (1993), 1–18.

(d) *External Affairs*

1. Aird, W.M. 'St Cuthbert, the Scots and the Normans', Bc69, 1–20.
2. Stalley, Roger. 'Ireland and Scotland in the later middle ages', Bc133, 108–17.
3. Webster, Bruce. 'The Declaration of Arbroath and Scottish national identity', *Medieval History* 3 (1993), 156–65.

(e) *Religion*

1. Adam, R.J. 'Meathie-Lour: a parish exploration', *Records of the Scottish Church History Soc.* xxv (1993), 40-67.
2. Anon. 'Appendix 2: concordance and measurements of the Govan stones', Bc111, 147–52.
3. Ash, Marinell; Broun, Dauvit. 'The adoption of St Andrew as patron saint of Scotland', Bc133, 16–24
4. Bailey, Richard N. 'Govan and Irish Sea sculpture', Bc111, 113–23.
5. Bailey, Richard N. 'The Ruthwell cross: a non-problem', *Antiquaries J.* 73 (1994 for 1993), 141–48.
6. Ballantyne, John. 'More on James Stewart of Jerusalem', *Innes R.* 45 (1994), 78–81.
7. Barrow, G.W.S. 'The medieval diocese of St Andrews', Bc133, 1–6.
8. Bryce, Ian B.D.; Roberts, Alasdair. 'Post-Reformation Catholic houses of north-east Scotland', *Soc. of Antiquaries of Scotland P.* 123 (1993), 363–72.

9. Cowan, Ian B. 'The monastic history of the diocese of St Andrews', Bc133, 7–15.
10. Cowan, Ian B.; Yellowlees, Michael J. 'The cathedral clergy of Dunkeld in the early sixteenth century', Bc83, 136–43.
11. Davies, Wendy. 'Ecclesiastical centres and secular society in the Brittonic world in the tenth and eleventh centuries', Bc111, 92–103.
12. Dilworth, Mark. *Scottish monasteries: monastic life in the 16th century.* Edinburgh; Edinburgh UP; 1994. Pp 240.
13. Dilworth, Mark. 'Canons regular and the Reformation', Bc83, 164–82.
14. Dilworth, Mark. 'Franco-Scottish efforts at monastic reform', *Records of the Scottish Church History Soc.* xxv (1994), 204–21.
15. Durkan, John. 'Heresy in Scotland: the second phase, 1546–1558', *Records of the Scottish Church History Soc.* xxiv (1992), 320-65.
16. Durkan, John. 'Scottish reformers: the less than golden legend', *Innes R.* 45 (1994), 1–28.
17. Fawcett, Richard. *Historic Scotland: Scottish abbeys and priories.* London; Batsford; 1994. Pp 144.
18. Fisher, Ian. 'Orphir church in its south Scandinavian context', Bc23, 375–80.
19. Ford, John D. 'The lawful bonds of Scottish society: the five articles of Perth, the negative confession and the national covenant', *Historical J.* 37 (1994), 45–64.
20. Halloran, Brian A. 'Spirited Scottish students: the Scots college, Paris, in 1639', *Innes R.* 45 (1994), 171–77.
21. Healey, Robert M. 'Waiting for Deborah: John Knox and four ruling queens', *Sixteenth Century J.* 25 (1994), 371–86.
22. Kirk, James. 'Iconoclasm and reform', *Records of the Scottish Church History Soc.* xxiv (1992), 366–83.
23. Leneman, Leah; Mitchison, Rosalind. 'Aquiescence in and defiance of Church discipline in early-modern Scotland', *Records of the Scottish Church History Soc.* xxv (1993), 19–39.
24. Lowe, Christopher E.; Craig, Derek; Dixon, Dianne. 'New light on the Anglian "minster" at Hoddom', *Dumfriesshire & Galloway Natural History & Antiquarian Soc. T.* 3rd ser. 66 (1991), 11–35.
25. Lynch, Michael. 'Preaching to the converted?: perspectives on the Scottish Reformation', Bc83, 301–43.
26. MacDonald, Alan R. 'The subscription crisis and church-state relations, 1584–1586', *Records of the Scottish Church History Soc.* xxv (1994), 222–55.
27. Mackey, James P. 'Was there (is there) a Celtic christianity?', *Records of the Scottish Church History Soc.* xxv (1993), 68–88.
28. MacQuarrie, Alan D. 'St Columba and his lay contemporaries in Scotland and Ireland', *Records of the Scottish Church History Soc.* xxv (1994), 188–203.
29. MacQuarrie, Alan D. 'The historical context of the Govan stones', Bc111, 27–33.
30. Rawles, Stephen. 'Glasgow university's copy of Robert Richardson's *Exegesis in canonem divi Augustini*', Bc83, 157–63.

31. Rendboe, Laurits. 'The Lord's prayer in Orkney and Shetland Norn', Bc23, 388–96.
32. Scott, J.G. 'Bishop John of Glasgow and the status of Hoddom', *Dumfriesshire & Galloway Natural History & Antiquarian Soc. T.* 3rd ser. 66 (1991), 37–45.
33. Thomas, Charles. 'Christianity at Govan: but when?', Bc111, 19–27.
34. Tolley, Thomas. 'Hugo van der Goes's altarpiece for Trinity college church in Edinburgh and Mary of Guelders, queen of Scotland', Bc133, 213–31.
35. Watt, Donald E.R. 'Abbot Walter Bower of Inchcolm and his *Scotichronicon*', *Records of the Scottish Church History Soc.* xxiv (1992), 286–304.
36. Watt, Donald E.R. 'Bishops in the Isles before 1203: bibliography and biographical lists', *Innes R.* 45 (1994), 99–119.
37. White, Allan. 'The Regent Morton's visitation: the Reformation of Aberdeen, 1574', Bc83, 246–63.

(f) *Economic Affairs*

1. Goodare, Julian. 'The long hundred in medieval and early modern Scotland', *Soc. of Antiquaries of Scotland P.* 123 (1993), 395–418.
2. Mahler, Ditlev L.D. 'Shielings and their role in the viking-age economy: new evidence from the Faroe Islands', Bc23, 487–505.
3. Murray, H.K.; Murray, J.C. 'Excavations at Rattray, Aberdeenshire: a Scottish deserted burgh', *Medieval Archaeology* 37 (1993), 109–219.
4. Reitemeier, Arnd. 'Das Handelsverbot der Hanse gegen Schottland (1412–1415/1418)' [The Hansa ban on trade with Scotland (1412–1415/1418)], *Hänsische Geschichtsblatter* 112 (1994), 161–236.
5. Stavert, Marion. L. (ed.). *The Perth guildry book, 1452, 1601*. Edinburgh; Scottish Record Soc.; 1993. Pp xxiv, 609.

(g) *Social Structure and Population*

1. Andersen, Per Sveaas. 'Den norske innvandringen til Hebridene i vikingtiden og den norrone bosetningens senere skjebne' [Norse immigration to the Hebrides in the viking period and the later fate of Norse settlement], *Historisk Tidskrift* 73 (1994), 265–85.
2. Arge, Símun V. 'On the landnam of the Faroe Islands', Bc23, 465–72.
3. Batey, Colleen E. 'The viking and late Norse graves of Caithness and Sutherland', Bc23, 148–72.
4. Bertelsen, Reidar; Lamb, Raymond G. 'Settlement mounds in the north Atlantic', Bc23, 544–54.
5. Bigelow, Gerald F. 'Archaeological and ethnohistoric evidence of a Norse island food custom', Bc23, 441–53.
6. Buckland, Paul C.; Sadler, Jon P.; Smith, David N. 'An insect's eye-view of the Norse farm', Bc23, 506–27.
7. Caldwell, David H.; Ewart, Gordon. 'Finlaggan and the lordship of the Isles: an archaeological approach', *Scottish Historical R.* 72 (1993), 146–66.

8. Debes, Hans Jacob. 'Problems concerning the earliest settlement in the Faroe Islands', Bc23, 454–64.
9. Fellows-Jensen, Gillian. 'Some Orkney personal names', Bc23, 397–407.
10. Gourlay, Robert B. 'Before the vikings: the pre-Norse background in Caithness', Bc23, 111–19.
11. Graham-Campbell, James A. 'The northern hoards of viking-age Scotland', Bc23, 173–86.
12. Hansen, Steffen Stummann. 'Viking-age Faroe Islands and their southern links in the light of recent finds at Toftanes, Leirvik', Bc23, 473–86.
13. Hunter, John R.; Bond, Julie M.; Smith, Andrea M. 'Some aspects of early viking settlement in Orkney', Bc23, 272–84.
14. Kaland, Sigrid H.H. 'The settlement of Westness, Rousay', Bc23, 308–17.
15. Lamb, Raymond G. 'Carolingian Orkney and its transformation', Bc23, 260-71.
16. Morris, Christopher D. 'Caithness: notes on field excursions', Bc23, 1–42.
17. Morris, Christopher D. 'Orkney: notes on field excursions', Bc23, 42–101.
18. Morris, Christopher D. 'The Birsay Bay project: a résumé', Bc23, 285–307.
19. Oram, Richard D. 'A family business?: colonization and settlement in twelfth- and thirteenth-century Galloway', *Scottish Historical R.* 72 (1993), 111–45.
20. Owen, Olwyn A. 'Tuquoy, Westray, Orkney: a challenge for the future', Bc23, 318–39.
21. Stewart, R. (ed.). *The people of Peterhead, 1696; taken from the list of pollable persons in the shires of Aberdeen, 1696*, vol. 1. Aberdeen; Aberdeen & North-East Scotland Family History Soc.; 1993. Pp 36.
22. Thomson, William P.L. 'Some settlement patterns in medieval Orkney', Bc23, 340-48.
23. Waugh, Doreen. 'Caithness: an onomastic frontier zone', Bc23, 120-28.

(h) *Naval and Military*

1. Hill, J. Michael. 'Killiecrankie and the evolution of highland warfare', *War in History* 1 (1994), 125–39.
2. MacIvor, Iain. *Blackness castle*. Edinburgh; Historic Scotland; 1993. Pp 24.
3. MacIvor, Iain. *Dumbarton castle*. Edinburgh; Historic Scotland; 1993. Pp 24.

(i) *Intellectual and Cultural*

1. Barnes, Michael P. 'The interpretation of the runic inscriptions of Maeshowe', Bc23, 349–69.
2. Batey, Colleen E. 'The sculptured stones in Glasgow museums', Bc111, 63–73.
3. Bawcutt, Priscilla. 'New light on Gavin Douglas', Bc83, 95–106.
4. Birrell, T.A. 'Some rare Scottish books in the old royal library', Bc83, 404–416.
5. Broadie, Alexander. 'James Liddell on concepts and signs', Bc83, 82–94.
6. Bruhn, Ole. 'Earl Rognvald and the rise of saga literature', Bc23, 240-47.

7. Bryce, Ian B.D. '"Castellum meum de Bucharin": a Banffshire hall-house of circa 1220', *Aberdeen University R.* 54/1 (1991), 13–22.
8. Bushnell, Rebecca W. 'George Buchanan, James VI and neo-classicism', Bc119, 91–112.
9. Caldwell, David H. 'Lead seal matrices of the 16th and early 17th century', *Soc. of Antiquaries of Scotland P.* 123 (1993), 373–80.
10. Caldwell, David H. 'The Beaton panels: Scottish carvings of the 1520s or 1530s', Bc133, 174–84.
11. Cameron, Neil. 'The Romanesque sculpture of Dumfermline abbey: Durham versus the Vicinal', Bc133, 118–23.
12. Cassidy, Brendan; Howlett, David R. 'Some eighteenth-century drawings of the Ruthwell cross', *Antiquaries J.* 72 (1994 for 1992), 102–17.
13. Cessford, Craig. 'The Borthwick Mains salmon', *Hawick Arch. Soc. T.* (1993), 26–29.
14. Chadburn, Ray. 'Appendix 1: the geology of the Govan stones, a preliminary assessment', Bc111, 145–6.
15. Clancy, Thomas Owen. 'The Drosten stone: a new reading', *Soc. of Antiquaries of Scotland P.* 123 (1993), 345–53.
16. Cosh, Mary. 'Organs, orrerys and others: a commission for Monymusk', Bc139, 69–90.
17. Craig, Derek. 'The early medieval sculpture of the Glasgow area', Bc111, 73–92.
18. Cramp, Rosemary A. 'Conclusions from the Govan conference', Bc111, 133–38.
19. Cramp, Rosemary A. 'The Govan recumbent cross-slabs', Bc111, 55–63.
20. Crawford, Barbara E. 'Earl William Sinclair and the building of Roslin collegiate church', Bc133, 99–107.
21. Crawford, Barbara E. 'The "Norse background" to the Govan hogbacks', Bc111, 103–13.
22. Cunningham, I.C. 'The Asloan manuscript', Bc83, 107–35.
23. Donaldson, Robert. '"M. Alex: Boyde.": the authorship of "Fra banc to banc"', Bc83, 344–66.
24. Dunbar, John G. 'Two late seventeenth-century designs for Kinnaird castle', Bc139, 43–52.
25. Elliot, Kenneth. '*Some helpes for young schollers*: a new source of early Scottish psalmody', Bc83, 264–75.
26. Evans, Godfrey. 'The mace of St Salvator's college', Bc133, 197–212.
27. Fawcett, Richard. *Architectural history of Scotland: accession of the Stewarts to the Reformation.* Edinburgh; Edinburgh UP; 1994. Pp 350.
28. Fawcett, Richard. 'Arbroath abbey', Bc133, 61–69.
29. Fenton, Alexander. 'Orkney Norn: a survey of "taboo" terms', Bc23, 381–87.
30. Fernie, E.C. 'The Romanesque churches of Dumfermline abbey', Bc133, 25–37.
31. Fisher, Ian. 'The Govan cross-shafts and early cross-slabs', Bc111, 47–55.
32. Gillies, William. 'The invention of tradition, highland-style', Bc83, 144–56.

33. Glenn, Jonathan A. (ed.). *The prose works of Sir Gilbert Hay, vol. 3: 'The buke of the ordre of knychthede' and 'The buke of the gouernaunce of princis'*. Edinburgh; Scottish Text Soc. (4th ser., vol. 21); 1993. Pp xiii, 174.
34. Graves, Pamela. 'Medieval stained and painted window glass in the diocese of St Andrews', Bc133, 124–36.
35. Gudmundsson, Finnbogi. 'On the writing of *Orkneyinga saga*', Bc23, 204–11.
36. Haggarty G.; Jennings, S. 'The imported pottery from Fast castle, near Dunbar, Scotland', *Medieval Ceramics* 16 (1992), 45–54.
37. Hagland, Jan Ragnar. 'The runic inscriptions from Orphir, Orkney', Bc23, 370-74.
38. Heyworth, Stephen. 'The church of St Rule, St Andrews', Bc133, 38–46.
39. Hillyard, Brian. '"Durkan & Ross" and beyond', Bc83, 367–84.
40. Hoey, Lawrence. 'Pier design in early Gothic architecture in east-central Scotland, *c.*1170-1250', Bc133, 84–98.
41. Horan, Martin. *Scottish castles*. Edinburgh; Chambers; 1994. Pp 128.
42. Howard, Deborah. 'Sir William Bruce's design for Hopetoun House and its forerunners', Bc139, 53–68.
43. Jesch, Judith. 'England and *Orkneyinga saga*', Bc23, 222–39.
44. Kelly, T.A. Davidson. 'The Govan collection in the context of local history', Bc111, 1–19.
45. Kelly, T.A. Davidson. 'The Partick hogback hunt', Bc111, 138–43.
46. Kirk, James. '"Melvillian" reform in the Scottish universities', Bc83, 276–300.
47. Kruse, Susan E. 'Silver storage and circulation in viking-age Scotland: the evidence of silver ingots', Bc23, 187–203.
48. Lang, James. 'The Govan hogbacks: a reappraisal', Bc111, 123–33.
49. Learmont, David. 'The restoration of the house of Dun', Bc139, 99–108.
50. Lyall, Roderick J. 'The literature of lowland Scotland, 1350-1700', Bc2, 77–98.
51. MacDonald, A.A. 'The Latin original of Robert Henryson's Annunciation lyric', Bc83, 45–65.
52. Macfarlane, Leslie J. 'William Elphinstone's library revisited', Bc83, 66–81.
53. Mackechnie, Aonghus. 'Design approaches in early post-Reformation Scots houses', Bc139, 15–34.
54. Mapstone, Sally. 'The Scots *Buke of phisnomy* and Sir Gilbert Hay', Bc83, 1–44.
55. Marshall, Rosalind K. '"Scarce a finer seat in Scotland: Kinneil castle and the 4th duke of Hamilton', Bc139, 35–42.
56. Maxwell-Irving, A.M.T. 'Lockerbie tower', *Dumfriesshire & Galloway Natural History & Antiquarian Soc. T.* 3rd ser. 67 (1992), 61–65.
57. McAleer, J. Philip. 'The west front of Arbroath abbey: its place in the evolution of the twin tower façade', Bc133, 70-83.
58. McKean, Charles. 'A plethora of palaces: some "castle-wise" country houses of early Renaissance Scotland', Bc139, 1–14.
59. Michael, Michael. 'The first maces of St Andrews university', Bc133, 185–96.

60. Mundal, Else. 'The Orkney earl and Scald Torf-Einarr and his poetry', Bc23, 248–59.
61. Norton, Christopher. 'Medieval floor tiles in Scotland', Bc133, 137–73.
62. O'Meadhra, Uaininn. 'Viking-age sketches and motif-pieces from the northern earldoms', Bc23, 423–40.
63. Salter, Mike. *The castles of south-west Scotland.* Malvern; Folly; 1993. Pp 152.
64. Salter, Mike. *The castles of the heartland of Scotland.* Malvern; Folly; 1994. Pp 140.
65. Salter, Mike. *The old parish churches of Scotland.* Malvern; Folly; 1994. Pp 124.
66. Sørensen, Preben Meulengracht. 'The sea, the flame and the wind: the legendary ancestors of the earls of Orkney', Bc23, 212–21.
67. Spearman, R.M. 'The Govan sarcophagus: an enigmatic monument', Bc111, 33–47.
68. Stoklund, Marie. 'Greenland runes: isolation or cultural contact?', Bc23, 528–43.
69. Thurlby, Malcolm. 'St Andrews cathedral-priory and the beginnings of Gothic architecture in northern Britain', Bc133, 47–60.
70. van Heijnsbergen, Theo. 'The interaction between literature and history in Queen Mary's Edinburgh: the Bannatyne manuscript and its prosopographical context', Bc83, 183–255.
71. Whitfield, Niamh; Graham-Campbell, James A. 'A mount with Hiberno-Saxon chipcarved animal ornament from Rerrick, near Dundrennan', *Dumfriesshire & Galloway Natural History & Antiquarian Soc. T.* 3rd ser. 67 (1992), 9–27.
72. Williamson, Arthur H. 'Number and national consciousness: the Edinburgh mathematicians and Scottish political culture at the union of the crowns', Bc119, 187–213.

L. IRELAND TO *c*.1640

See also Aa61,87,104,110,157,b3,25,30,50,c28,60,84; Fg5; Jd1; Ke4,28

(a) *General*

1. Bourke, Edward J. *Shipwrecks of the Irish coast, 1105–1993.* Dublin; Bourke; 1994. Pp 240.
2. Casey, Christopher; Rowan, Alistair. *North Leinster.* London; Penguin (The buildings of Ireland); 1993. Pp 544.
3. Cullen, Clara; Henchy, Monica; Ward-Perkins, Sarah. *Writings on Irish history.* Dublin; Irish Commission of Historical Sciences; 1994. Pp 110.
4. Gosling, Paul. *Archaeological inventory of county Galway, 1: West Galway (including Connemara and the Aran Islands).* Dublin; Stationery Office; 1993. Pp xii, 242.
5. Power, Denis. *Archaeological inventory of county Cork, 1: West Cork.* Dublin; Stationery Office; 1992. Pp 478.

(b) *Politics*

1. Hill, J. Michael. 'The rift within clan Ian Mor: the Antrim and Dunyreg MacDonnells, 1590-1603', *Sixteenth Century J.* 24 (1993), 865–79.
2. Lennon, Colm. *Sixteenth-century Ireland: the incomplete conquest.* Dublin; Gill & Macmillan; 1994. Pp 390.
3. McCavitt, John. 'The flight of the earls, 1607', *Irish Historical Studies* 29 (1994), 159–73.
4. Morgan, Hiram. 'A booke of questions and answers concerning the wars or rebellions of the kingdome of Ireland', *Analecta Hibernica* 36 (1994), 93–156.

(c) *Constitution, Administration and Law*

1. Connolly, Philomena. 'An attempted escape from Dublin castle: the trial of William and Walter de Bermingham, 1332', *Irish Historical Studies* 29 (1994), 100-08.
2. McCavitt, John. '"Good planets in their several speares"—the establishment of the assize circuits in early seventeenth-century Ireland', *Irish Jurist* ns 24 (1993 for 1989), 248–78.

(d) *External Affairs*

1. Eastwood, Bruce Stansfield. 'The astronomy of Macrobius in Carolingian Europe: Dungal's letter of 811 to Charles the Great', *Early Medieval Europe* 3 (1994), 117–34.
2. Hudson, Benjamin. 'William the conqueror and Ireland', *Irish Historical Studies* 29 (1994), 145–58.

(e) *Religion*

1. Bhreathnach, Edel. 'Killeshin: an Irish monastery surveyed', *Cambrian Medieval Celtic Studies* 27 (1994), 33–47.
2. Bieler, Ledwig (ed.). *Clavis Patricii, II. Libri epistolarum sancti Patricii episcopi: introduction, text and commentary.* Dublin; Royal Irish Academy; 1994. Pp 294.
3. Caball, Marc. 'Providence and exile in early seventeenth-century Ireland', *Irish Historical Studies* 29 (1994), 174–88.
4. Cotter, Francis J. *The friars minor in Ireland, from their arrival to 1400.* New York; Franciscan Inst., St Bonaventure University; 1994. Pp 289.
5. Cunningham, Bernadette. 'A view of religious affiliation and practice in Thomond, 1591', *Archivium Hibernicum* 48 (1994), 13–24.
6. Howlett, David R. (ed.). *Liber epistolarum sancti Patricii episcopi: the book of letters of saint Patrick the bishop.* Dublin; Four Courts; 1994. Pp 172.
7. Lennon, Colm. 'The foundation charter of St Sythe's guild, Dublin, 1476', *Archivium Hibernicum* 48 (1994), 3–12.
8. Leslie, James B. (comp.). *Clergy of Connor: from patrician times to the present day.* Belfast; Ulster Historical Foundation; 1993. Pp xi, 672.

9. Lyons, Mary Ann. 'Sidelights on the Kildare ascendancy: a survey of Geraldine involvement in the Church, *c.*1470-*c.*1520', *Archivium Hibernicum* 48 (1994), 73–87.
10. McCafferty, John. 'Defamation and the church courts in early sixteenth-century Armagh', *Archivium Hibernicum* 48 (1994), 88–106.
11. Ó Catháin, Séamas. 'Hearth-prayers and other traditions of Brigit: Celtic goddess and holy woman', *Royal Soc. of Antiquaries of Ireland J.* 122 (1992), 12–34.
12. Ó Cúiv, Brian (ed.). *Aibidil Gaoidheilge agus Caiticiosma [Irish alphabet and catechism]: Seán Cearnaigh's Irish primer of religion, published in 1571.* Dublin; Dublin Inst. of Advanced Studies; 1994. Pp 246.
13. Ó Riain, Pádraig (ed.). *Beatha Bharra: saint Finbarr of Cork—the complete life.* London; Irish Texts Soc.; 1994. Pp 322.
14. O'Floinn, Raghnall. *Irish shrines and reliquaries of the middle ages.* Dublin; Town House & Country House; 1994. Pp 46.
15. O'Sullivan, William. 'Two Clogher constitutions', Bc102, 351–72.

(f) *Economic Affairs*

1. Cleary, R.M. 'Medieval town wall at Emmet Street, Clonmel', *Tipperary Historical J.* (1993) 194–98.
2. Hickey, Elizabeth. *Skryne and the early Normans.* Tara (Meath); Meath Arch. & Historical Soc.; 1994. Pp 218.
3. Simms, Angret (ed.). *Irish towns.* Cork; Mercier; 1994. Pp 192.
4. Smith, Brendan. 'The de Pitchford family in thirteenth-century Ireland', *Studia Hibernica* 27 (1993), 29–43.
5. Wallace, Patrick F. 'The archaeological identity of the Hiberno-Norse town', *Royal Soc. of Antiquaries of Ireland J.* 122 (1992), 35–66.

(g) *Social Structure and Population*

1. Brady, Niall. 'Labour and agriculture in early medieval Ireland: evidence from the sources', Bc149, 125–45.
2. King, H.A. 'Archaeological investigations in the environs of a motte and bailey at Ballinaclogh Lower, near Timanhoe, co. Laois', *J. of the County Kildare Arch. Soc.* 18 (1992–93), 8–19.
3. McNeill, Hugh. *The annals of the parish of Derrykeighan: from AD 453 to AD 1890.* Ballymena; Mid-Antrim Historical Group; 1993. Pp 81.
4. Patterson, Nerys. *Cattle lords and clansmen: the social structure of early Ireland.* Notre Dame (In); Notre Dame UP; 1994. Pp 424.

(h) *Naval and Military*

1. McNeill, Tom E. 'The outer gate at Dunamase castle, co. Laois', *Medieval Archaeology* 37 (1993), 236–39.
2. Stradling, R.A. *Spanish monarchy and Irish mercenaries, 1618–1668: the wild geese in Spain.* Blackrock (Dub); Irish Academic; 1994. Pp 220.
3. Sweetman, P.D. 'Wiggins rebuked', *Archaeology of Ireland* 7/4 (1993), 19–20.

4. Taaff, S. 'The evidence for town walling at Athy, co. Kildare', *J. of the Arch. Soc. of University College, Dublin* 4 (1993), 26–31.
5. Wiggins, K. 'King John's castle (1976) ... revisited', *Archaeology of Ireland* 7/3 (1993), 26–28.

(i) *Intellectual and Cultural*

1. Carroll, Clare. 'Representations of women in some early modern English tracts on the colonization of Ireland', *Albion* 25 (1993), 379–93.
2. de Barra, Seamas. '*Pairlement Chloinne Tomais*: léirmhiniu, datai agus udar' ["Pairlement Chloinne Tomais": intention, dates and author], *Studia Hibernica* 26 (1991/2), 107–46.
3. Edwards, Nancy. 'The iconography of Irish high crosses: Carolingian influence in Ireland in the ninth century' [review article], *Early Medieval Europe* 3 (1994), 63–71.
4. Fanning, Thomas. *Viking age ringed pins from Dublin*. Dublin; Royal Irish Academy; 1994. Pp 140.
5. Hadfield, Andrew; McVeath, John (ed.). *Strangers to that land: British perceptions of Ireland from the Reformation to the famine*. Gerards Cross; Smythe; 1994. Pp 315.
6. Kelly, Dorothy. 'The high crosses of Ireland', *Royal Soc. of Antiquaries of Ireland J.* 122 (1992), 67–78.
7. Lysaght, Patricia. 'Women, milk and magic at the boundary festival of May', Bc30, 208–29.
8. Maley, Willy. '"The supplication of the blood of the English most lamentably murdered in Ireland cryeng out of the yearth for revenge (1598)"', *Analecta Hibernica* 36 (1994), 3–92.
9. Teahan, John. *Irish furniture and woodcraft*. Dublin; Town House & Country House; 1994. Pp 46.

M. IRELAND SINCE c.1640

See also Aa8,41,80,95,99,102,114,119,128,141,152,157,b1–2,25,43,64,c14,19; Ba72, c7,84,140; Gi85; Hg14

(a) *General*

1. Robinson, Philip. *The plantation of Ulster*. Belfast; Ulster Historical Foundation; 1994. Pp xxix, 254.

(b) *Politics*

1. Belchem, John. 'Republican spirit and military science: the "Irish brigade" and Irish-American nationalism in 1848', *Irish Historical Studies* 29 (1994), 44–64.
2. Bew, Paul. *Ideology and the Irish question: Ulster Unionism and Irish nationalism, 1912–1916*. Oxford; Oxford UP; 1994. Pp xix, 165.

3. Broderick, Eugene. 'The corporate labour policy of Fine Gael, 1934', *Irish Historical Studies* 29 (1994), 88–99.
4. Connolly, Sean J. 'Late eighteenth-century Irish politics', *Parliamentary History* 13 (1994), 227–35.
5. Coogan, Tim Pat. *De Valera: long fellow, long shadow*. London; Hutchinson; 1993. Pp xii, 772.
6. Curtin, Nancy J. *The United Irishmen: popular politics in Ulster and Dublin, 1791–1798*. Oxford; Oxford UP; 1994. Pp ix, 317.
7. Durey, Michael. 'The Dublin Society of United Irishmen and the politics of the Carey-Drennan dispute, 1792–1794', *Historical J.* 37 (1994), 89–111.
8. English, Richard. *Radicals and the republic: socialist republicanism in the Irish Free State, 1925–1937*. Oxford; Oxford UP; 1994. Pp vi, 309.
9. English, Richard. '"Paying no heed to public clamor": Irish republican solipsism in the 1930s', *Irish Historical Studies* 28 (1993), 426–39.
10. Graham, B.J.; Hood, Susan. 'Town tenant protest in late nineteenth- and early twentieth-century Ireland', *Irish Economic & Social History* 21 (1994), 39–57.
11. Hughes, Michael. *Ireland divided*. Cardiff; Wales UP; 1994. Pp xiii, 143.
12. Jackson, Alvin. *Sir Edward Carson*. Dundalk; Dundalgan; 1993. Pp 74.
13. Jupp, Peter J.; Royle, Stephen A. 'The social geography of Cork city elections, 1801–1830', *Irish Historical Studies* 29 (1994), 13–43.
14. Keenan, Joe (ed.). *'The Labour opposition of Northern Ireland': a complete reprint of Northern Ireland's first Labour newspaper, 1925–1926*. Belfast; Athol; 1992. Pp 212.
15. Kelly, James. *Henry Grattan*. Dundalk; Dundalgan; 1993. Pp 54.
16. Lindley, Keith. 'Irish adventurers and godly militants in the 1640s', *Irish Historical Studies* 29 (1994), 1–12.
17. McKenny, Kevin. 'Charles II's Irish cavaliers: the 1649 officers and the Restoration land settlement', *Irish Historical Studies* 28 (1993), 409–25.
18. Moran, Gerard. *Father Patrick Lavelle: the rise and fall of an Irish nationalist, 1825–1886, a radical priest in Mayo*. Dublin; Four Courts; 1994. Pp xii, 226.
19. O'Callaghan, Margaret. *British high politics and a nationalist Ireland: criminality, land and the law under Foster and Balfour*. Cork; Cork UP; 1994. Pp x, 223.
20. Ohlmeyer, Jane H. 'The "Antrim plot" of 1641: a rejoinder', *Historical J.* 37 (1994), 431–37.
21. O'Sullivan, Michael. *Sean Lemass: a biography*. Dublin; Blackwater; 1994. Pp viii, 216.
22. Owen, Arwel Ellis. *The Anglo-Irish agreement: the first three years*. Cardiff; Wales UP; 1994. Pp x, 263.
23. Perceval-Maxwell, M. *The outbreak of the Irish rebellion of 1641*. Dublin; Gill & Macmillan; 1994. Pp xvii, 390.
24. Perceval-Maxwell, M. 'The "Antrim plot" of 1641—a myth?: a response', *Historical J.* 37 (1994), 421–30.
25. Phoenix, Eamon. *Northern nationalism: nationalist politics, partition and the Catholic minority in Northern Ireland, 1890-1940*. Belfast; Ulster Historical Foundation; 1994. Pp xix, 485.

26. Quinlivan, Patrick J. 'Hunting the Fenians: problems in the historiography of a secret organisation', Bc7, 133–53.

(c) *Constitution, Administration and Law*

1. Barnard, Toby C. 'Lawyers and the law in later seventeenth-century Ireland', *Irish Historical Studies* 28 (1993), 256–82.
2. Campbell, Colm. *Emergency law in Ireland, 1918–1925*. Oxford; Oxford UP; 1994. Pp xxiii, 429.
3. Crossman, Virginia. *Local government in nineteenth-century Ireland*. Belfast; Inst. of Irish Studies, Queen's University; 1994. Pp 1, 115.
4. d'Arcy, Fergus. 'The Ranger of the Curragh of Kildare: the case of a curious office, 1687–1961', *Royal Soc. of Antiquaries of Ireland J.* 122 (1992), 5–11.
5. Duggan, John P. *The Irish Free State diplomatic service, 1922–1937*. Blackrock (Dub); Irish Academic; 1994. Pp 208.
6. Elvert, Jurgen. 'Der "Republic of Ireland Bill" vom 18. April 1949: einige Überlegungen zur irischen Sezession vom British Commonwealth of Nations' [The Republic of Ireland bill of 18th April 1949: some considerations on the Irish secession from the British Commonwealth of Nations], *Historische Mitteilungen der Ranke-Gesellschaft* 5 (1992), 114–29.
7. Kotsonouris, Mary. *Retreat from revolution: Dail courts, 1920-1924*. Blackrock (Dub); Irish Academic; 1994. Pp 176.

(d) *External Affairs*

1. Hopkinson, Michael. 'President Woodrow Wilson and the Irish question', *Studia Hibernica* 27 (1993), 89–111.

(e) *Religion*

1. Dunne, Eamonn. 'Action and reaction: Catholic lay organisations in Dublin in the 1920s and 1930s', *Archivium Hibernicum* 48 (1994), 107–18.
2. Forristal, Desmond. *Edel Quinn, 1907–1944*. Dublin; Dominican; 1994. Pp 226.
3. Kerr, Donal A. *"A nation of beggars?": priests, people and politics in famine Ireland, 1846–1852*. Oxford; Oxford UP; 1994. Pp xiv, 338.
4. Kilroy, Phil. *Protestant dissent and controversy in Ireland, 1660-1714*. Cork; Cork UP; 1994. Pp ix, 300.
5. Leighton, C.D.A. *Catholicism in a Protestant kingdom: a study of the Irish 'ancien regime'*. Basingstoke; Macmillan; 1994. Pp x, 218.
6. Lenihan, Pádraig. 'Catholicism and the Irish confederate armies: for God or king?', *Recusant History* 22 (1994), 182–98.
7. Macaulay, Ambrose. *William Crolly, archbishop of Armagh, 1835–1849*. Dublin; Four Courts; 1994. Pp 492.
8. Ó Ferghail, Fearghus. 'An Ossory instruction of 1773', *Archivium Hibernicum* 48 (1994), 25–36.
9. O'Flaherty, Eamon. 'Clerical indiscipline and ecclesiastical authority in Ireland, 1690-1750', *Studia Hibernica* 26 (1991/2), 7–29.

(f) *Economic Affairs*

1. Bourke, Joanna. 'Avoiding poverty: strategies for women in rural Ireland, 1880-1914', Bc151, 292–311.
2. Crawford, W. *The handloom weavers and the Ulster linen industry*. Belfast; Ulster Historical Foundation; 1994. Pp viii, 95.
3. Humphreys, Madeleine. 'An issue of confidence: the decline of the Irish whiskey industry in independent Ireland, 1922–1952', *J. of European Economic History* 23 (1994), 93–113.
4. Ireland, John de Courcy. 'An analysis of the history of fish exportation from Ireland', Bc127, 15–37.
5. MacDonnell, Hector. 'A seventeenth-century inventory from Dunlace castle, county Antrim', *Royal Soc. of Antiquaries of Ireland J.* 122 (1992), 109–27.
6. Mangan, James J. (ed.). *Robert Whyte's 1847 famine ship diary: the journey of an Irish coffin ship*. Cork; Mercier; 1994. Pp 124.
7. Ó Gráda, Cormac. *Ireland: a new economic history, 1780-1939*. Oxford; Oxford UP; 1994. Pp xvi, 536.
8. O'Rourke, Kevin. 'The repeal of the Corn laws and Irish emigration', *Explorations in Economic History* 31 (1994), 120-38.
9. O'Rourke, Kevin; Polak, Ben. 'Property transactions in Ireland, 1708–1988: an introduction', *Irish Economic & Social History* 21 (1994), 58–71.
10. Rhodes, Rita M. *Women and the family in post-famine Ireland: status and opportunity in a patriarchical society*. New York; Garland; 1992. Pp xix, 366.
11. Somerville, Alexander. *Letters from Ireland during the famine of 1847*, ed. D.K.M. Snell. Dublin; Irish Academic; 1994. Pp 219.
12. Takei, Akihio. 'The first Irish linen mills, 1800-1824', *Irish Economic & Social History* 21 (1994), 28–38.
13. Vaughan, W.E. *Landlords and tenants in mid-Victorian Ireland*. Oxford; Oxford UP; 1994. Pp xxiii, 339.
14. Williamson, Jeffrey G. 'Economic convergence: placing post-famine Ireland in comparative perspective', *Irish Economic & Social History* 21 (1994), 5–27.

(g) *Social Structure and Population*

1. Clark, Mary. 'Dublin city pipe water accounts, 1704/5', *Irish Genealogist* 9 (1994), 76–88.
2. Cohen, Marilyn. 'Religion and social inequality in Ireland', *J. of Interdisciplinary History* 25 (1994), 1–21.
3. Finnegan, Ruth; Collins, Brenda. 'Staying and moving: links between migration and community', Bc116, 162–80.
4. Kelly, James. 'Harvests and hardship: famine and scarcity in Ireland in the late 1720s', *Studia Hibernica* 26 (1991/2), 65–105.
5. Lane, Pádraig G. 'Agricultural labourers and rural violence, 1850-1914', *Studia Hibernica* 27 (1993), 77–87.
6. Maguire, Martin. 'The organisation and activism of Dublin's Protestant working class, 1883–1935', *Irish Historical Studies* 29 (1994), 65–87.

7. McKernan, Anne. 'War, gender and industrial innovation: recruiting women weavers in early nineteenth-century Ireland', *J. of Social History* 28 (1994), 109–24.
8. Miller, Kerby A.; Wagner, Paul. *Out of Ireland: the story of Irish emigration to America*. London; Aurum; 1994. Pp 132.
9. O'Dowd, Anne. 'Women in rural Ireland in the nineteenth and early twentieth centuries: how the daughters, wives and sisters of small farmers and landless labourers fared', *Rural History* 5 (1994), 171–83.

(h) *Naval and Military*

1. McCracken, Donal P. *The Irish pro-Boers, 1877–1902*. Johannesburg; Persekor; 1989. Pp xx, 189.
2. McIvor, Aidan. *A history of the Irish naval service*. Blackrock (Dub); Irish Academic; 1994. Pp 256.
3. Perry, Nicholas. 'Nationality in the Irish infantry regiments in the first world war', *War & Soc.* 12 (1994), 65–95.

(i) *Intellectual and Cultural*

1. Anderson, William K. *James Connolly and the Irish left*. Dublin; Irish Academic; 1994. Pp 200.
2. Barnard, Toby C. 'Protestants and the Irish language, *c.*1675–1725', *J. of Ecclesiastical History* 44 (1993), 243–72.
3. Edwards, Owen Dudley. 'The stage Irish', Bc7, 83–114.
4. Gaughan, John Anthony. *Alfred O'Rahilly, vol. 3: controversialist, part 1: social reformer*. Dublin; Kingdom; 1992. Pp 306.
5. Hyland, Áine; Milne, Kenneth (ed.). *Irish educational documents*, vol. 2. Dublin; Church of Ireland College of Education; 1992. Pp xxiv, 571.
6. Maume, Patrick. 'Daniel Corkery: a reassessment', *Studia Hibernica* 26 (1991/2), 147–66.
7. McCartney, Donal. *W.E.H. Lecky: historian and politician, 1838–1903*. Dublin; Lilliput; 1993. Pp 271.
8. Ó Buachalla, Breandán. 'Seacaibiteachas Thaidhg Ui Neachtain' [The Jacobitism of Thaidg Ui Neachtain], *Studia Hibernica* 26 (1991/2), 32–64.
9. Ó Tuathaigh, M.A.G. 'Irish historical "revisionism": state of the art or ideological project? ', Bc84, 306–26.
10. O'Brien, Conor Cruise. *Ancestral voices: religion and nationalism in Ireland*. Dublin; Poolbeg; 1994. Pp 197.
11. Williams, Jeremy. *Architecture in Ireland, 1837–1971: companion guide*. Dublin; Irish Academic; 1994. Pp x, 424.

(j) *Local History*

1. Barnard, Toby C. 'Athlone, 1685; Limerick, 1710: religious riots or charivaris?', *Studia Hibernica* 27 (1993), 61–75.
2. Jordan, Donald E. *Land and popular politics in Ireland: county Mayo from the plantation to the land war*. Cambridge; Cambridge UP; 1994. Pp xiv, 369.

(k) *Science and Medicine*

1. Burke, Helen. *The royal hospital, Donnybrook: a heritage of caring, 1743–1993*. Dublin; Royal Hospital Donnybrook; 1993. Pp x, 378.

N. EMPIRE AND COMMONWEALTH POST 1783

See also Aa69,c5; Ha15,e33,44,f14,i22; If87,i2,8,45,86,j58–59

(a) *General*

1. Ackrill, Margaret. *British imperialism in microcosm: the annexation of the Cocos (Keeling) Islands*. London; London School of Economics (working papers in economic history 18); 1994. Pp 32.
2. Beinart, William. *Twentieth-century South Africa*. Oxford; Oxford UP; 1994. Pp xvi, 293.
3. Buckner, Phillip; Reid, John. *The Atlantic region to confederation: a history*. Toronto; Toronto UP; 1994. Pp xvii, 491.
4. Connell, John. 'Britain's Caribbean colonies: the end of the era of decolonization', *J. of Commonwealth & Comparative Politics* 32 (1994), 87–106.
5. Conrad, Margaret; Finkel, Alvin; Jaenen, Cornelius. *A history of the Canadian peoples, vol. 1: beginnings to 1867*. Mississauga (On); Copp Clark Pitman; 1993. Pp xxiii, 631.
6. Cooper, Frederick. 'Conflict and connection: rethinking colonial African history', *American Historical R.* 99 (1994), 1515–45.
7. Davidson, Basil. *The search for Africa: a history in the making*. London; Currey; 1994. Pp x, 374.
8. Dawson, Graham. *Soldier heroes: British adventure, empire and the imagining of masculinities*. London; Routledge; 1994. Pp 356.
9. Fletcher, Brian. 'Australian history: the imperial context, 1880s-1939', *Australian J. of Politics & History* 40 (1993), 1–15.
10. French, Patrick. *Younghusband: last great imperial adventurer*. London; HarperCollins; 1994. Pp xxii, 440.
11. Goldsworthy, David (ed.). *British documents on the end of empire, series A, vol. 3: the Conservative government and the end of empire, 1951–1957, part I: international relations*. London; HMSO; 1994. Pp 422.
12. Goldsworthy, David (ed.). *British documents on the end of empire, series A, vol. 3: the Conservative government and the end of empire, 1951–1957, part II: politics and administration*. London; HMSO; 1994. Pp 410.
13. Goldsworthy, David (ed.). *British documents on the end of empire, series A, vol. 3: the Conservative government and the end of empire, 1951–1957, part III: economic and social policies*. London; HMSO; 1994. Pp 422.
14. Grimshaw, Patricia; Lake, Marilyn; McGrath, Ann; Quartly, Marian. *Creating a nation* [Australia]. Melbourne; McPhee Gribble; 1994. Pp 360.

15. James, Lawrence. *The rise and fall of the British empire*. London; Little, Brown; 1994. Pp xvi, 704.
16. Lari, Suhail Zaheer. *A history of Sindh*. Oxford; ; 1994. Pp 212.
17. Louis, William Roger; Robinson, Ronald E. 'The imperialism of decolonization', *J. of Imperial & Commonwealth History* 22 (1994), 462–511.
18. Owram, Doug. *Canadian history: a reader's guide, vol. 2: confederation to the present* Toronto; Toronto UP; 1994. Pp 437.
19. Porter, Andrew N. *European imperialism, 1860-1914*. Basingstoke; Macmillan; 1994. Pp xiii, 119.
20. Ross, Robert. *Beyond the pale: essays on the history of colonial South Africa*. Hanover (South Africa)/London; Wesleyan UP; 1993. Pp xii, 270.
21. Semmel, Bernard. *The liberal ideal and the demons of empire: themes of imperialism from Adam Smith to Lenin*. Baltimore (Md); Johns Hopkins UP; 1993. Pp xxi, 223.
22. Shillington, Kevin. *African adventure: a brief life of Cecil Rhodes*. Bishops Storford; Rhodes Memorial Museum and Commonwealth Centre; 1992. Pp 64.
23. Stern, Robert W. *Changing India: bourgeois revolution on the subcontinent*. Cambridge; Cambridge UP; 1993. Pp 267.
24. Stockwell, Anthony J. 'Southeast Asia in war and peace: the end of European colonial empires', Bc45, 325–85.
25. Sutcliffe, Bob. *Imperialism*. London; Tauris; 1994. Pp 256.
26. Taylor, M. Brook. *Canadian history: a reader's guide, vol. 1: beginnings to confederation*. Toronto; Toronto UP; 1994. Pp 525.
27. Taylor, Stephen. *Shaka's children: a history of the Zulu people*. London; HarperCollins; 1994. Pp 432.
28. Walvin, James. *The life and times of Henry Clarke of Jamaica, 1828–1907*. London; Cass; 1994. Pp 155.
29. Winks, Robin W. *The imperial revolution: yesterday and tomorrow*. Oxford; Oxford UP; 1994. Pp 28.
30. Winsnes, Selena Axelrod (ed.). *Letters on West Africa and the slave trade: Paul Erdmann Isert's "Journey to Guinea and the Caribbean islands in Columbia" (1788)*. Oxford; British Academy and Oxford UP; 1992. Pp x, 278.

(b) *Politics*

1. Allen, D.F. 'New light on Malta during the peace of Amiens, 1801–1803', *British Library J.* 20 (1994), 174–83.
2. Bassett, Michael. *Sir Joseph Ward: a political biography*. Auckland; Auckland UP; 1993. Pp xii, 330.
3. Bell, Coral. *Dependent ally: a study in Australian foreign policy*. Sydney; Allen & Unwin; 1993. Pp vii, 205.
4. Buckley, Ken; Dale, Barbara; Reynolds, Wayne. *Doc. Evatt: patriot, internationalist, fighter and scholar*. Melbourne; Longman Cheshire; 1994. Pp 433.
5. Crockett, Peter. *Evatt: a life*. Oxford; Oxford UP; 1994. Pp iv, 388.
6. Dockrill, Michael. 'The British empire and the peace conferences, 1919–1923', *Historian* 42 (1994), 3–8.

7. Eldredge, Elizabeth A. *A south African kingdom: the pursuit of security in nineteenth-century Lesotho*. Cambridge; Cambridge UP; 1993. Pp xv, 250.
8. English, John. *The worldly years: the life of Lester Pearson*, vol. 2. New York; Knopf Canada; 1992. Pp 473.
9. Fisher, Michael H. (ed.). *The politics of the British annexation of India, 1757–1857*. Oxford; Oxford UP; 1994. Pp xvi, 313.
10. Fraser, Cary. *Ambivalent anti-colonialism: the United States and the genesis of West Indian independence, 1940-1964*. London; Greenwood; 1994. Pp 240.
11. Furedi, Frank. *Colonial wars and the politics of third world nationalism*. London; Tauris; 1994. Pp viii, 338.
12. Gill, Sadiq Ali. *Anglo-American diplomacy and the emergence of Pakistan, 1940-1947*. Lahore (Pak); Research Soc. of Pakistan, University of the Punjab; 1990. Pp xiii, 362.
13. Hack, K. 'Decolonization and the Pergau-dam affair: Anglo-Malaysian relations, past and present', *History Today* 44/11 (1994), 9–11.
14. Horner, David. *High command: Australia's struggle for an independent war strategy, 1939–1945*. Sydney; Allen & Unwin; 1992. Pp xxv, 556.
15. Ingham, Kenneth. *Obote: a political biography*. London; Routledge; 1994. Pp ix, 228.
16. Kratoska, Paul; Batson, Ben. 'Nationalism and modernist reform', Bc45, 249–324.
17. Kwan, Elizabeth. 'The Australian flag: ambiguous symbol of nationality in Melbourne and Sydney, 1920-1921', *Australian Historical Studies* 26 (1994), 280-303.
18. Lowe, John. *The great powers, imperialism and the German problem, 1865–1925*. London; Routledge; 1994. Pp 240.
19. Lwanda, John Lloyd. *Kamuzu Banda of Malawi: a study in promise, power and paralysis*. Glasgow; Nudu Nsomba; 1993. Pp 319.
20. Martin, A.W.; Hardy, Patsy (ed.). *Dark and hurrying days: Menzies' 1941 diary*. Canberra; National Library of Australia; 1993. Pp x, 177.
21. Mostert, Noel. *Frontiers: the epic of South Africa's creation and the tragedy of the Xhosa people*. London; Cape; 1992. Pp xxix, 1355.
22. Pearson, Geoffrey A.H. *Seize the day: Lester B. Pearson and crisis diplomacy*. Ottawa; Carleton UP; 1994. Pp 190.
23. Phillips, Walter. *James Jefferis: prophet of federation*. Melbourne; Australian Scholarly; 1993. Pp xviii, 341.
24. Schoeman, Karel (ed.). *The British presence in the Transorange, 1845–1854*. Cape Town; Human & Rousseau; 1992. Pp 147.
25. Silverfarb, Daniel. *Twilight of British ascendancy in the middle east: case study of Iraq, 1941–1950*. London; Macmillan; 1994. Pp 306.
26. Singh, D.S. Ranjit. *Brunei, 1839–1983: the problems of political survival*. Singapore; Oxford UP; 1994. Pp xii, 260.
27. Tang, James H. 'From empire defence to imperial retreat: Britain's post-war China policy and the decolonization of Hong Kong', *Modern Asian Studies* 28 (1994), 317–37.
28. Tarling, Nicholas. 'The establishment of the colonial régimes', Bc45, 5–78.
29. Templeton, Malcolm. *Ties of blood and empire: New Zealand's involvement*

in middle east defence and the Suez crisis, 1947–1957. Auckland; Auckland UP; 1994. Pp 288.
30. Trainor, Luke. *British imperialism and Australian nationalism: manipulation, conflict and compromise in the late nineteenth century*. Melbourne; Cambridge UP; 1994. Pp xi, 213.
31. Wainwright, A. Martin. *Inheritance of empire: Britain, India, and the balance of power in Asia, 1938–1955*. Westport (Ct); Praeger; 1994. Pp xiv, 237.
32. Waite, P.B. *The loner: three sketches of the personal life and ideas of R.B. Bennett, 1870-1914*. Toronto; Toronto UP; 1992. Pp xiv, 110.
33. Weiss, Ruth. *Sir Garfield Todd and the making of Zimbabwe*. London; Tauris; 1994. Pp 224.
34. Wylie, D. 'Shaka and the modern Zulu state', *History Today* 44/5 (1994), 8–11.
35. Young, Thomas-Durell. *Australian, New Zealand, and United States security relations, 1951–1986*. Boulder (Co); Westview; 1992. Pp xxvi, 284.

(c) *Constitution, Administration and Law*

1. Ahmed, Akbar. 'Jinnah and the quest for Muslim identity', *History Today* 44/9 (1994), 34–40.
2. Anderson, David M. 'Policing and communal conflict: the Cyprus emergency, 1954–1960', Bc42, 177–207.
3. Arnold, John; Spearitt, Peter; Walker, David. *Out of empire: the British dominion of Australia*. Melbourne; Mandarin; 1993. Pp ix, 342.
4. Baines, Gary. 'The contradictions of community politics: the African petty bourgeoisie and the New Brighton advisory board, *c*.1937–1952', *J. of African History* 35 (1994), 79–97.
5. Baker, Colin. *Development governor: a biography of Sir Geoffrey Colby*. London; British Academic; 1994. Pp xiv, 407.
6. Blake, Raymond B. *Canadians at last: the integration of Newfoundland as a province*. Toronto; Toronto UP; 1994. Pp 272.
7. Burroughs, Peter. 'The law, the citizen, and the state in nineteenth-century Australia', *J. of Imperial & Commonwealth history* 22 (1994), 542–54.
8. Calligas, Eleni. 'Lord Seaton's reforms in the Ionian Islands, 1843–1848: a race with time', *European History Q.* 24 (1994), 7–29.
9. Carmichael, John. *African Eldorado: Ghana from the Gold Coast to independence*. London; Duckworth; 1993. Pp viii, 224.
10. Chakrabarty, Bidyut. 'The 1947 United-Bengal Movement: the examination of the short-lived proposal to create three, not two, sovereign polities—India, Pakistan and United Bengal: a thesis with a synthesis', *Indian Economic & Social History R.* 30 (1993), 467–88.
11. Chatterji, Joya. *Bengal divided: Hindu communalism and partition, 1932–1947*. Cambridge; Cambridge UP; 1994. Pp 300.
12. Clark, Aylwin. *An enlightened Scot: Hugh Cleghorn, 1752–1837*. Duns; Black Ace; 1993. Pp xvii, 333.
13. Cowen, M.P.; Shenton, R.W. 'British neo-Hegelian idealism and official colonial practice in Africa: the Oluwa land case of 1921', *J. of Imperial & Commonwealth History* 22 (1994), 217–50.

14. Crisswell, Colin N. *The end of the Brooke Raj in Sarawak*. Stirling; Kiscadale; 1994. Pp 197.
15. Dalton, Dennis. *Mahatma Gandhi's non-violent power in action*. New York; Columbia UP; 1993. Pp xii, 279.
16. Daly, M.W. (ed.). *On trek in Kordofan: the diaries of a British district officer in the Sudan, 1931–1933*. Oxford; Oxford UP; 1994. Pp ix, 304.
17. Daramy, Sheikh Batu. *Constitutional developments in the post-colonial state of Sierra Leone, 1961–1984*. Lewiston/Lampeter; Mellen; 1993. Pp 303.
18. Darwin, John. 'The Central African emergency, 1959', Bc42, 217–34.
19. Farnsworth, Susan H. *The evolution of British imperial policy during the mid nineteenth-century: a study of the Peelite contribution, 1846–1874*. London; Garland; 1992. Pp x, 438.
20. Flanagan, Thomas. *Louis Riel*. Ottawa; Canadian Historical Association; 1992. Pp 27.
21. Furedi, Frank. 'Creating a breathing space: the political management of colonial emergencies', Bc42, 89–106.
22. Heehs, Peter. *The bomb in Bengal: the rise of revolutionary terrorism in India, 1900-1910*. Delhi; Oxford UP; 1994. Pp xii, 324.
23. Heehs, Peter. 'Foreign influences in Bengali revolutionary terrorism, 1902–1908', *Modern Asian Studies* 28 (1994), 533–56.
24. Heuman, Gad. *The killing time: the Morant Bay rebellion in Jamaica*. London; Macmillan; 1994. Pp 240.
25. Holland, Robert. 'Never, never land: British colonial policy and the roots of violence in Cyprus, 1950-1954', Bc42, 148–76.
26. Jeffery, Robin. 'Democracy in south Asia', *History Today* 44/5 (1994), 43–49.
27. Jumare, Ibrahim M. 'The late treatment of slavery in Sokoto: background and consequences of the 1936 proclamation', *International J. of African Historical Studies* 27 (1994), 303–22.
28. Kaul, Chandrika. 'England and India: the Ilbert bill, 1883. A case study of the metropolitan press', *Indian Economic & Social History R.* 30 (1993), 413–36.
29. Kolinsky, Martin. *Law, order and riots in mandatory Palestine, 1928–1935*. Basingstoke; Macmillan; 1993. Pp xviii, 295.
30. Lancaster, J.C.S. *The governorship of Sir Benjamin d'Urban at the Cape of Good Hope, 1834–1838*. Pretoria; Government Printer; 1991. Pp xiii, 237.
31. Llewelyn-Jones, Rosie. *A very ingenious man: Claude Martin in early colonial India*. New York; Oxford UP; 1993. Pp xxi, 241.
32. Loo, Tina. *Making the law: order and authority in British Columbia, 1821–1871*. Toronto; Toronto UP; 1994. Pp 352.
33. Madden, A. Frederick. *The Dominions and India since 1900: select documents on the constitutional history of the British empire and commonwealth, vol. 6*. London; Greenwood; 1994. Pp xxix, 873.
34. Maloba, Wunyabari. *Mau Mau and Kenya: an analysis of a peasant revolt*. Bloomington (In); Indiana UP; 1993. Pp x, 228.
35. Marshall, P.J. 'Parliament and property rights in the late eighteenth-century British empire', Bc150, 530-44.

36. Martin, Ged W. 'Identity and interaction: a defence of Natal history', *J. of Imperial & Commonwealth History* 22 (1994), 317–31.
37. Matson, Albert Thomas. *Nandi resistance to British rule, vol. 2: the volcano erupts*. Cambridge; African Studies Centre, Cambridge University; 1993. Pp viii, 202.
38. Maxon, Robert M. *The struggle for Kenya: the loss and reassertion of imperial initiative, 1912–1923*. London; Associated UP; 1993. Pp 351.
39. Mazumdar, A. *The Indian press and freedom struggle, 1937–1942*. London; Sangam; 1993. Pp 357.
40. McCracken, John Leslie. *New light at the Cape of Good Hope: William Porter, the father of Cape liberalism*. Belfast; Ulster Historical Foundation; 1993. Pp 160.
41. Menon, Dilip M. *Caste, nationalism and communism in south India: Malabar, 1900-1948*. Cambridge; Cambridge UP; 1993. Pp xv, 228.
42. Miller, Carman. *Painting the map red: Canada and the South African war, 1899–1902*. Kingston; McGill-Queen's UP; 1993. Pp xvi, 544.
43. Moleah, Alfred Tokollo. *South Africa: colonialism, apartheid and African dispossession*. Wilmington (De); Disa; 1993. Pp xiii, 505.
44. Mookerjee, Syama Prasad. *Leaves from a diary*. Oxford; Oxford UP; 1993. Pp xx, 240.
45. Morley, John Arthur Elwell. *Colonial postscript: diary of a district officer, 1935–1956*. London; Radcliffe; 1992. Pp xii, 212.
46. Mukonoweshuro, Eliphas G. *Colonialism, class formation, and underdevelopment in Sierra Leone*. Lanham (Md); UP of America; 1993. Pp 257.
47. O'Regan, John. *From empire to commonwealth: reflections on a career in Britain's overseas service*. London; Radcliffe; 1994. Pp xi, 196.
48. Parekh, Bhikhu. *Communalism, tradition and reform: an analysis of Gandhi's political discourse*. New Delhi; Sage; 1989. Pp 288.
49. Parekh, Bhikhu. *Gandhi's political philosophy: a critical examination*. Basingstoke; Macmillan; 1989. Pp 248.
50. Pratt, John. *Punishment in a perfect society: the New Zealand penal system, 1840-1939*. Wellington (NZ); Victoria UP; 1992. Pp 288.
51. Rai, Vijai Shankar. *The last phase of the transfer of power in India*. New Delhi; Arnold; 1990. Pp 175.
52. Rathbone, Richard. 'Police intelligence in Ghana in the late 1940s and 1950s', Bc42, 107–28.
53. Russell, B.J.F. *Kenya, beyond the Marich pass: a district officer's story*. London; Radcliffe; 1994. Pp xii, 161.
54. Sapire, Hilary. 'Apartheid's "testing ground": urban "native policy" and African politics in Brakpan, South Africa, 1943–1948', *J. of African History* 35 (1994), 99–123.
55. Sharma, Archana. *British policy towards Malaysia, 1957–1967*. London; Sangam; 1994. Pp xix, 197.
56. Sheppard, George. *Plunder, profit and paroles: a social history of the war of 1812 in upper Canada*. Kingston; McGill-Queen's UP; 1994. Pp 368.
57. Singh, Braj Kishore. *The Indian national congress and the partition of India, 1937–1947*. Delhi; Capital; 1990. Pp iii, 224.

58. Singh, Kewal. *Partition and aftermath: memoirs of an ambassador.* New Delhi; Vikas; 1991. Pp 412.
59. Smith, Simon C. 'The rise, decline and survival of the Malay rulers during the colonial period, 1874–1957', *J. of Imperial & Commonwealth History* 22 (1994), 84–108.
60. Snelson, Peter. *To independence and beyond: memoirs of a colonial and commonwealth civil servant.* London; Radcliffe; 1993. Pp 342.
61. Stockwell, Anthony J. '"A widespread and long-concocted plot to overthrow government in Malaya"?: the origins of the Malayan emergency', Bc42, 66–88.
62. Switzer, Les. *Power and resistance in an African society: the Ciskei Xhosa and the making of South Africa.* Pietermaritzburg; Natal UP; 1993. Pp 464.
63. Tabili, Laura. '"Keeping the natives under control": race segregation and the domestic dimensions of empire, 1920-1939', *International Labor & Working-Class History* 44 (1993), 64–78.
64. Trench, Charles Chevenix. *Men who ruled Kenya: the Kenya administration, 1892–1963.* London; Radcliffe; 1993. Pp 316.
65. Trocki, Carl A. 'Political structure in the nineteenth and early twentieth centuries', Bc45, 79–130.
66. Tsokhas, Kosmas. 'Dedominionization: the Anglo-Australian experience, 1939–1945', *Historical J.* 37 (1994), 861–83.
67. Twaddle, Michael. *Kakungulu and the creation of Uganda, 1868–1928.* London; Currey; 1993. Pp ix, 304.
68. Verney, David (ed.). *In vice-regal India, 1916–1921: the letters of Ralph Verney.* Padstow; Tabb House; 1994. Vol. 2. Pp xii, 211.
69. Visram, Rozina. *Women in India and Pakistan: the struggle for independence from British rule.* Cambridge; Cambridge UP; 1992. Pp 64.
70. Vyas, R.P. *British policy towards the princely states of India.* Jodhpur; Rajasthan-Vidya Prakashan; 1991. Pp xxiv, 224.
71. Webb, Patrick. 'Guests of the crown: convicts and liberated slaves on McCarthy Island, the Gambia', *Geographical J.* 160 (1994), 136–42.
72. White, Nicholas J. 'Government and business divided: Malaya, 1945–1957', *J. of Imperial & Commonwealth History* 22 (1994), 251–74.
73. Whitehead, John. *Thangliena: the life of T.H. Lewin, amongst wild tribes of India's north-east frontier.* Stirling; Kiscadale; 1992. Pp xv, 437.
74. Willis, Justin. 'Killing bwana: peasant revenge and political panic in early colonial Ankole', *J. of African History* 35 (1994), 379–400.
75. Worden, Nigel. *The making of modern South Africa: conquest, segregation and apartheid.* Oxford; Blackwell; 1993. Pp 166.

(d) *Religion*

1. Adamson, Christopher. 'God's continent: politics and religion in upper Canada and the northern and western United States, 1775 to 1841', *Comparative Studies in Soc. & History* 36 (1994), 417–46.
2. Bebbington, David W. 'Evangelicalism in its settings: the British and American movements since 1940', Bc91, 365–88.

3. Bugge, Henriette. *Mission and Tamil society: social and religious change in south India, 1840-1900.* Richmond; Curzon; 1994. Pp 220.
4. Busch, Briton C. 'Whalemen, missionaries, and the practice of Christianity in the nineteenth-century Pacific', *Hawaiian J. of History* 27 (1993), 91–118.
5. Collins, T. *The Baptist mission in Nigeria, 1850-1993: a history of the southern Baptist convention missionary work in Nigeria.* Ibadan; Y-Books; 1993. Pp 96.
6. Dart, Martha. *Marjorie Sykes, Quaker-Gandhian: a narrative biography.* York; Sessions; 1993. Pp xii, 159.
7. Downs, Frederick S. *The history of Christianity in India, vol. 5: north-east India in the nineteenth and twentieth centuries.* Bangalore; Church History Association of India; 1992. Pp xv, 236.
8. Glen, Robert M. *Mission and Moko: aspects of the work of the Church Missionary Society in New Zealand, 1814–1882.* Christchurch (NZ); Latiner Fellowship of New Zealand; 1992. Pp 245.
9. Ileto, Reynaldo. 'Religion and anti-colonial movements', Bc45, 197–248.
10. Ipenburg, At. *All good men: the development of Lubwa mission, Chinsali, Zambia, 1905–1967.* New York; Lang; 1992. Pp 345.
11. Manthadam, J. 'The state of Indian missions in 1862', *Indian Theological Studies* 31 (1994), 26–43.
12. Nwosu, L. Ugwuanya. 'The missionary factor in the development of nationalism in colonial Africa: 1891–1900 revisited', *Vidyajyoti* 58 (1994), 134–50.
13. Piggin, Stuart. 'The American and British contributions to evangelicalism in Australia', Bc91, 290-309.
14. Porter, Andrew N. 'War, colonialism, and the British experience: the redefinition of Christian missionary policy, 1938–1952', *Kirchliche Zeitgeschichte* 5 (1992), 269–88.
15. Powell, Avril Ann. *Muslims and missionaries in pre-mutiny India.* London; Curzon; 1993. Pp ix, 339.
16. Rawlyk, George A. *The Canada fire: radical evangelicalism in British north America, 1775–1812.* Kingston; McGill-Queen's UP; 1994. Pp 272.
17. Schoeman, Karel (ed.). *The Wesleyan mission in the OFS, 1833–1854.* Cape Town; Human & Rousseau; 1991. Pp 128.
18. Tull, Barbara Mitchell. *Affectionately Rachel: letters from India, 1860-1884.* Kent (Oh); Kent State UP; 1992. Pp xv, 351.
19. Tyrrell, Alex. *A sphere of benevolence: the life of Joseph Orton, Wesleyan Methodist missionary (1795–1842).* Melbourne; State Library of Victoria; 1993. Pp xiv, 265.
20. van Die, Marguerite. '"The double vision": evangelical piety as derivative and indigenous in Victorian English Canada', Bc91, 253–72.
21. Walls, Andrew. 'The evangelical revival, the missionary movement and Africa', Bc91, 310-30.
22. Wells, David. 'On being evangelical: some theological differences and similarities', Bc91, 389–410.

(e) *Economic Affairs*

1. Agbodeka, Francis. *An economic history of Ghana from the earliest times.* Accra; Ghana UP; 1992. Pp 182.

2. Ally, Russell. *Gold and empire: the Bank of England and South Africa's gold producers, 1886–1926*. Johannesburg; Witwatersrand UP; 1994. Pp 229.
3. Balachandran, G. 'Towards a "Hindu marriage": Anglo-Indian monetary relations in interwar India, 1917–1935', *Modern Asian Studies* 28 (1994), 615–47.
4. Brown, Nathan J. 'Who abolished corvée labour in Egypt and why?', *Past & Present* 144 (1994), 116–37.
5. Bryant, Raymond L. 'Shifting the cultivator: the politics of teak regeneration in colonial Burma', *Modern Asian studies* 28 (1994), 225–50.
6. Butlin, Noel G (ed.). *Economics and the dreamtime: a hypothetical history*. Cambridge; Cambridge UP; 1993. Pp 252.
7. Elson, Robert E. 'International commerce, the state and society: economic and social change', Bc45, 131–95.
8. Fieldhouse, David K. *Merchant capital and economic decolonization: the United Africa Company, 1929–1987*. Oxford; Oxford UP; 1994. Pp xxviii, 832.
9. Fieldhouse, David K. 'Gentlemen, capitalists, and the British empire', *J. of Imperial & Commonwealth History* 22 (1994), 531–41.
10. Graves, Adrian. *Cane and labour: the political economy of the Queensland sugar industry, 1862–1906*. Edinburgh; Edinburgh UP; 1993. Pp xxi, 282.
11. Gregory, Robert G. *South Asians in east Africa: an economic and social history, 1890-1990*. Boulder (Co); Westview; 1993. Pp xiii, 410.
12. Kaberuka, Will. *The political economy of Uganda, 1890-1979: a case study of colonialism and underdevelopment*. New York; Vantage; 1993. Pp xxv, 309.
13. Laurence, Keith Ormiston. *A question of labour: indentured immigration into Trinidad and British Guiana, 1875–1917*. London; Currey; 1994. Pp 648.
14. Livingston, K.T. 'Anticipating federation: the federalising of telecommunications in Australia', *Australian Historical Studies* 102 (1994), 97–117.
15. Look Lai, Walton. *Indentured labor, Caribbean sugar: migrants to the British West Indies, 1838–1918*. Baltimore (Md); Johns Hopkins UP; 1993. Pp xxviii, 370.
16. McCalla, Douglas. *Planting the province: the economic history of upper Canada, 1784–1870*. Buffalo (NY); Toronto UP; 1993; Pp xviii, 446.
17. Moore, Henrietta L.; Vaughan, Megan. *Cutting down trees: gender, nutrition and agricultural change in the northern province of Zambia, 1890-1990*. London; Currey; 1994. Pp xxvi, 278.
18. Morgenrood, Pierre. 'Public sector capital expenditure and its funding in the Cape Colony' [part 1], *Q. B. of the South African Library* 48 (1994), 97–108.
19. Murray, Colin. *Black mountain: land, class and power in the eastern Orange Free State, 1880s-1980s*. Edinburgh; Edinburgh UP; 1992. Pp 340.
20. Overton, John. *Colonial green revolution?: food, irrigation and the state in colonial Malaya*. Wallingford; C.A.B. International; 1994. Pp 256.
21. Owen, J.M. Kalinga. 'The Master Farmers' scheme in Nyasaland, 1950-1962: a study of a failed attempt to create a yeoman class', *African Affairs* 92 (1993), 367–87.
22. Oxley, Deborah. 'Packing her (economic) bags: convict women workers', *Australian Historical Studies* 102 (1994), 57–76.

23. Pirie, G.H. 'Railway ganging in southern Africa, *c.*1900-1937', *J. of Transport History* 3rd ser. 14 (1993), 64–76.
24. Reddock, Rhoda E. *Women, labour and politics in Trinidad and Tobago: a history*. London; Zed; 1994. Pp 346.
25. Rothermund, Dietmar. *India in the great depression, 1929–1939*. New Delhi; Manohar; 1992. Pp xii, 292.
26. Roy, Tirthankar. *Artisans and industrialization: Indian weaving in the twentieth century*. Delhi; Oxford UP; 1994. Pp 254.
27. Schenk, Catherine R. 'Closing the Hong Kong gap: the Hong Kong free dollar market in the 1950s', *Economic History R.* 2nd ser. 47 (1994), 335–53.
28. Tsokhas, Kosmas. 'The Matson line: Australian shipping policy and imperial relations, 1935–1939', *Australian J. of Politics & History* 40 (1994), 364–78.
29. Wadley, Susan S. *Struggling with destiny in Karimpur, 1925–1984*. Berkeley (Ca); California UP; 1994. Pp 319.
30. Ward, J.R. 'The industrial revolution and British imperialism, 1750-1850', *Economic History R.* 2nd ser. 47 (1994), 44–65.
31. Warren, D. *Merchants, commissioners and wardmasters: municipal and colonial politics in Cape Town, 1840-1854*. Pretoria; Government Printer; 1992. Pp 575.
32. Washbrook, David. 'The commercialization of agriculture in colonial India: production, subsistence and reproduction in the "dry south", *c.*1870-1930', *Modern Asian Studies* 28 (1994), 129–44.
33. West, Michael O. 'Pan-Africanism, capitalism and racial uplift: the rhetoric of African business formation in colonial Zimbabwe', *African Affairs* 92 (1993), 263–83.
34. Wingent, Peter. *Movements of aircraft on Imperial Airways' African route, 1931–1939*. Winchester; Wingent; 1991. Pp vii, 211.

(f) *Social Structure and Population*

1. Akenson, Donald H. *Occasional papers on the Irish in South Africa*. Grahamstown (South Africa); Inst. of Economic & Social Research, Rhodes University; 1991. Pp 95.
2. Arnold, David. *Colonizing the body: state medicine and epidemic disease in nineteenth-century India*. Berkeley (Ca); California UP; 1993. Pp xii, 354.
3. Bush, Julia. '"The right sort of woman": female emigrators and emigration to the British empire, 1890-1910', *Women's History R.* 3 (1994), 385–409.
4. Fleras, Augie; Elliot, Jean Leonard. *The "nations within": Aboriginal-state relations in Canada, the United States and New Zealand*. Toronto; Oxford UP; 1992. Pp xii, 267.
5. Harrison, Mark. *Public health and prevention medicine in British India, 1859–1914*. Cambridge; Cambridge UP; 1994. Pp 344.
6. Hassam, Andrew. *Sailing to Australia: shipboard diaries by nineteenth-century British emigrants*. Manchester; Manchester UP; 1994. Pp x, 235.
7. Jackson, R.V. 'Theory and evidence: Bentham, Collins, and the New South Wales penal settlement', *Australian J. of Politics & History* 39 (1993), 318–29.

8. James, Ronald M. 'Defining the group: nineteenth-century Cornish on the North American mining frontier', Bc135, 32–47.
9. Lewandoski, Ken. 'A new transportation for the penitentiary era: some *Household Words* on free emigration', *Victorian Periodicals R.* 26 (1993), 8–18.
10. McCracken, Donal P. *The Irish in southern Africa, 1795–1910.* Durban; UP of Durban-Westville; 1992. Pp 290.
11. Summers, Carol. *From civilization to segregation: social ideals and social control in Southern Rhodesia.* Athens (Oh); Ohio UP; 1994. Pp 280.
12. Thom, G. 'Rutland convicts transported to Australia', *Rutland Record* (1993), 136–40.

(g) *Naval and Military*

1. Barthorp, Michael. *The British troops in the Indian mutiny, 1857–1859.* London; Osprey; 1994. Pp 48.
2. Barua, P. 'Military developments in India, 1750-1850', *J. of Military History* 58 (1994), 599–616.
3. Beckett, Ian F.W. 'The Indian expeditionary force on Malta and Cyprus, 1878', *Soldiers of the Queen* 76 (1994), 6–11.
4. Bishop, Arthur. *The splendid hundred: the true story of Canadians who flew in the greatest air battle of world war II.* Whitby (On); McGraw-Hill Ryerson; 1994. Pp 192.
5. Cain, Frank. *The wobblies at war: a history of the IWW and the great war in Australia.* Melbourne; Spectrum; 1993. Pp viii, 300.
6. Castle, Ian; Knight, Ian J. *Fearful hard times: the siege and relief of Eshowe, 1879.* London; Greenhill; 1994. Pp 256.
7. Cock, Jacklynn. *Women and war in South Africa.* London; Open Letters; 1992. Pp 254.
8. Cooper, Anthony. 'The Australian historiography of the first world war: who is deluded?', *Australian J. of Politics & History* 40 (1993), 16–35.
9. Cowman, Ian. 'Main fleet to Singapore? Churchill, the admiralty and force Z', *J. of Strategic Studies* 17 (1994), 179–193.
10. English, Barbara. 'The Kanpur massacres in India in the revolt of 1857', *Past & Present* 142 (1994), 169–78.
11. Flynn, Michael C. *The second fleet: Britain's grim convict armada of 1790.* Sydney; Library of Australian History; 1993. Pp 787.
12. Gordon, Andrew. 'The Admiralty and imperial overstretch, 1902–1941', *J. of Strategic Studies* 17 (1994), 63–85.
13. Greenhous, Brereton; Harris, Stephen J.; Johnston, William C.; Rawling, William G.P. *The crucible of war, 1939–1945: the official history of the Royal Canadian Air Force*, vol. 3. Toronto; Toronto UP; 1994. Pp 900.
14. Hamlin, John F. *Military aviation in Malta GC, 1915–1993.* Peterborough; GMS Enterprises; 1994. Pp 256.
15. Heathcote, T.A. *The military in British India: the development of British land forces in south Asia, 1600-1947.* Manchester; Manchester UP; 1994. Pp 240.
16. Heathcote, T.A. 'The army of British India', Bc109, 376–402.

17. Hirst, J.B. 'Australian defence and conscription: a reassessment, part 1', *Australian Historical Studies* 101 (1993), 608–27.
18. Hirst, J.B. 'Australian defence and conscription: a reassessment, part 2', *Australian Historical Studies* 102 (1994), 39–56.
19. Kent, John. 'The Egyptian base and the defence of the middle east, 1945–1954', Bc42, 45–65.
20. Killingray, David. 'The "rod of empire": the debate over corporal punishment in the British African colonial forces, 1888–1946', *J. of African History* 35 (1994), 201–16.
21. Lambert, Andrew D. '*Dalhousie*: the last flagship of the Indian Navy and the question of imperial defence', *Warship* (1993), 9–18.
22. Lee, Cecil. *Sunset of the raj: the fall of Singapore, 1942*. Edinburgh; Pentland; 1994. Pp xxxii, 292.
23. Milner, Marc. *The U-boat hunters: the Royal Canadian Navy and the offensive against Germany's submarines*. Annapolis (Md); Naval Inst.; 1994. Pp 280.
24. Moreman, T.R. 'The arms trade and the north-west frontier Pathan tribes, 1890-1914', *J. of Imperial & Commonwealth History* 22 (1994), 187–216.
25. Morton, Desmond. *When your number's up*. Toronto; Random House of Canada; 1994. Pp 508.
26. Mukherjee, Rudrangshu. 'The Kanpur massacres in India in the revolt of 1857: a reply', *Past & Present* 142 (1994), 178–89.
27. Omissi, David. *The sepoy and the raj: the Indian Army, 1860-1940*. Basingstoke; Macmillan; 1994. Pp xx, 313.
28. Savona-Ventura, C. 'Malta and the British navy: the medical connection during the nineteenth century, part 1: the establishment of the naval hospital at Bighi', *J. of the Royal Navy Medical Service* 78/1 (1992), 171–76.
29. Seegers, Annette. *The military and the making of modern South Africa*. London; Tauris; 1994. Pp 228.
30. Sibbald, Raymond. *War correspondents: the Boer war*. Stroud; Sutton; 1993. Pp 256.
31. Thompson, P.S. 'Isandlwana to Mome: Zulu experience in overt resistance to colonial rule', *Soldiers of the Queen* 76 (1994), 11–15.

(h) *Intellectual and Cultural*

1. Aiken, S. Robert. *Imperial belvederes: hill stations of Malaya*. Malaysia; Penerbit Fajar Bakti; 1994. Pp xi, 84.
2. Allman, Jean. 'Making mothers: missionaries, medical officers and women's work in colonial Asante, 1924–1945', *History Workshop* 38 (1994), 23–47.
3. Atkins, Keletso E. *The moon is dead! Give us our money!: cultural origins of an African work ethic in Natal, c.1843–1900*. London; Currey; 1994. Pp 288.
4. Buchert, Lene. *Education in the development of Tanzania, 1919–1990*. London; Currey; 1994. Pp xv, 192.
5. Chaudhuri, Nupur; Strobel, Margaret. *Western women and imperialism: complicity and resistance*. Bloomington (In); Indiana UP; 1992. Pp vi, 276.
6. Choko, Marc H. *Canadian war posters: 1914–1918*. Quebec; Editions du Meridien; 1994. Pp 199.

7. Evans, Dorothy. *Schooling in the south Atlantic islands, 1661–1992.* Oswestry; Nelson; 1994. Pp 414.
8. Francis, Mark. 'Anthropology and social Darwinism in the British empire: 1870-1900', *Australian J. of Politics & History* 40 special issue (1994), 203–15.
9. Gregory, Robert G. *The rise and fall of philanthropy in east Africa: the Asian contribution.* New Brunswick (NJ); Transaction; 1992. Pp x, 251.
10. Harries, Patrick. *Work, culture and identity: migrant labourers in Mozambique and southern Africa, c.1860-1910.* London; Currey; 1994. Pp 320.
11. Irschick, Eugene F. *Dialogue and history: constructing south India, 1795–1895.* Berkeley (Ca); California UP; 1994. Pp 278.
12. Israel, Milton. *Communications and power: propaganda and the press in the Indian nationalist struggle, 1920-1947.* Cambridge; Cambridge UP; 1994. Pp xiii, 336.
13. Jeater, Diana. *Marriage, perversion and power: the construction of a moral discourse in Southern Rhodesia, 1894–1930.* Oxford; Oxford UP; 1993. Pp viii, 281.
14. Kaul, Chandrika. 'A new angle of vision: the London press, governmental information management and the Indian empire, 1900-1922', *Contemporary Record* 8 (1994), 213–41.
15. Kumar, Radha. *The history of doing: an illustrated account of movements for women's rights and feminism in India, 1800-1990.* London; Verso; 1993. Pp 220.
16. Levere, Trevor H. *Science and the Canadian Arctic: a century of exploration, 1818–1918.* Cambridge; Cambridge UP; 1993. Pp xiv, 438.
17. MacDonald, Robert H. *The language of empire: myths and metaphors of popular imperialism, 1880-1918.* Manchester; Manchester UP; 1994. Pp xii, 268.
18. Maddocks, R.J.; Bratzel, M.P. *The postmarks and postal history of the Cameroons under British administration, 1916–1961.* Windsor (On); MPB Canada; 1993. Pp 349.
19. Marks, Shula; Engels, Dagmar. *Contesting colonial hegemony: state and society in Africa and India.* London; British Academic; 1994. Pp viii, 349.
20. McCracken, John. 'African history in British universities: past, present and future', *African Affairs* 92 (1993), 239–53.
21. Mitter, Partha. *Art and nationalism in colonial India, 1850-1922.* Cambridge; Cambridge UP; 1994. Pp 520.
22. Moir, Geoffrey D. *Falklands Islands history through philately.* Croydon; Moir; 1993. Pp 51.
23. Molloy, Frank. '"The sigh of thy harp shall be sent o'er the deep": the influence of Thomas Moore in Australia', Bc7, 115–32.
24. Mungazi, Dickson A. *Colonial education for Africans: George Stark's policy in Zimbabwe.* London; Praeger; 1992. Pp xxii, 154.
25. Nayder, Lillian. 'Class consciousness and the Indian mutiny in Dickens' *The perils of certain English prisoners*', *Studies in English Literature* 32 (1992), 690-705.

26. Peers, Douglas M. 'Soldiers, scholars, and the Scottish Enlightenment: militarism in early nineteenth-century India', *International History R.* 16 (1994), 441–65.
27. Richards, Thomas. *The imperial archive: knowledge and the fantasy of empire*. London; Verso; 1993. Pp viii, 179.
28. Riffenburgh, Beau. *The myth of the explorer: the press, sensationalism and geographical discovery*. London; Bellhaven; 1993. Pp 226.
29. Strobel, Margaret. *Gender, sex and empire*. Washington (DC); American Historical Association; 1993. Pp 31.
30. Sweeting, Anthony. *A phoenix transformed: the reconstruction of education in post-war Hong Kong*. Oxford; Oxford UP; 1993. Pp 294.
31. Thomas, Nicholas. *Colonialism's culture: anthroplogy, travel and government*. Cambridge; Polity; 1994. Pp xi, 238.
32. Young, David. 'East-End street names and British imperialism', *Local Historian* 22 (1992), 84–88.
33. Youngs, Tim. *Travellers in Africa: British travelogues, 1850-1900*. Manchester; Manchester UP; 1994. Pp 220.
34. Zachernuk, Philip S. 'Of origins and colonial order: southern Nigerian historians and the "Hamitic hypothesis", *c.*1870-1970', *J. of African History* 35 (1994), 427–55.
35. Zastoupil, Lynn. *John Stuart Mill and India*. Cambridge; Cambridge UP; 1994. Pp 280.

INDEX OF AUTHORS

Index of Authors

Index of Authors

Index of Authors

Index of Authors

Index of Authors

Index of Authors

Index of Authors

Index of Authors

Index of Authors

Index of Authors

Index of Authors

Index of Authors

Index of Authors

Index of Authors

Index of Authors

Index of Authors

Index of Authors

Index of Authors

Index of Authors

Index of Authors

Index of Authors

Index of Authors

Index of Authors

Index of Authors

Index of Authors

Index of Authors

Index of Authors

Index of Authors

Index of Authors

Index of Authors

Index of Authors

Index of Authors

Index of Authors

Index of Authors

Index of Authors

Index of Authors

Index of Authors

Index of Authors

INDEX OF PERSONAL NAMES

Index of Personal Names

Index of Personal Names

Index of Personal Names

Index of Personal Names

Index of Personal Names

Index of Personal Names

Index of Personal Names

Index of Personal Names

Index of Personal Names

COUNTY ABBREVIATIONS

ENGLAND

Bedfordshire	Bdf	Berkshire	Brk
Buckinghamshire	Bkm	Cambridgeshire	Cam
Cheshire	Chs	Cornwall	Con
Cumberland	Cul	Derbyshire	Dby
Devon	Dev	Dorset	Dor
Durham	Dur	Essex	Ess
Gloucestershire	Gls	Hampshire	Ham
Herefordshire	Hef	Hertfordshire	Hrt
Huntingdonshire	Hun	Isle of Wight	Iow
Isles of Scilly	Ios	Kent	Ken
Lancashire	Lan	Leicestershire	Lec
Lincolnshire	Lin	Middlesex	Mdx
Norfolk	Nfk	Northamptonshire	Nth
Northumberland	Nbl	Nottinghamshire	Ntt
Oxfordshire	Oxf	Rutland	Rut
Shropshire	Shr	Somerset	Som
Staffordshire	Sts	Suffolk	Sfk
Surrey	Sry	Sussex	Ssx
Warwickshire	War	Westmorland	Wes
Wiltshire	Wil	Worcestershire	Wor
Yorkshire	Yks		
Channel Islands	CI	Isles of Scilly	IoS
Isle of Man	IoM		

SCOTLAND

Aberdeen	Abd	Angus	Ans
Argyll	Arl	Ayr	Ayr
Banff	Ban	Berwick	Bew
Bute	But	Caithness	Cai
Clakmannan	Clk	Dumfries	Dfs
Dumbarton	Dmb	East Lothian	Eln
Fife	Fif	Hebrides	Wis
Inverness	Inv	Kinross	Krs
Kincardine	Kcd	Kirkcudbright	Kkd
Lanark	Lks	Midlothian	Mln
Moray	Mor	Nairn	Nai
Orkney	Ork	Peebles	Pee
Perth	Per	Renfrew	Rfw
Ross & Cromarty	Roc	Roxburgh	Rox
Selkirk	Sel	Shetland	Zet
Stirling	Sti	Sutherland	Sut
West Lothian	Wln	Wigtown	Wig

County Abbreviations

WALES

Anglesey	Agy	Brecknock	Bre
Caernarvon	Cae	Cardigan	Cgn
Carmarthen	Cmn	Denbigh	Den
Flint	Fln	Glamorgan	Gla
Merioneth	Mer	Monmouth	Mon
Montgomery	Mgy	Pembroke	Pem
Radnorshire	Rad		

IRELAND

Carlow	Car	Cavan	Cav
Clare	Cla	Cork	Cor
Donegal	Don	Dublin	Dub
Galway	Gal	Kerry	Ker
Kildare	Kid	Kilkenny	Kik
Laois	Las	Leitrim	Let
Limerick	Lim	Longford	Log
Louth	Lou	Mayo	May
Meath	Mea	Monaghan	Mog
Offaly	Off	Roscommon	Ros
Sligo	Sli	Tipperary	Tip
Waterford	Wat	Westmeath	Wem
Wexford	Wex	Wicklow	Wic

NORTHERN IRELAND

Antrim	Ant	Armagh	Arm
Derry	Dry	Down	Dow
Fermanagh	Fer	Tyrone	Tyr

INDEX OF PLACES

Index of Places

Index of Places

Index of Places

Index of Places

Index of Places

Index of Places

Index of Places

Index of Places

INDEX OF SUBJECTS

Index of Subjects

Index of Subjects

Index of Subjects

Index of Subjects

Index of Subjects

Index of Subjects

Index of Subjects

Index of Subjects

Index of Subjects